Angelo J. Polvere

The American Civil Engineer
1852-1974

The American Civil Engineer 1852-1974

The history, traditions and development of the

By William H. Wisely
Hon. M. ASCE
Executive Director Emeritus, ASCE

American Society of Civil Engineers
345 East 47th Street
New York, N.Y. 10017

Library of Congress Catalog Card Number: 74-17792
ISBN: 0-87262-000-X

CONTENTS

PREFACE

When the Bicentennial of the United States of America is commemorated in 1976, the American Society of Civil Engineers will have been a witness to 124 years of those two centuries. This historical account, hopefully, will help to relate the American Civil Engineer to the development of his country and his profession through this period.

A "Historical Sketch" of ASCE, authored by Secretary Charles Warren Hunt, was published as a book in 1897. Mr. Hunt followed this with a sequel in 1917 that was published in *Transactions*, Vol. 82, under the title "The Activities of the American Society of Civil Engineers During the Past Twenty-Five Years." In 1947 Professor Edward C. Thoma, M. ASCE, of Purdue University wrote a short review on the "Rise and Growth of the American Society of Civil Engineers," which was made available only in a limited edition of mimeographed copies.

The chief aim of this book is to provide an accurate documentary reference to the broad range of ASCE activity since the founding of the Society. If the record is more detailed than may be desired by the casual reader, it has been made so in the interest of completeness.

The book is not, however, a chronological recital of events as they occurred. The various programs and service areas of the Society are treated independently, each as a story in itself. This is intended to enhance readability and to provide better topical access for reference purposes.

Except for the appendices, the use of names of living persons has been almost completely avoided. It is deemed paramount that the thrust of the Society as an entity be unobstructed even by the most impartial effort to identify and judge the roles of the host of individuals who have contributed so much to the work of ASCE. These evaluations and interpretations are left to other historians.

In a further effort to improve readability, specific source references have been limited to a few major items. The primary sources of documentation were the published *Proceedings* and *Transactions*, the annual report of the Society, *Civil Engineering*, and the minutes of the meeting of the Board of Direction. It will not be difficult, therefore, for the reader to pursue a more detailed inquiry in the time frame in which an event is reported.

The hope that the book will be a useful historical reference has been noted. An even more important purpose would be served, however, if it might help both the student and the practitioner to orient themselves to the rich heritage of the civil engineering profession. Present and future leaders of the Society may also benefit from the guidance given here concerning past successes — and failures — in policy and administration. During my work on the book I have had frequent occasion to regret that the research could not have been undertaken before my own service on the staff began in 1955.

The project has been infinitely absorbing and stimulating, and I am deeply obliged to Executive Director Eugene Zwoyer for his continuing interest and encouragement. Many other members of the headquarters staff have responded most generously to my frequent and oft-times impatient requests for data and assistance. Public Information Services Manager Herbert R. Hands and Miss Mary E. Jessup, formerly News Editor of *Civil Engineering*, have been especially helpful in the exhaustive research task that has been necessary. The cooperation of Paul A. Parisi, Director of Publication Services, and Irving Amron, Editor of Information Services — who indexed the book — is noted gratefully.

The Book, *Engineering and American Society, 1850-75*, (University Press of Kentucky) was a particularly useful reference, remarkably perceptive in its analysis of the motivation of the early American civil engineer. Its author, Dr. Raymond H. Merritt*, was kind enough to review this manuscript in detail, and his thoughtful and scholarly criticism was a major contribution.

Highly competent reviews were provided also by Past-President G. Brooks Earnest and Secretary Emeritus William N. Carey, both of whom have been deeply involved with the management of the Society for many years.

Acknowledgment is made to the Department of Civil and Coastal Engineering of the University of Florida — Dr. James H. Schaub, Head, and the office staff collectively — for their tolerance and cheerful cooperation. The staff of the Physics and Engineering Library of the University was ever ready to assist, often under trying conditions.

Grateful appreciation is extended to Professor Daniel H. Calhoun, author, and to the MIT Press, publisher, of the book *The American Civil Engineer — His Origins and Conflicts*, for their permission to use a portion of that title for this book.

As she has done so many times in the past 44 years, my dear wife has again indulged me in a professional adventure that has imposed heavily upon my family and homelife. This time she has not only graciously foregone much inconvenience and interference with our retirement activities, but she has also served as my reviewer and critic. I dedicate this book to her.

October, 1974 W.H.W.

*Director of Cultural and Technological Studies,
University of Wisconsin — Milwaukee

CHAPTER I

EMERGENCE OF AMERICAN ENGINEERING AS A PROFESSION

As was the case with technology itself, the emergence of the American Society of Civil Engineers as a significant influence in the development of the United States of America was a sporadic and sometimes frustrating process. The series of events resulting in the creation of the first national professional engineering organization in America covered a period of 28 years. It encompassed several false starts in the 1830's, a resumption of interest some years later leading to the formal organization of the Society in 1852, a second hiatus through the Civil War years, and the ultimate resurgence of ASCE in 1867 as a modest but viable identity of the new profession of engineering.

An overview of the state of the nation and its need for technology during this period reveals the importance of a professional relationship among those providing that technology. Considering the boundless number and magnitude of public works needed, however, and the thin ranks of those in any way capable of building them, it is not surprising that organizational endeavors had to be subordinated.

The first half of the 19th century was a crucial period in the youth of America. Migration to the west from the eastern seaboard extended to the Mississippi by 1820. This development was largely rural in character, with 90 percent of the national population of 10 million living in settlements of less than 2,500. The fertile western farms were devoted to food crops, while agriculture in the South was dedicated to the production of cotton. Manufacturing was making a modest beginning in New England and some larger western cities.

Facilities for communication and transport were sorely needed to integrate these regional elements into a national structure for trade and commerce. Fortunately, these services were to be forthcoming through many bold and imaginative enterprises in which engineers played important roles.

The earliest non-military engineering art was applied mainly in the domain of transportation. The 1787 law establishing the Northwest Territory decreed that 2 percent of revenues from land sales would be allocated to construction of the National Road or "Cumberland Pike" from Washington to the Ohio River. Begun in 1806, the first 130 miles to Wheeling, Virginia (now West Virginia), were opened in 1820, and the road was eventually extended to Columbus and St. Louis. The project was significant in that it represented the first use of federal funds for major civil works construction.

The 9,000 miles of rock-and gravel-surfaced roads in 1820 grew to 88,000 miles in the next four decades.

Equally significant was the Canal Era, even though it was to be shortlived. Beginning with the South Hadley and Middlesex Canals in Massachusetts (1793 and 1804), the period was highlighted by the greatly successful Erie Canal in New York (1825) and followed by the Union Canal in Pennsylvania (1829), the Morris Canal in New Jersey (1831), and the Chesapeake and Ohio Canal (1851). By mid-century the steam engine was in general use for water and land transportation. About 3,000 miles of canals were in operation in 1840, by which time the waterway was rapidly yielding its leading role in transportation to the railroad.

The South Carolina Railroad was the first in the United States. It was soon followed by the Baltimore and Ohio; the Pennsylvania; the Pittsburgh, Fort Wayne and Chicago; the Rock Island; and the Erie railroads. It is noteworthy that the

Erie was to be a training ground for seven engineers who were later to become presidents of ASCE.*

By 1852, the experimental and early development phase of the American railroad industry was accomplished, with about 9,000 miles of operating trackage. This increased ten-fold in the next thirty years.

The railroad brought new emphasis to the art of bridge building, which until the early 19th century was largely confined to timber and stone masonry. The middle years of the century, however, brought a renaissance in bridge building, both in the transition from wood to cast and wrought iron as well as in the advancement from the early empirical trussed arches to the proprietary trusses of Howe, Pratt, Warren, Whipple, Bollman, Fink and others. The systematic analysis of stress in the truss members and early efforts toward rational design soon followed. The first metal truss bridge (cast and wrought iron) was designed and built on the Reading Railroad in 1845. By mid-century wrought iron was in general use as a structural material, and the arrival of the Bessemer converter in 1856 introduced the Steel Age.

In 1850, there were 83 municipal water supplies in the United States, thanks largely to pioneering efforts in Boston, New York City and Bethlehem, Pennsylvania. The supplies provided untreated water from wells, spring and surface sources. The earliest American work on water-treatment was initiated at St. Louis in 1866, when slow sand filters were utilized.

Although some sewer lines in Boston were in operation by 1800, the earliest public system of sewers was built in Chicago in 1855 under the direction of City Engineer E.S. Chesbrough. By 1860, public sewer systems were serving about a million people in ten of the largest cities. Sewage was dis-

*James P. Kirkwood (1868), William J. McAlpine (1869), Horatio Allen (1872), Julius W. Adams (1875), William E. Worthen (1887), Octave Chanute (1891) and George S. Morison (1895).

charged without treatment to the nearest watercourse.

What manner of men were these "civil engineers" who were assuming such a significant leadership role in planning, building and managing the bold public works projects that were shaping America in the mid-19th century? The term "Civil Engineer" was first adopted by John Smeaton, builder of early roads, structures and canals in England, who about 1782 signed himself under that title in presenting expert testimony in the courts.

In America the Continental Congress legislated the appointment of engineer officers in the army, most of these positions being filled by Europeans. Although this "Corps of Engineers" was disbanded at the end of the Revolutionary War, it was reconstituted in 1794 and has prevailed ever since. The United States Military Academy was created by act of Congress in 1802, with the intent that it would function as an arm of the Corps of Engineers. This administration prevailed for more than 60 years. In 1821 the Congress enacted legislation directing the Corps of Engineers to make surveys of major roads and canals, and prescribed that this work be performed by office and field parties under the direction of a supervisory board, all to be jointly constituted of "engineer officers" and "civil engineers." This is certainly one of the earlier distinctions in America between the military engineer and the civilian or "civil" engineer.

The importance of the USMA as a source of trained engineers is attested to by the fact that of 572 graduates in the period 1802-29, 49 had been appointed chief or resident engineers on railroad or canal projects by 1840.

Thus, most of the formally educated engineers of this period were graduates of West Point, the remainder having supplemented their general academic training by scientific study and field experience. A large segment of engineers in the mid-1800's, however, had little or no formal education, acquiring their technical knowledge through self-study and apprenticeship, often as axmen or rodmen in surveying parties. The roads, canals and railroads on which they worked served as their "universities."

Engineering education in America began in the United

States Military Academy, which produced its first graduate in 1802. The curriculum was greatly strengthened during the tenure of Sylvanus Thayer as Superintendent from 1817-33. The first civil engineering course outside of West Point was offered in 1821 by the American Literary, Scientific and Military Academy, later to be known as Lewis College and then, in 1834, renamed Norwich University. The first civil engineering degree was conferred by Rensselaer Polytechnic Institute in 1835. By mid-century engineering courses were being offered by Union College (1845), Harvard College (1846), and Yale College (1846). In the next twenty years about seventy institutions of higher learning initiated engineering programs.

The national census of 1850 counted only 512 civil engineers, of whom two-thirds were resident in the states of Massachusetts (68), New York (62), Ohio (59), Pennsylvania (55), Connecticut (46) and Wisconsin (44). Although few in numbers, they were an elite group in stature. Regardless of the pathway of their careers, the civil engineers of this era were supremely confident of their capabilities, and they jealously guarded their independence of professional decision and action. As champions of the public interest, they were outspoken in their criticism of questionable political and industrial management. Their performance earned them prestige, public respect and financial remuneration to a degree unexcelled by any other profession at the time.

America in the mid-19th century was a brash, lusty young nation, rich in natural resources, blessed with a virile blend of rugged frontiersmen and immigrant artisans in its population of 24 million, and fostering in its government structure unprecedented encouragement to personal initiative, innovation and private enterprise. The Singer sewing machine, Goodyear rubber, Colt firearms and the McCormick reaper — all to become giants of American industry — were among the foremost contributors to the London Exposition in 1851. The stage was well-set for the "Industrial Revolution," as this era was later to become known.

Abroad, engineering had attained professional status with the creation in 1818 of the British Institution of Civil Engineers, culminating organizational efforts going back at least to 1771. This movement established a pattern for the world to follow. National engineering organizations were formed in Holland (1843), Belgium (1847), Germany (1847) and France (1848).

The Society for Promoting the Improvements of Roads and Inland Navigation, reported to have been formed in 1789 by citizens of Pennsylvania, may be the earliest organization in America to be interested in public works engineering. Civil Engineer William Strickland visited England on behalf of this society in 1825. His report, published in 1826, described canals, locks, bridges, roads, tunnels, tramways, breakwaters, harbor works and industrial installations.

It was against this backdrop* that the leaders among America's professional builders began to think and talk about an organization that would provide a facility for their communication and concerted action in matters of technical and professional concern. The time could not have been more propitious!

Early Organizational Disappointments

When the Franklin Institute was established in Philadelphia in 1824 it was the professional home of many engineers, and for some years its journal carried much of the American and foreign engineering literature. The Institute welcomed such use of its facilities.

In a classic paper (*Transactions*, Vol. 24, 1891), Dr. J. Elfreth Watkins mentions the earliest effort in America toward the formation, in 1836, of a "National Society of Civil Engineers" by a group of engineers connected with the Charleston and Cincinnati Railroad. The venture "did not meet with the encouragement expected, and the society was short-lived."

Late in 1838, "a highly respectable meeting of members of the profession in Augusta, Georgia," resulted in a call for a

*An excellent review of the status of technology in this period will be found in the book, *Engineering in American Society: 1850-1875*, by Raymond H. Merritt (1969, University Press of Kentucky, Lexington, Kentucky).

convention of civil engineers in the United States to be held in Barnum's Hotel in Baltimore on February 11, 1839. The Baltimore convention was held, "forty gentlemen of the profession being present, from the States of Massachusetts, New York, New Jersey, Pennsylvania, Illinois, Maryland, Virginia, Missouri, North Carolina, Georgia and Louisiana." Benjamin H. Latrobe, Jr. was elected President of the convention, and John Frederick Houston was appointed Secretary.

The attendance of 40 practitioners at this convention is noteworthy, as this number must have represented at least 10 percent of all the civil engineers then resident in the United States.

An early action of the assembly was to adopt the following resolution:

> "Resolved, that the convention now proceed to the election of a committee of seventeen, to prepare and adopt a constitution and form a Society of Civil Engineers of the United States."

In accord with a decision that, "the different portions of the Union may be represented" in the committee, those elected to serve were:

Benjamin Wright	New York
William S. Campbell	Florida
Claude Crozet	Virginia
W. M. C. Fairfax	Virginia
C. B. Fisk	Maryland
Edward F. Gay	Pennsylvania
Walter Gwynn	North Carolina
J. B. Jervis	New York
Jonathan Knight	Maryland
Benjamin H. Latrobe, Jr.	Maryland
W. G. McNeill	South Carolina
Edward Miller	Pennsylvania
Moncure Robinson	Virginia
J. Edgar Thomson	Georgia
Isaac Trimble	Maryland
Sylvester Welsh	Kentucky
G. W. Whistler	Connecticut

The convention received a most generous offer from the Franklin Institute of Philadelphia which would have provided accommodations for headquarters, meetings and library, secretarial service, space for publication of papers in the Institute's Journal and even editorial staff.

Another resolution of the Baltimore convention authorized appointment of "a committee of five to draft an address to the Civil Engineers of the United States, and to superintend the publication of such portions of the proceedings of this convention as they may deem expedient." Appointees to this committee were C. B. Fisk, Isaac Trimble, J. B. Jervis, G. W. Whistler (later replaced by Edward Miller) and S. W. Roberts of Pennsylvania.

The Committee of Five promptly convened on March 20, 1839, to draft the statement for circulation to all civil engineers in the Union, in fulfillment of its charge. Several portions of the "address" are revealing of professional circumstances and attitudes of the time:

> ". . . Public works are now so extended in our country, and the mass of experimental knowledge to be gained from those in use is so great and so peculiarly applicable to our circumstances, that it is even more valuable to the American Engineer than what he can learn in Europe, where larger means have permitted greater expenditures. In this country, it is of paramount importance to obtain the greatest amount of useful effect at the smallest cost; and of attempts to attain this end, the Union now contains a multitude of instructive examples. Some have been eminently successful, and others less so; but of either kind, the student, or the more advanced Engineer, too often seeks in vain for any satisfactory written or printed description, and is unable to obtain anything more than vague, doubtful, and incorrect information. This evil can only be removed by the exertions of the Engineers themselves.
>
> "They are now established as a distinct class, and have long felt the want of such an association as that proposed, but it has hitherto been supposed that the proper time for its organization had not yet arrived.
>
> "The success that has attended the labours of the London Institution of Civil Engineers, its high standing and great

usefulness, prove that such societies may be of great public utility, when properly conducted, and are incentives to induce us to imitate so excellent an example.

"It is admitted, however, that a society in this country must differ somewhat in its plan of operations from the British Institution, which can readily give utterance to its opinions elicited after frequent and full discussion, since a large portion of its members during the winter, have their residences within the limits of London. Here, however, owing to the vast extent of territory over which are scattered the members of our profession, the usefulness of the Society must (for the present at least) depend more upon the facts and experience of its members, made known in written communications, than upon their opinions orally expressed in public discussions.

"The very fact that our improvements are so widely spread, that few, if any, members are able to give even the most important of them a personal examination, affords, perhaps, the strongest argument in favour of a society that shall, by a concert of action, bring the experience of the whole country within the reach of each member.

"The difficulty of meeting at any one point, caused by the time and expense required in traveling from distant portions of so extensive a country as the United States, is a serious obstacle, but it has been much diminished by the facilities afforded by the rail-roads already in use, which are among the valuable results of the labours of our Civil Engineers. . .

"The standing of the profession in our country is, fortunately, such that its importance need not be dilated upon; it is, therefore, the more necessary that every thing in the power of the members should be done to add to its respectability and increase its usefulness. We look forward to the formation of the Society as a valuable means of advancing these desirable ends.

"The Committee will close this address by a quotation from the inaugural address of the distinguished Thomas Telford, the first President of the London Institution, which appears to them peculiarly appropriate:

" 'In foreign countries similar establishments are instituted by government, and their members and proceedings are under its control, but here a different course being adopted, it becomes incumbent on each individual member to feel that the very existence and prosperity of the institution depends, in no small degree, on his personal conduct and exertions; and the merely mentioning the circumstance will, I am convinced, be sufficient to command the best efforts of the present and future members, always keeping in mind that talents and respectability are preferable to numbers, and that from too easy and promiscuous admissions, unavoidable and not unfrequent incurable inconveniences, perplex most societies.' "

When the Committee of Seventeen met on its appointed date of April 10, 1839, only four of its members attended — Messrs. Wright, Campbell, Fisk and Miller. This group proceeded to draft a constitution for "The American Society of Civil Engineers," to be "instituted for the collection and diffusion of professional knowledge, the advancement of mechanical philosophy, and the elevation of the character and standing of the Civil Engineers of the United States."

Although the Member grade was to be limited to persons "who are or have been engaged in the practice of a Civil Engineer," an Associate grade would have accommodated architects and "eminent Machinists, and others, whose pursuits constitute branches of Engineering, but who are not engineers by profession." Another interesting provision imposed a fine of $10 against any member who failed "to produce to the Society at least one unpublished communication in each year, or present a scientific book, map, plan or model, not already in the possession of the Society. . ."

The limited participation of members in the meeting of the Committee of Seventeen was a foreboding omen. Only four other members of the committee (Robinson, Welsh, Latrobe and Jervis) endorsed the draft constitution as produced by its four authors. The constitution was never submitted to another convention for adoption, and the offer of support made by the Franklin Institute went begging.

The Secretary of the Committee of Seventeen, Edward Miller, gave his explanation for the collapse of the movement in a communication published in the *American Railroad Journal* in February 1840. In his opinion, the Committee of Seventeen was too large for its purpose, in addition to "the facts that most of those appointed were ignorant of their appointment, several absolutely indifferent or hostile to the formation of any institution; and that many were unknown to each other and so scattered as to render a meeting difficult. . ." He noted further that "under these conditions there can hardly be a necessity for pointing out the local views, partialities and jealousies which influenced in some measure the result."

Mr. Miller also concluded that jealousy and discontent must result from the necessity of vesting the management in a few who must reside near the point at which the Society's hall is located. Four independent regional societies, each with its own headquarters, was his answer to this provincialism. Although Mr. Miller's recommendations were never followed, time has proven him to be a perceptive and astute analyst. In the years to come there were several manifestations of regional and intra-professional bias that had to be dealt with in various ways.

Mr. Miller's disappointment at the collapse of the 1839 organizational movement comes through clearly in the following letter that he addressed to the members of the Committee of Seventeen:

Philadelphia July 15, 1839

"Sir:

"I have the honour to inform you that the form of Constitution proposed for the Society of Civil Engineers by that portion of the Committee of Seventeen which met in Philadelphia on the 10th of April, agreeably to their appointment, is rejected.

"The votes are as follows:

Approving— Benjamin Wright; Wm. S. Campbell; Charles B. Fisk; and M. Robinson. Edw. F. Gay also approves of the Constitution but declines becoming a member of the Society under any circumstances.

Disapproving—W.M.C. Fairfax; Walter Gwynn; John B. Jervis; Jonathan Knight; B.H. Latrobe, Jr., W.G. McNeill; J. Edgar Thomson. C. Crozet expresses no opinion on the subject, but declines his appointment as one of the Committee of Seventeen.

"Isaac Trimble; Sylvester Welsh; and G.W. Whistler, have made no reply to my circular letter, and I am not acquainted with their views.

"From the tenor of the letters received from the different

members of the Committee, I have been convinced that a *National Society* on a broad and useful basis, can not be formed by gentlemen holding such discordant opinions, unless they will take the pains to meet together, and give the subject a fair discussion. I am also of opinion that a subject of such importance should not be decided by a meagre majority, and therefore (although still believing the Constitution to be a good one) I add my vote to the negatives, which will make the names against the proposed measure eight. Mr. Crozet having withdrawn from the Committee, the proposed Constitution is of course *rejected*, even though all the remaining gentlemen to be heard from should vote in its favor.

"The Adjournment of the Committee in April was "*sine die*," and those portions of Sections 15 and 16, which fix a meeting time, fall to the ground with the body of the instrument. Matters must consequently commence *de novo*, and additional correspondence will be necessary before a day of meeting can be selected.

"I have hitherto cheerfully attended to the duties which the Baltimore Convention, and subsequently the Committee imposed upon me, but must confess that I now see no prospect of a beneficial result to the profession. So I have no leisure for useless correspondence, I respectfully decline acting longer as the organ of the Committee, and will hand over the papers and correspondence, in my hands, to any one whom they may designate.

"Very respectfully,
Edw. Miller"

Thus did the progressive discussions of "the highly respectable meeting of members of the profession in Augusta, Georgia," come to a disappointing end.

The Watkins paper also refers to an abortive move in Albany, New York, in 1841 to establish "The American Institute of Civil Engineers." Then, in 1848, local activity resulted in the formation of the New York Institute of Civil Engineers and the Boston Society of Civil Engineers. The New York group published a Transactions volume, but ceased activity in 1850 for lack of continuing support. The mobility of its membership, making it difficult to attract sufficient attendance at meetings, was suggested as the likely cause for the demise of the Institute.

The Boston Society of Civil Engineers was to become the first permanent engineering organization in America. Several factors contributed to its success: The concentration of engineers in Massachusetts and Connecticut; the rigid admission requirements that limited membership to mature, less mobile and more prestigious individuals; the immediate establishment of a headquarters for meetings and a library; and the focus of activities upon social as well as professional functions. Whatever the reasons, the BSCE survived to serve its regional membership with distinction for 126 years. In 1974 it affiliated with ASCE (page 50).

James Laurie, one of the founders of the Boston Society of Civil Engineers, was later to become a Founder and the first President of ASCE.

The American Society of Civil Engineers and Architects — 1852

It all began with the following notice sent to practitioners of civil engineering in and near New York City:

"New York City, October 23rd, 1852

"Dear Sir:

"A meeting will be held in the office of the Croton Aqueduct Department, Rotunda Park, on Friday, November 5th, at 9 o'clock P.M. for the purpose of making arrangements for the organization, in the city of New York, of a Society of Civil Engineers and Architects.

"Should the object of the meeting obtain your approval, you are respectfully invited to attend.

Wm. H. Morell — Wm. H. Sidell
J.W. Adams — A.W. Craven
James Laurie — James P. Kirkwood

and others."

There were twelve respondents to this invitation, who gathered at the appointed time in the office of Alfred W. Craven, chief engineer of the Croton Aqueduct Department: Julius W. Adams, J.W. Ayres, Alfred W. Craven, Thomas A. Emmet, Edward Gardiner, Robert B. Gorsuch, James Laurie, W.H. Morell and W.H. Sidell of New York; G.S. Greene of Albany; S.S. Post of Oswego and W.H. Talcott of New Jersey.

With Mr. Craven presiding, the group resolved to incorporate an American Society of Civil Engineers and Architects, and designated Messrs. Laurie, Adams and Sidell to draft a Constitution. The committee retired and, surprisingly, returned after only a short deliberation to present a draft Constitution that was discussed, amended and adopted forthwith! This instrument remained in force until 1868.

James Laurie, First President of ASCE

The first officers to be elected were: James Laurie, President; Edward Gardiner and Charles W. Copeland, Vice Presidents; J.W. Adams, A.W. Craven, James P. Kirkwood, William H. Morell and William H. Sidell, Directors; and Robert B. Gorsuch, Secretary and Treasurer. Mr. Kirkwood was elected in absentia.

It appears logical that only the twelve individuals who gathered in Rotunda Park on November 5, 1852, merit the appellation of "Founder." This, at least was the vigorous contention of an unidentified member of the original twelve in a communication to the *Engineering News* of March 8, 1890. The letter reviewed accurately the organization of ASCE and observed:

> ". . . that no one could be considered a 'founder' of the Society, who, in view of the published call to engineers, had virtually declined, by neglecting either to wish us 'God-speed,' or to personally participate in the honors attending the leading of what was so generally regarded as a forlorn hope."

The editor of *Engineering News* admitted the validity of the premise, and apologized for past inaccuracies in referring to "founders" of the Society.

The announcement of the death of Colonel Julius Walker Adams in the December 16, 1899, issue of the *Engineering Record* indicates that his role in the formation of ASCE was noteworthy. Referring to Colonel Adams' service as editor of Appletons' *Mechanic's Magazine and Engineer's Journal*, the notice states:

> ". . . It was while he held this office that the famous Wacamahaga, a half-social, half-technical club, was formed by Mr. Adams, Henry R. Worthington, Charles W. Copeland, James O. Morse, James How, C.M. Guild and others, who later organized what is now the American Society of Civil Engineers."

It appears that the name "WA-CA-MA-HA-GA" was derived from the first initials of the names of some of its members. During 1852 the proceedings of the club were published by the *Mechanic's Magazine and Engineer's Journal.* Although the name of C.M. Guild is not found in any ASCE membership list, Colonel Adams was a Founder of the Society, and Copeland and Morse were Charter Members, How became a member in 1868 and Worthington in 1876. At least one other Founder of ASCE, Alfred W. Craven, was affiliated with the WA-CA-MA-HA-GA.

A further issue from this extraordinarily productive organization meeting was the adoption of an "address" to be distributed to all engineers and architects in the United States considered eligible for membership, to invite their affiliation. This interesting and informative statement is reproduced herewith:

Rooms of
The American Society of Civil Engineers and Architects,
New York, November 10th, 1852

"It has been for some time under advisement to form in the city of New York a Society of Civil Engineers, embracing also the kindred professions, with a view to their mutual improvement and the public good. Accordingly a meeting was called on the evening of the 5th of November of such professional men as were accessible and were supposed to be favorably inclined to such an association. The objects of the contemplated Society were laid before this meeting, as also the means by which it was proposed to accomplish the end in view. A Constitution was drawn up, discussed in detail, and finally, after much labor, approved and accepted as the basis for the government of the 'American Society of Civil Engineers and Architects.' Officers were elected in accordance with the provisions of the Constitution, and the Society was duly organized.

"It becomes our duty, in conformity to a resolution of said Society, to address such members of the respective professions as are known to us throughout the country, and, laying before them in brief the result of their deliberation, invite them to co-operate in a furtherance of the aim and objects of the Society, so far as they may be found to accord with their individual views. Such gentlemen only as receive this circular are eligible as members of the Society by a bare notice of their desire to become such and a compliance with the accompanying forms (on or before December 1st, 1852). All others will be elected by ballot in conformity with the requirements of the Constitution.

"It will be admitted that no point in our country offers the facilities for rendering such a society of practical benefit to the public as well as to its own members as the city of New York, and so long as this city retains its present commercial importance, so long it will be a center around and within which there will accumulate by a natural law practical commercial and professional information not elsewhere to be sought, and which, embodying the elements of successful enterprise, may be regarded as a fund of valuable data, equally valuable to the man of science and to the political economist. Much of this information is, from its nature, unwritten, but is entirely accessible, and, under the auspices of a society formed for that purpose, could be rendered available to such members as may desire the benefit of it.

"The Constitution of the Society declares that it has for its object:

"The professional improvement of its members, the encouragement of social intercourse among men of practical science, the advancement of engineering in its several branches, and of architecture, and the establishment of a central point of reference and union for its members.

"Among the means to be employed for attaining these ends shall be periodical meetings for the reading of professional papers, and the discussion of scientific subjects, the foundation of a library, the collection of maps, drawings and models, and the publication of such parts of the proceedings as may be deemed expedient.

"Civil, geological, mining and mechanical engineers, architects and other persons who, by profession, are interested in the advancement of science, shall be eligible as members.

"It is anticipated that the union of the three branches of civil and mechanical engineering and architecture will be attended by the happiest results, not with a view to the fusion of the three professions in one; but as in our country, from necessity, a member of one profession is liable at times to be called upon to practice to a greater or lesser extent in the others, and as the line between them cannot be drawn with precision, it behooves each, if possible, to be grounded in the practice of the others; and the bond of union established by membership in the same Society, seeking the same end, and by the same means, will, it is hoped, do much to quiet the unworthy jealousies which have tended to diminish the usefulness of distinct societies formed heretofore by the several professions for their individual benefit. . .

"It is scarcely necessary in this place, even did space permit, to enlarge on the many advantages which may be anticipated to flow from the judicious management of such a society as the one we have organized.

"In the formation of a constitution for our Society it has been the endeavor, so far as possible, to profit by the experience of similar societies which have preceded us in the same field, and we may be permitted to hope, that in framing that paper, we have succeeded in embodying those features from among them best calculated to ensure the result at which we aim.

"In reference to the revenue of the Society, it has been considered, that as members not residing in this city will, or may, under the arrangements which we propose, benefit by the professional material collected to nearly the same extent as resident members, the initiation fee might, with propriety, be fixed as the same for all members; while the yearly contribution or assessment, will, in the case of the resident members, be double that of the non-resident. The rates were accordingly fixed at $10 initiation fee for all members, and $10 yearly assessment on each member residing within 50 miles of the city of New York, and $5 yearly assessment on all members residing beyond that distance. . ."

By the time of the next meeting, on December 1, 1852, ten more members had responded to the invitation to join. At that meeting Bylaws were read, discussed, amended and adopted, and the formal organization of the American Society of Civil Engineers and Architects was consummated.

Following is the list of ten men who accepted membership and paid dues on or before the designated date of December 1, 1852, thus acquiring claim to the designation "Charter Member" together with the twelve founders:

John F. Winslow	McRee Swift
John A. Roebling	Robert A. Brown
E. French	H.A. Gardner
Archibald Kennedy	James B. Francis
I.C. Chesbrough	George M. Dexter

All of the 22 Charter members were engineers except Mr. Dexter, a Boston architect.

There is a significant difference between the 1839 organization movement and the successful effort in 1852. In the latter instance, the participants were few in number, all located in or near New York City, and probably well known to each other. The facility with which so much important business was accomplished at the November 5, 1852, meeting would indicate that there had been extensive preparation and strong leadership.

While these factors may have expedited the organization process, they also may have been at least partly responsible for the problems that plagued the Society in its early years. Some time was to elapse before the interest and support of a majority of the prominent civil engineers of the Union were to be forthcoming.

Action toward the objectives of the Society was initiated in the meeting of January 5, 1853, with eight members present, when a circular was authorized soliciting the donation of reports, maps, plans and other data from those in charge of public works functions. Under the title, "The Relief of Broadway," President James Laurie presented a proposal for placing railway tracks above the level of the street.

Eight meetings were held in 1853 and, while President Laurie set a fine example by his perfect attendance, the average total attendance was only six. In addition to the Broadway Elevated Street Railway proposal, topics of discussion included projects to improve Church and Mercer Streets in New York City, "The Use and Abuse of Iron as Applied to Building Purposes," and a presentation by J.W. Adams on the Lexington and Danville Railroad suspension bridge over the Kentucky River.

The first Annual Report of the Board of Direction, dated October 10, 1853, recorded a total membership of 55, receipts of $700 and expenditures of $115.12. The precarious state of the Society is documented by the following excerpt:

> "In view of the limited number of resident members, and the uncertainty whether it would be practicable to establish the Society on a basis that would be attended with beneficial

results, the policy of the Board during the past year has been to husband the resources of theSociety, to make no expenditures that could well be avoided, so that in case of failure the funds collected might be returned to the members.

"For these reasons no steps have been taken towards the formation of a library, or for renting rooms for the use of the Society.

"The Board regrets that they cannot speak in more flattering terms of the success of the Society, or with more confidence of its future prospects, but, believing that such an institution is much wanted, and that it rests with and is entirely within the power of those eligible as members to make it eminently useful, they recommend that the organization be kept up, and that renewed efforts be made to obtain additional members who are residents of the city or vicinity, and can attend the meetings. And, meanwhile, that the same course of policy with respect to the funds of the Society as indicated above be continued."

The first Annual Report also listed the roster of members:

HONORARY MEMBERS

John James Abert
Alexander Dallas Bache
Henry Burden
Dennis Hart Mahan
Moncure Robinson
Joseph G. Totten

CORRESPONDING MEMBER

T.S. Brown

MEMBERS

Julius Walker Adams
James Barnes
E.L. Berthoud
Robert N. Brown
I.C. Chesbrough
Stephen Chester
Charles W. Copeland
Alfred Wingate Craven
Matthias Oliver Davidson
George M. Dexter
Thomas Addis Emmet
Theodore D. Judah
Archibald W. Kennedy
James P. Kirkwood
James Laurie
Isaiah William Penn Lewis
William Jarvis McAlpine
John McRae
Thomas C. Meyer
J.F. Miller
D. Mitchel, Jr.
James E. Montgomery

James K. Ford
James Bicheno Francis
Edmund French
Edward Gardiner
Henry A. Gardner
Robert B. Gorsuch
H. Grassau
George Sears Greene
Daniel L. Harris
Waldo Higginson
George E. Hoffman
Josiah Hunt
M.B. Inches
William H. Morell
William W. Morris
James Otis Morse
Thomas S. O'Sullivan
William D. Picket
Simeon S. Post
John Augustus Roebling
William H. Sidell
Israel Smith, Jr.
McRee Swift
William H. Talcott
William Wallace
John F. Winslow

Average attendance of the six meetings in 1854 was even lower than in the year before. The Annual Report for that year showed but 54 members, despite the fact that dues for residents had been reduced from $10 to $5 and for non-residents from $5 to $3. There were only ten resident members.

In reducing the dues "to render membership less onerous, and with the hope that by so doing new members might be included to seek a connection with the Society," it was also noted that "the labors of the members more than their money is wanted to make the Society useful."

Among subjects discussed in the 1854 meetings were the replacement of an aqueduct on the Morris Canal, evaluation of materials used for conveyance of water, the comparative economy of inclined planes and steam locomotives on railroads, and "Ball's indestructible water pipe."

The struggle to arouse membership interest and participation continued into 1855 when the first formal paper, "Results of Some Experiments on the Strength of Cast Iron," was presented by W.H. Talcott. The record of another meeting noted that "a desultory conversation ensued, including in its range the marvelous execution of the 'Minie Ball' as shown on the Siege of Sebastopol." Other meeting topics included "Blasting Rocks by a New Process," comparison of English and American wire rope, and "Recent Inventions for Economizing Fuel in Generating Steam."

An interesting incident concerns James O. Morse, who was elected in December, 1854, to the combined offices of Secretary and Treasurer. On January 5, 1855, he resigned the secretaryship, effective with the appointment of his successor. His replacement did not take place until 1867, twelve years later!

By this time, it was evident that the Society was suffering from the lack of two important resources: A publication for its transactions and for communication with its non-resident members, and a permanent headquarters and meeting place. Heretofore all meetings had been held in the offices of the Croton Aqueduct Department opposite Rotunda Park in New York City.

The need for rooms to house offices, meetings, a library and museum was frequently the subject of discussion and committee study. On March 2, 1855, the following committee report was presented to the Board of Direction:

"The Committee to whom was referred at the last meeting of the Society the question of whether it is advisable, in the present circumstances of the Society, to rent a room for its business purposes, submit the following report:

" 'The desirableness of the Society possessing rooms of its own being on all hands admitted, the question is now as to the expediency of encroaching at this stage upon its limited means.

" 'We have no place now where country members can find the Society except once a month, and no place where any papers, reports, or maps, which we have collected, can be seen or consulted except at the monthly meetings.

" 'As country members may be frequently in the city between the monthly meetings, and can but rarely suit their visits to the time of these meetings, and as we believe that a sufficient room can be obtained at an expense not inconsistent with the means of the Society, we think that it would be well to secure such a room, and arrange to have it open daily to all the members, so that those who are unable to attend the monthly meetings of the Society can yet find and have access to any information on its files.

" 'To such a room we would by advertisement invite all persons having models of patented or proposed improvements to exhibit, to send them occasionally, for the inspection of such engineers or architects as might be in the city. We would endeavor to make the room in this way a point of interest for engineers, and for all inventors or others who have anything to communicate or explain to the profession.

" 'We are inclined to believe that unless some such step is taken by which those members who cannot attend our meetings may understand our disposition and desire to consult as well their convenience and profit, our receipts from country subscribers will fall off very sensibly.

" 'We suggest that the experiment be tried for one year, and to that end recommend that a committee be appointed to procure a room convenient, if possible, to the office of some member who may be willing to take supervision of it, and that a desk and other necessary furniture be procured, provided that the rent of the room do not exceed $250.00, and the cost of furnishing it do not exceed $150.00.' "

The Board considered the report in executive session, and the record shows only that "no motion was made, and the Society adjourned."

Concerning this action, Charles Warren Hunt, a former Secretary and earlier historian* of the Society, observes:

"It is to be regretted that no report of this discussion has been found, as it is difficult to understand why the experiment of giving to the Society a local habitation was not tried, it having been clearly demonstrated during its two years of life that without one success was impossible."

The next meeting of the Society took place twelve and a half years later. The Civil War (1861-65) was undoubtedly partly responsible for the duration of the hiatus. Certainly the complex tensions and emotions associated with the war would have militated against any movement toward national unity. It appears, however, that not more than a dozen of the members in 1855 were active in the war in a military capacity.

**Historical Sketch of the American Society of Civil Engineers*, by Charles Warren Hunt, 1897, American Society of Civil Engineers, New York, N.Y.

The war notwithstanding, Messrs. Laurie, Craven, McAlpine, Morse, Adams, Swift, Talcott and others found a wide assortment of engineering problems to engage their interest while the Society was inactive. It must be assumed that these leaders had been sorely discouraged by the early lack of interest in the Society. Obviously, any desire for professional association among engineers was not at a high order of priority at this time.

The length of the gap in ASCE activity may to some extent have been dependent upon the engagements occupying James Laurie in that period. After a three-year study of bridges for the State of New York, he spent two years in Nova Scotia in evaluating railroads and planning extensions. From 1860-66 he was chief engineer of the New Haven, Hartford and Springfield Railroad, during which time he designed and built an iron and lattice bridge to replace a wooden structure carrying a single railroad track — without interruption of traffic on the line! It is significant that steps were not taken to revive ASCE until James Laurie returned to New York City in 1867!

A New Beginning

Upon call by President James Laurie, a special meeting of the Society was held on October 2, 1867, in the office of C.W. Copeland, 171 Broadway. Others present in addition to Messrs. Laurie and Copeland were J.W. Adams, James K. Ford, W.J. McAlpine, James O. Morse (still Secretary and Treasurer), Israel Smith, McRee Swift and W.H. Talcott. Action to accept the minutes of the meeting was followed by unanimous endorsement of the objective of the special meeting, i.e., "to take such steps as might be necessary to resuscitate the Society." Messrs. Copeland, McAlpine and Morse were appointed a committee to prepare "a plan for the revival of meetings, said plan to be so arranged as not to call for the expenditure of more than $1,200 for the coming year." The meeting was adjourned to October 9.

In the absence of President Laurie, Vice President Copeland presided over the adjourned meeting, which was also attended by Messrs. Adams, McAlpine, Morse, Swift, Talcott and James How. The committee report reproduced below was received, its recommendations adopted, and the Board of Direction authorized to implement them:

"October 8th, 1867

"*To the American Society of Civil Engineers and Architects:*

"The Committee appointed at a meeting of the Society held October 2d, for the purpose of proposing a plan for the more permanent establishment of the Society, beg leave to report:

"It is assumed that whatever is to be done in the way of reviving and re-establishing the Society must be done mainly by the few members who were present at the last meeting, and who are residents of New York and the immediate vicinity.

"Without meaning to discuss the causes that led to the suspension for twelve years of all meetings of the Society, it may not yet be out of place to suggest that in the future we strive to avoid whatever we may have found in the past to have been detrimental to the life and prosperity of the association.

"It is, we think, true, that all societies similar to ours, that have been successful and grown to greatness, have had their beginning in a small way, but in those beginnings the social element has always been cultivated, and out of the frequent and pleasant meetings, in a social way, of a few men of kindred tastes and pursuits, have grown most, if not all, of the permanent associations of the world, that, like ours, are devoted to science and to art.

"A large majority of the future members of our Society will be non-residents of New York, but most, if not all, will at times visit here. As members of the Society they will wish to meet with some of its representatives here, and to know for themselves that the Society is a fixed institution.

"Heretofore we have not been able to offer any such welcome to our members, or to do anything to keep alive a feeling of interest in the Association.

"But the argument need not be further gone into. We will confine ourselves to the plan we have to propose.

"We have offered to us, on the corner of William and Cedar Streets, two blocks from Broadway, and two short blocks from

Wall Street, two rooms in the third story, directly over the rooms of the Chamber of Commerce. The building is elegant, its entrance and stairway are commodious, and there is altogether an air of quiet respectability about the place, in every way suited to our wants. Together with these advantages, there is one which may contribute a good deal to the fitness of the place for our wants. We allude to a private restaurant in the upper story of the building, kept by the janitor of the building (an aged old-time negro servant) for the use of the occupants of the building, and many of the gentlemen connected with the banks and insurance companies of the neighborhood.

"The rooms are connected by sliding doors. The larger one is about 20 x 14, the other about 14 x 10; both have grates, and the larger one has fixtures for Croton water. Gas outlets are provided suitable to our wants, but there are no gas fixtures.

"The owners will paint and clean the rooms, and intimated that they would give us a new handsome grate and mantel. The rent is, in our opinion, remarkably low, it being but $400 for one year. We can take the rooms for a term of years, or for one year.

"We recommend that said rooms be taken by the Society, that they be suitably fitted up, and that the next annual meeting be held there.

"To this end we recommend that a committee be appointed, charged with the duty of engaging the rooms, of furnishing the same at a cost not exceeding $600, and of preparing such additions to the By-Laws of the Society as may be deemed necessary to serve for the proper care and management of the rooms."

The tenuous thread by which the Society clung to life at the time of its renaissance is manifest from the following extract from the published memoir of Secretary-Treasurer James O. Morse:

"Mr. Morse will be remembered as the first Treasurer of our Society, who, after the war, was instrumental in its revival. The Society had languished for several years, in fact, was considered as having died out entirely—no meeting having been held for some time—when it was proposed, by a few of the old members, to endeavor to organize a new society; but the lack of funds was an obstacle to its immediate success—nothing considered to be available from the wreck of the old organization, which was regarded as having died intestate. It was then that Mr. Morse stepped forward and presented an account in detail, showing every dollar that the Society had received and expended, and allowing compound interest for the difference, showed a very respectable balance to the credit of the Society, and upon this the reorganization was based which has resulted so successfully."

The financial resources of the Society were modest but adequate. The cash balance held by the Treasurer in a savings bank had increased from $266.93 to $497.57 during the 1855-67 suspension of activity. In addition, Mr. Laurie presented a check for $558.25, representing the dividends at compound interest that he had collected as President on five shares of New York Central Railroad stock held by the Society. The total assets of the Society at its rebirth came to almost $1,600, more than enough to meet all needs.

The New York Central Railroad investment occupies considerable space in the record. A letter from Mr. Laurie to Mr. Morse on November 20, 1867, indicates that the entire funds of the Society were lost through imprudent speculation in 1854 or earlier, when it was decided that the principal portion of funds would thereafter be invested in stocks. The New York Central stock certificate, however, was somehow misplaced during the 1855-67 inactive period, and the Company withheld payment of dividends on that account. It was not until 1878 that the value of the original shares plus accumulated dividends, totalling $1,475.79, was recovered.

The Annual Meeting on November 6, 1867, marked a new era for the Society. It was the first to be held in the new headquarters in the Chamber of Commerce Building, 63 William St. Fifty-four new members were admitted at the next meeting, and the custom of regular fortnightly meetings was initiated. The collection of library material was begun. After the first issue of *Transactions* a few years later, there was no longer doubt concerning the stability and promising future of the new organization.

In his presidential address at the 1870 Annual Convention,

Alfred W. Craven reflects both the discouragement of the first years and the confidence that pervaded the regeneration of the Society:

"Our Society, as you all know, was first formally organized in the year 1852. At that time there was scarcely enough members within immediate vicinity of New York to supply incumbents for the offices alone. Still we met, and strove our best to nurse the infant in whose health and growth we all had so deep an interest. We met, as our early records show, to the number of seven, five, sometimes three on an evening, and every man present was an official. We could not afford to hire a room, and so our 'corporal's guard,' or rather our guard of corporals, met at night in one of the vacant rooms of the Croton Aqueduct Department in this city. There we may be said to have worked hard; for even if we did nothing else, to courageously and persistently attend such meetings was to work hard. The struggle lasted three years, until at last, in March, 1855 we were obliged to succumb. Our failure seemed to be so complete, that with some of us there was no hope of recuperation, and had there been any means of returning to absent members their respective portions of the money remaining in our hands, the majority of the active members would have counseled such return. Some however, still retained their courage, and urged the careful investment of our funds and the safe storage of our books and papers until hoped-for brighter days. The result has proved the value of that courage and the wisdom of that advice. Our syncope lasted twelve years; and when, resolved upon one more strong effort for life, we came together in October, 1867, we started with a fund which added materially to our strength and spirits. We took our present rooms, and, so far as our limited means permitted, we tried to make them attractive. We were no longer vagrants. We had at last established an actual habitation of our own; a point to which each member might come as to his own house, and at which he might meet those who, with him, were interested in common enterprise, and so we all felt renewed hope. That hope has, in my opinion, been justified by events; for, although we are still far behind what we all desire to be, and though some impatient hearts are still unsatisfied, we have, considering the character of our work, and the age of our organization, made what may fairly be called large strides. Look at the facts: At the time of regathering for this work, the members on our old register, known to be alive and thought to be willing to aid in the attempt to revive the Society, numbered in all only twenty-eight. Of those who were present in the city to speak positively to that point, there were but thirteen. This was in October 1867, not yet three years ago. Now we count upon our rolls 179 regular members and the number is constantly increasing."

At the organization meeting of the Society in 1852 it had been resolved to incorporate under the name "American Society of Civil Engineers and Architects." The incorporation procedure was somehow overlooked, so this name never acquired legal status. The American Institute of Architects having been organized in 1857, the resolution adopted by vote of 17 to 4 on March 4, 1868, to change the name to "American Society of Civil Engineers" was not surprising. Again, however, no immediate action was taken to incorporate, and it was not until April 1877 that the present name of the Society became legally registered.

No chronicle of the birth of the American Society of Civil Engineers would be complete without acknowledgment of the professional conviction, leadership, and steadfast perseverance of James Laurie. It was he who insisted that a national engineering society was necessary and feasible, despite the failure of previous attempts to organize. He was a leader in the discussion which lead to the calling of the November 5, 1852, meeting. On that occasion he was elected the first President of the Society, and he labored diligently to carry it through those difficult early years. Perhaps his greatest contribution was his direction of the revival effort in 1867, which finally brought success. The following resolution was adopted at the 1867 Annual Meeting; it will be earnestly and wholeheartedly endorsed by all members of the Society now and in the future:

"*Resolved*, that we tender our thanks to Mr. James Laurie for his faithful services as our President, for his efforts to re-establish and reorganize this Society on a basis which gives promise of a successful and useful continuance, and particularly for his care of our funds, to which we are greatly indebted for our present unencumbered and hopeful condition."

By 1875 the membership of the Society included 6 Honorary Members, 362 Members, 17 Associates, and 23 Juniors, a total of 408. About 70 percent of these were Non-Resident Members, residing beyond 50 miles of New York City. It was no longer true that ASCE was a "New York City gentleman's club," although such might have been an apt description in the beginning.

The professional accomplishments of its leadership and members also gave stature to the Society as a national entity. These men were dominant figures in the design, construction and management of the major railroads, canals, roads, municipal services and other public works projects of the day. Some of their important engagements are documented in Chapters V and VI; they are recounted in more detail in the memoirs of early members as published in *Transactions*.

* * *

Thus did the evolution of ASCE to a state of soundness and permanence extend through a period of a quarter of a century. The process was a part of the maturing of the engineering profession, which by this time had made countless contributions to the development of America. If it should appear from this record that a profession founded upon decisive action and efficiency should have been able to organize itself with less difficulty, let it be remembered that these men were so preoccupied with building a nation that they were unable to give adequate attention to their own desires and needs. This is a manifestation of the service ethic—the hallmark of the true profession.

CHAPTER II

THE MANAGEMENT AND RESOURCES OF ASCE

Before exploring historically the growth, development, activities, services and accomplishments of America's first national engineering society, it will be advantageous first to review the functional mechanism and resources available for its operations. This will enhance understanding of the sometimes slow and ponderous procedures, and the complex administrative involvements that were not uncommon. It must also be remembered that the manpower and fiscal resources of the Society—almost entirely contributed by its members — were indeed modest in consideration of its professional goals.

The management structure of the Society was prescribed in the Constitution, supervised by the Board of Direction, and executed by the staff. For many years virtually all the actual work of the Society was performed by the membership. Only since about 1940 have an expanded staff and outside specialists been utilized in the realization of programs.

This evolutionary review will include the Constitution and subordinate regulations, the Board of Direction, the staff, the organizational structure (national, regional and local), the headquarters facility, the financial resource, and the most important resource of all—the membership.

The Constitution and Bylaws

The maintenance of the Constitution and Bylaws has engaged more time in the business meetings of the Society and its Board of Direction than any other single topic. The Constitution was amended 44 times between 1852 and 1974 (Appendix VIII). More than half of the amendments dealt only with membership requirements or admission procedures, or with procedures for nomination and election of officers.

The Constitution was extensively overhauled in 1891, 1921 and 1950, and significant amendments dealing with the composition and authority of the Board of Direction were also enacted in 1897, 1908, 1930, 1966 and 1970. The amendments establishing geographical representation districts (1894) and extending full membership privileges to Juniors (1947 and 1950) were also especially significant.

Discussion of proposed constitutional amendments was sometimes highly controversial, reaching a high in this respect in 1921 when ex-officio membership of the Secretary and Treasurer in the Board of Direction was discontinued. Differences of opinion on that issue reached such a degree of bitterness that a portion of the record was ordered expunged by the Board.

The first Bylaws were a combination of rules of order for meetings, and elementary operational regulations. Amendments were so prolific as to defy detailed reporting. With the 1921 redraft of the Constitution, the Bylaws became essential to the administration of the increasingly complex membership and election procedures and the growing number of committees. This trend continued apace, and in October 1951, a third level of regulation was created in the Rules of Policy and Procedure, which thereafter supplemented the Bylaws. By 1974, amendment of the Bylaws and the Rules was invariably a part of every Board of Direction meeting.

There is no apparent reason for the overwhelming concern of the Board of Direction with organizational and procedural detail, which has persisted throughout the life of the Society. Not only did this propensity draw upon the time and stamina of the Board members, but it also sometimes diverted attention from business more relevant to the objectives of the Society.

The Board of Direction

The changing composition of the Board of Direction is shown in the tabulation below. From 1891 to 1897, when all Past-Presidents served, the size of the Board varied from 35 to 39 members.

COMPOSITION OF BOARD OF DIRECTION

Year	*President*	*President Elect*	*Past President*	*Vice President*	*Secretary*	*Treasurer*	*Directors*	*Total*
1852	1	—	—	2	1	—	5	9
1891	1	—	12*	4	1	1	18	37
1897	1	—	5	4	1	1	18	30
1921	1	—	2	4	—	—	18	25
1930	1	—	2	4	—	—	19	26
1966	1	—	1	4	—	—	19	25
1970	1	1	1	4	—	—	21	28

*All Past-Presidents were members of the Board of Direction.

Initially, the authority of the Board was limited to "general care of the affairs of the Society" and management of its funds. All major decisions were made by vote of the membership, either in general business meeting or by letter ballot. Special committees could be authorized only by majority vote of the Society. Membership admissions were subject to letter ballot of the Society, with three negative votes excluding.

Not until 1908 was the Board empowered to elect members and to transfer them in grade. Special committees could be appointed by the Board, however, only when authorized by a general business meeting. Full authority to appoint special committees was not forthcoming until 1921.

In 1970 the Board of Direction was delegated complete management power when the schedules for annual dues and entrance fees were transferred from the Constitution—which could be amended only by membership referendum—to the Bylaws, which could be amended by Board action alone. This action gave the Board complete management authority over the financial affairs of the Society.

In 1964 about 29% of the ASCE membership were engaged in the area of private practice, 40% in public practice, 7% in education and 24% in all other areas. Representation in the Board of Direction, as summarized below, has not followed this pattern, although it has been reasonably compatible with the geographical distribution of the membership:

AREA OF PRACTICE OF BOARD MEMBERS

	Year				
	1875	*1900*	*1925*	*1950*	*1974*
Private practice	5	11	8	12	10
Government service	3	8	8	5	9
Railroad executive	2	7	3	1	—
Other industry	—	1	6	4	2
Education	—	1	2	3	7

The field of private practice has always been strongly represented in the Board of Direction, and education is gradually assuming a stronger role in the management of the Society. The declining role of the railroad executive is interesting.

The public-practice sector has consistently been underrepresented, probably because working conditions in government are not always conducive to the undertaking of outside activity. More liberal policies on time for professional activities, travel reimbursement for attendance of meetings, and recognition of professional development must be forthcoming for engineers in public agencies in order for this imbalance to be corrected.

Awareness of the importance of the leadership of the Society has been manifest in the ever-present concern with the procedures for nominating and electing officers. The President, Vice Presidents and Directors served one-year terms from 1852 to 1891; thereafter, the President served for one year, the Vice Presidents for two years and the Directors for three years. The Secretary and Treasurer were elected annually by the membership until 1894, and after that by the Board; both were members of the Board until 1921.

From 1852 to 1891 all officers were elected by the membership at the Annual Meeting, a small Nominating Committee being provided in 1878 to expedite the process. The new Constitution in 1891 called for a 19-member Nominating Committee with election by letter ballot of all Corporate Members. This system prevailed until 1921. When the Districts were first created in 1895 it was specified that the Board would include a Vice President and six Directors representing District 1 (the New York City area).

In 1921 the Nominating Committee was discontinued and all officers were nominated by a two-ballot system, but six years later a Nominating Committee comprising the Directors and Past-Presidents was authorized to designate the "Official Nominee" for the office of President. A single nomination ballot was adopted in 1950 to determine official nominees for the offices of Vice President and Director.

A significant change was the provision made in 1966 for a President-elect, to advance automatically into the presidency after a year in the post. This insured familiarity with current Society affairs by the President as he assumed office.

The nomination of Vice Presidents and Directors by mail ballot of the membership, never a popular process, became increasingly awkward and expensive as the Society grew. These procedures were under review in 1974, and return to one or more nominating committees appeared to be imminent.

Because the Society was organized in New York City, it is not surprising that there was sensitivity to domination by "New York" even as ASCE grew nationally. Indeed, there was strong justification for such concern in the constitutional requirement as late as 1921 that one Vice President and six of the eighteen Directors in the Board of Direction be elected from District 1. Since 1921 the number of Directors has been based upon membership distribution, and the representation from District 1 has gradually diminished to two of the nineteen

Directors in 1974. Nevertheless, when the fund-raising campaign for the United Engineering Center was under way in 1958, there was still some feeling against the "ghost" of the New York hierarchy.

Actually, while the influence of New York City in the Board of Direction was declining, a new concentration of power was quietly growing in California. In 1974, the four California Sections were included in District 11, which was allocated four of the nineteen Directors in the Board.

A roster of the Presidents of the Society is given in Appendix I, and a listing of all officers with their terms of service is provided in Appendix II.

The Headquarters Facility

The first home of ASCE was a two-room suite in the Chamber of Commerce Building, 63 William St., New York City, where the first Annual Meeting was held on November 6, 1867. According to the October 8, 1867, report:

> "The building is elegant, its entrance and stairway commodious, and there is altogether an air of quiet respectability about the place, in every way suited to our wants."

An additional room was taken four years later, but the space was outgrown by 1875, when new rooms were taken at 4 East 23rd St.

In 1877 a dwelling house at 104 East 20th St. was rented, but proved unsatisfactory in several respects. On January 1, 1881, a circular letter to the membership stated that the Society "needs a more permanent home, properly arranged for its use, where the business of the Society and that of its non-resident members on a visit to New York can be carried on." Contributions to a Building Fund were invited, the goal being $25,000 to $30,000.

In April 1881, there was an opportunity to purchase a desirable property at 127 East 23rd St. for $30,000, with a $5,000 down-payment. This sum was raised by $500 advances on the part of ten members, to be repaid from subsequent subscriptions to the Building Fund. This house accommodated the library and operations of the Society until 1897, by which time it had been improved to a value of $50,000. It was here that the American Institute of Electrical Engineers came into being on May 13, 1884.

Growth of the Society brought on an ambitious project in 1895, when two lots were purchased at 218-220 West 57th St. for $80,000. With a fine new building costing $90,000, it was necessary to raise $60,000 by Building Fund subscriptions to supplement available resources and a $60,000 mortgage to cover the total cost of $170,000.

In compliance with action requiring ". . . that the construction and architecture of the new Society House be entrusted to Members of the American Society of Civil Engineers and to none others," plans were prepared by Joseph M. Wilson, ASCE Vice President (1894-95).

At the 1896 Annual Meeting it was decided, however, to hold an architectural competition, open to a selected list of architects who were not members of ASCE. One of these, C.L.W. Eidlitz, was the successful competitor.

The ground floor of the new "House" comprised a large foyer, three office rooms and a spacious lounge for informal and social gatherings. The second floor featured a Reading Room and 500-seat auditorium. The third floor was devoted entirely to office space and the fourth to library storage for 150,000 volumes. The facade of the building was attractive and gracious in style, and years later it was to be a 57th St. landmark.

The building was formally opened on November 24, 1897. Additional land was purchased and a 50 percent expansion of the building was constructed in 1905, bringing the total investment in the property to $360,000.

From the beginning there was disappointment in the acoustics of the new auditorium; this was to be the subject of much subsequent discussion. A special committee in 1900 recommended carpeting, certain wall hangings and other treatment.

The problem was still unsolved thirteen years later, when a long-suffering member made a stirring appeal for relief:

> ". . . I think it is a reproach to a Society as scientific as ours . . . that it continue to maintain an auditorium with such uncommonly poor acoustic properties . . .
>
> "I remember the first meeting we had in this room. . . It was soon perceived that members were changing their seats to hear what was being said . . . and somebody asked whether something could not be done to improve these acoustic defects. Some member suggested that wire might be hung across the room, and one man misunderstood the meaning of the suggestion and approved it on the theory that it was a motion to hang the architect!"

This plea elicited action to "employ the best talent available" to remedy the situation. And the architect escaped unscathed!

57th Street Headquarters

While ASCE was in the midst of this ambitious building project the following communication aroused great interest in the engineering profession:

"2 East 91st Street
New York, February 14, 1903

"Gentlemen of the
American Society of Civil Engineers
American Society of Mechanical Engineers
American Institute of Mining Engineers
American Institute of Electrical Engineers
and the Engineers' Club:

"It will give me great pleasure to give, say, one million dollars to erect a suitable Union Building for you all, as the same may be needed.

"With best wishes,

"Truly yours,
"Andrew Carnegie"

ASCE withdrew from co-trusteeship of the Carnegie gift in accord with the mandate of a membership referendum, the vote being 1,139 to 662 against participation. Nevertheless, ASME, AIME and AIEE proceeded to form the United Engineering Society to administer conversion of the Carnegie gift into the Engineering Societies Building at 33 West 39th St., New York City (page 308).

Although ASCE, unlike AIME, ASME and AIEE, was already housed in a handsome, new, well-financed headquarters, it might have opted to join with the other societies in accepting the Carnegie gift had it not been for an apparent misunderstanding. From the extensive discussions on the matter, it seems that some members thought that a merger of the societies was intended rather than a cooperative venture to own and manage a common building. This engendered concern for the identity, autonomy and high membership standards of ASCE. There was particular reluctance to accept affiliation with the New York Engineers' Club, a private social organization that was included in the Carnegie offer. Such ASCE notables as Octave Chanute, Oberlin Smith, Rudolph Hering and Charles Macdonald favored ASCE participation in the joint building project, but the proposal may not have been clearly understood by the voting membership.

Engineering Societies Building, 33 West 39th St.

By 1914, the Engineering Societies Building had been cleared of debt, but it was felt that the venture was incomplete without the participation of the oldest of the national societies. In 1915, ASCE was offered full status as a Founder Society for $250,000, of which $225,000 would cover the cost of adding two floors to the Engineering Societies Building for ASCE occupancy.

This time the mandate of the ASCE membership, by a vote of 2,500 to 300, was in favor of ASCE affiliation with United Engineering Societies—a strong endorsement of the furtherance of engineering unity that was implicit in the move. The height of the Engineering Societies Building was raised by two floors—it being necessary to carry the load to foundation by an independent structure—and ASCE became an occupant and the fourth Founder with appropriate pomp and fanfare on December 17, 1917. The actual cost came to $262,500.

The 57th St. property was leased after it was vacated by the Society, and became a highly profitable investment. It was sold in 1966 for $850,000.

**United Engineering Center,
345 East 47th St.**

By 1928, some facilities of the Engineering Societies Building were outgrown, and by 1950 active study of larger quarters was under way. This led in 1957 to the acquisition of land on First Avenue facing the United Nations complex, upon which a 20-story tower was to be erected at a cost of $12.8 million. The engineering societies were asked to raise $3,787,000 by contributions from their members, of which the ASCE quota was $800,000.

The fund-raising campaign proved to be an exciting experience in professional collaboration. Only AIChE, with a much smaller quota ($300,000) and its membership largely concentrated in a few very large corporations, reached its quota before ASCE. Under the slogan, "Let's Get the Job Done," the Society achieved that goal in just 33 months. Success was largely a result of efforts by the Local Sections, the first five to reach their apportioned quotas being the Kentucky, Lehigh Valley, Nashville, Cincinnati and Columbia Sections, in that order.

Simply by being the first Society to ask for the space, ASCE was assigned the top two floors of the new United Engineering Center, which were occupied on September 5, 1961. In 1974 growth of the staff required occupancy of a third floor by the Society. At that time some of the other societies occupying the Center were becoming restive because of high costs and a shortage of competent office personnel in New York City. A few of them vacated all or part of their space in the Center and moved to other locations, and this was a source of some concern to United Engineering Trustees (UET).

The United Engineering Center had proved to be an efficient and functional headquarters, despite the urban problems plaguing New York City. The location of the Center—in the prestigious United Nations complex—gave a measure of status that was valuable to the engineering profession. So long as the property was exempted from New York City taxes the economic factor would be favorable to the occupant societies. Unless there was to be some major complication it seemed likely, in 1974, that ASCE would support continuation of the Center in its present location.

The Library

Among the first actions after the founding of ASCE, in January 1853, was the issuance of a circular to "All men in charge of public works, asking for printed reports, maps, plans, etc., in order to start an Engineering Library in connection with the Society." It was not until the first headquarters were occupied in 1867, however, that this actually was initiated.

As funds were not available, all acquisitions were by donation. Particularly important were the gifts of extensive personal collections, such as those by William Y. Arthur, M. ASCE, in 1872 and by Past-President William J. McAlpine in 1873. By 1873, the library contained 3,433 items. A special committee was created at that time to plan and draft policy for the library.

The 1885 attempt to establish a joint library for ASCE,

AIME, ASME and AIEE was one of the earliest efforts toward intersociety cooperation. Early indications that a workable plan could be devised proved unfounded, and the effort was abandoned three years later.

By 1892 the ASCE library held 16,000 accessions, and the collection had grown to 22,000 by 1897, when the Society moved into its new 57th St. building.

When the library was merged into the Engineering Societies Library on October 1, 1916, there were more than 89,000 accessions, of which 77,000 were unduplicated in the ESL. The remaining 12,000 volumes were presented to the Cleveland Association of Members.

In the years to follow, the Engineering Societies Library was to acquire recognition as one of the outstanding engineering libraries in the world. Its search and translation services were available by mail across the nation. Funding depended, however, upon a per capita assessment against the constituent bodies of United Engineering Societies, later called UET. During the "Information Explosion" of the 1960's there arose a need for means of information storage and retrieval that exceeded the resources of the ESL, and some of the supporting societies evidenced reluctance to continue the membership assessment.

The Staff

Until 1872 the Society staff comprised the Secretary alone, who was elected by the membership and served without compensation. Five incumbents during that period served but one year, the exception being James P. Morse, who resigned on January 5, 1855, but was not replaced until 1869, following the Society's 12-year hiatus. Mr. Morse also served as Treasurer for 21 years.

In June 1872, it was "the unanimous sense of the membership . . . that it is injurious to the dignity of the Society to longer accept the services of any gentleman acting as permanent Secretary without compensation." Gabriel Leverich became the first paid Secretary, being required "to devote all time necessary to the thorough development of the Society's interests" for an annual salary of $3,000. He was empowered to hire any necessary assistants, but they had to be paid from his specified salary. Duties of the Secretary included primary responsibility for correspondence, meetings, development of membership and professional functions, and service as Librarian.

This arrangement continued generally until 1892, except for relaxation in 1885 of the requirement that the Secretary pay his assistants from his $3,000 salary. The new Constitution in 1891 provided Secretary Francis Collingwood with an Assistant Secretary. The post was filled by Charles Warren Hunt, who also assumed the duties of Librarian. Mr. Hunt was later to serve the Society as Secretary for a quarter century, and became its first official historian.

The staff in 1892 consisted of the Secretary, Assistant Secretary, Auditor, Assistant Librarian, one clerk and two stenographers, a total of seven full-time employees in addition to a janitor and two office boys. Part-time service was rendered by the Treasurer and one stenographer. The total payroll was $12,722.

By 1900, Charles Warren Hunt had been Secretary for five years, and the staff had grown to nine full-time members in addition to janitors, office boys and part-time help. All salaries totalled $16,283 at this time.

When the Society moved into the Engineering Societies Building in 1916, before transferring the library to United Engineering Societies, there were 22 staff members exclusive of maintenance personnel. Operations of the staff were limited to administration, publications, meetings and the library.

The 1930 Functional Expansion Program revolutionized the range of Society activity, and for the first time staff members were assigned to full-time service in professional development functions. Regrettably, details are lacking in the records. Of the 56 members of the staff in 1940 at least four were engaged with public affairs, education, employment conditions and other professional activities.

Unusual emphasis was devoted in this period to decentralization of the staff function. In 1935 a staff post of Field Secretary was created "to bring to Local Sections a view of their new responsibilities and opportunities." The Washington (Eastern) Field Office was opened mainly as a wartime facility in 1941 (page 247), but continued into the postwar years. A West Coast Field Secretary in Los Angeles was authorized in 1944, and a Mid-West Regional Office in Chicago followed two years later. Apparently the benefit-cost ratios were unfavorable, however, as the Chicago and Los Angeles operations were closed out at the end of 1948. The Washington office was suspended in 1955, but was reestablished in 1972.

The evolution of the headquarters staff operation, with the category "Professional Development" including all technical, professional and research activities, is summarized below:

STAFF PERSONNEL

Year	*General Services*	*Publications*	*Professional Development*	*Total*
1852	1	—	—	1
1875	1	—	—	1
1900	7	2	—	9
1916	NR	NR	NR	16
1940	NR	NR	NR	56
1950	50	21	6	72
1974	52	38	19	109

The complete roster of executive officers of the Society with their periods of service is given in Appendix III. First in tenure was Charles Warren Hunt, who served with dedication for 25 years. He appears to have been an excellent manager, albeit occasionally controversial, and was highly regarded by the membership.

Though his period of service was brief, only from 1922-24, John H. Dunlap directed the secretariat with distinction until his tragic death as the result of a train accident following an Annual Convention (page 203).

George T. Seabury, second in tenure with 20 years of service, was an effective executive during a critical period of the

Society's development. He maintained a high order of rapport with both the leadership and the membership of the Society. He died a week before his retirement would have become effective on June 1, 1945, and among the many tributes to his memory was the observation that "Twenty fruitful years of the history of the American Society of Civil Engineers constitutes his monument."

In 1974, there were only 1.5 staff members per 1,000 members of the Society. This figure is well below the ratio prevailing in the other Founder Societies, despite the broader range of professional activitiy in ASCE, while maintaining a comparable level of membership services in other areas.

Organizational Structure

Concern for a form of organization that would be most effective in achieving national recognition for the civil engineering profession dates back to 1874. The President and Vice Presidents at that time studied the feasibility of a federal charter as a means of gaining such stature, and recommended that an application be made to Congress "for a national charter, upon the most favorable terms which may consistently be granted, keeping in view mutual advantages between the Society and the country." The Board of Direction was empowered to execute this directive, but no further action is reported.

In 1922, Past-President Clemens Herschel again urged that the Society seek a federal charter as a means of acquiring a higher order of public recognition. This movement also foundered on the basis of the following conclusion to a complicated opinion by legal counsel of the Society:

> ". . . Assuming that the Federal corporation is authorized to operate with the same powers as those possessed by the existing corporation, it will have no advantages of power, privileges or immunities over those possessed by the ASCE. No one can say whether Congress would or would not confer special or extraordinary privileges over the new corporation, but presumably it would not."

As early as 1873 there was restiveness on the part of Non-Resident Members concerning the concentration of Society meetings and management in New York City. Considerable attention was given to this situation at the Louisville Convention that year, and something of a crisis was averted by extension of the voting privilege to Non-Resident Members by letter ballot.

Ten years later a proposed constitutional amendment would have authorized the formation of Sections of the Society "for the advancement of a special Branch of Engineering." From the discussion it was apparent that the proposal had two purposes: (1) A means for organizing engineers engaged in various fields of technical specialty and (2) a facility for engineers in any geographical location to come together for the presentation and discussion of papers. In support of the first purpose was the possibility that a new "Society of River and Harbor Engineers" would be formed if ASCE did not act. There was much more interest, however, in the creation of local geographical units. In any event, the amendment failed of adoption.

At the 1885 Annual Convention Arthur M. Wellington, M. ASCE, presented a resolution calling for a task committee "to consider the matter of making such changes in the organization of the Society as may be desirable in connection with the subject of local engineering societies or clubs, and of sections or chapters of the Society; also to take into consideration the future policy of the Society in relation to the admission of branches of engineering not now generally represented . . ." Only the Western Society of Engineers (Chicago) and the Civil Engineers Club of Cleveland were mentioned as local bodies in discussion of this proposal. The committee was formed, but was discharged a year later when its members failed to reach agreement on a report.

The Board of Direction was then requested to invite comments from Society members and from local engineering clubs on the subject of national organization of the engineering profession. Only eleven members responded, in addition to the Engineers Club of St. Louis, the Boston Society of Civil Engineers and the Western Society of Engineers. The views expressed were so diverse that the Board concluded, in 1887, "that there is at present no desire for changes in the organization of the Society."

These efforts did, nevertheless, have an effect. As a result of the deliberations in St. Louis, sixteen ASCE members joined together on February 29, 1888, as the St. Louis Association of Members, for discussion of Society affairs and to debate the merits of local projects. This was the first local organization of ASCE members, although it was not officially a segment of the Society until 1914.

A move in 1890-91 to codify and update the ASCE Constitution included exhaustive efforts to develop local identity of the Society either by merging with existing local clubs and associations, or by creating new branches of present members. A compromise provision that would have permitted affiliation of the 27 local clubs then extant was defeated by a narrow margin.

Early in 1905 associations of ASCE members similar to that in St. Louis were formed in Kansas City (April 8) and in San Francisco (April 28). The trend was noted at the Annual Convention, and it was decided to recommend that:

> ". . . wherever possible, steps be taken calling attention of the membership to this matter (of local organizations), and that such Associations be formed."

A model set of Bylaws was adopted for the guidance of proposed local associations. Indications of interest were reported in several cities.

By 1915 there were 17 Associations of Members (Kansas City, San Francisco, Memphis, Colorado, Atlanta, Philadelphia, Seattle, Portland, Los Angeles, Texas, Spokane, Louisiana, Baltimore, St. Louis, Northwestern, Cleveland and San Diego, in the order of their formation). The presidents of fourteen of these associations were brought together at the 1915 Annual Meeting, for a conference that produced the recommendation that:

". . . the membership of the Society be divided geographically into Districts; every member residing in the territory covered by that District to become a part of the District Organization without payment of further dues. Each District to have a President, or Chairman, and Secretary, and existing Local Associations in that District to become Local Sections reporting to the management of the Society through the District Organization; each District to elect its own representative on the Board of Direction of the Society."

To explore further the ramifications of this revolutionary plan a general meeting was held on April 19, 1916, with representatives of AIME, ASME and AIEE on the subject, "What relations should exist between the National Engineering Societies and the local sections or associations of their members, and, in the interests of the Profession, what should be the attitude of both of the above to other local engineering societies or clubs?" The principal product of the occasion was a simple plan (not adopted) to advance the "Solidarity of the Engineering Profession" through a Joint National Conference Committee and a cadre of Joint Local Conference Committees, set forth by ASCE President Elmer L. Corthell, a truly cosmopolitan engineer.

A step in the direction of a regional structure had been taken in 1894, when seven Districts were designated simply as a means of achieving a measure of geographical representation in the Board of Direction. A change from seven to thirteen of these representation Districts in 1915 proved to be the only regional development actually to derive from the recommendation of the Conference of Local Association Presidents earlier that year.

In April 1917, the task committee studying the matter reported to the Board, in part as follows:

"The Local Associations of the American Society of Civil Engineers are in many cases responding to a sentiment that there should be more cooperation with other Engineering Organizations, more influence exerted in local communities, and more intimate relations established between the Society

> and its non-resident members. With this end in view, the plan of district representation was approved by the conference of Presidents of fourteen Associations in 1915, but, at the request of the Secretary, in February, for an expression of opinion, only four Associations responded, and only one of these advocated the districts plan . . . All the Associations which did offer suggestions, however, agreed that they wanted few restrictions and freedom to deal with local affairs.
>
> "Your Committee is convinced that the future of the American Society of Civil Engineers, as well as the welfare of its members, is in many ways dependent on the success of its Local Associations, and they should be encouraged to widen their field and strengthen themselves in all possible ways . . ."

These views prefaced an excellent statement of policies and rules providing for relationships of the local units to the national Society, to other engineering organizations and to the public. The report was adopted, and became the guideline for further organization of local Associations of Members, to a total of 23 in 1920.

The 1921 Annual Register (issued in February) for the first time listed "Local Sections" instead of Associations of Members, notwithstanding the fact that it was not until October of that year that a new Constitution was adopted giving formal and official status to Local Sections. From 30 in 1921 their number grew to 79 in 1974, with 109 Branches, 14 Associate Member Forums and 163 Technical Groups (Appendix VI).

A significant event at the local-organization level took place in April 1974, when the prestigious Boston Society of Civil Engineers, organized in 1848 (see page 13), merged with the Massachusetts Section of ASCE to form the Boston Society of Civil Engineers Section, ASCE. Such a consolidation had been under study for more than fifty years, and the manner of its accomplishment appropriately preserved the identity and traditions of the Boston society.

The original constitutional authorization permitted a section to be formed "in any locality," although organization on state boundaries with internal branches was favored from the standpoint of action in public affairs at the state and local levels. The Texas Section, with 13 Branches in 1974, was so organized. Some states, such as Ohio and New York, each with six Sections, found it necessary to set up State Councils in order to coordinate consideration and action on matters of statewide import.

A trend toward regional structuring reappeared in 1948 when the Sections in District 9 and in the Pacific Southwest area formed Councils primarily to expedite the nomination of Directors and Vice Presidents of the Society. Fourteen such Councils were in existence in 1974. The meetings of most Councils were limited to delegates of the Sections in consideration of Society affairs, but several of them also offered professional programs open to the membership.

There has never been a successful move nationally to formalize cooperation among the local units of the various engineering societies, although this has been accomplished independently in several states and localities. ASCE participation in a Founder Society movement toward this end was authorized in 1924, and a policy statement favoring coordination of Local Section functions was adopted in June of that year. In only three or four cities were the policy recommendations implemented with regard to consolidation of Local Section headquarters services, meetings, employment services and public affairs action.

In 1931 the American Institute of Electrical Engineers proposed that the four Founder Societies explore the feasibility of joint state engineering councils, which would represent the engineering profession "in legislation and other nontechnical matters affecting the status of the profession and the welfare of the public." ASCE promptly endorsed and supported the venture, but it failed to survive, apparently for lack of interest on the part of AIME and ASME.

The Illinois Engineering Council was very effective in coordinating engineering action at the state level for at least 35 years, and a state legislative council in California has also enjoyed success. An apparent need to exercise vigilance with regard to engineering registration laws in the mid-1960's led ASCE to bring the Consulting Engineers Council and the National Society of Professional Engineers together to set up tripartite "Joint Engineering Action Groups" at the state level. Several JEAGs were formed and at least one of them—in Tennessee—was still active in 1974.

Since 1959 formation of new Sections has been restricted to the United States (see page 328). Overseas Units existed in Australia and West Pakistan in 1974 (see page 329).

The creation of a Women's Auxiliary was considered by the Board of Direction in 1924 after this had been done by ASME and AIME. After study it was concluded that the establishment of such auxiliaries was best left to the option and execution of the sections. Specific authorization for the sections to set up subsidiary bodies was not formalized in the Bylaws until 1950.

In 1931, however, the Tacoma Section formed its "Wives of ASCE" club, and was followed by "ASCE Wives" groups in the Dallas and Fort Worth Branches of the Texas Section in 1938 and 1940. Other clubs were organized in Cleveland (1944), Seattle (1947) and Indiana (1950). In 1974 sixteen Wives Clubs were functioning.

FIGURE 1 — ASCE FINANCIAL

INCOME

Year	*Fees and Dues*	*Publications*	*Advertising*	*Other*	*Total*
1853	$700	—	—	—	$700
1875	7,858	$233	$605	$70	8,765
1900	42,888	2,394	2,221	4,037	51,540
1925	242,448	6,916	—	52,546	301,910
1950	463,111	66,580	148,254	93,046	770,991
1973	2,368,120	988,486	782,101	404,732	4,543,439

*Estimated

The shift toward sectionalization and localization of the Society, which began in the early 1900's, may have been the result of a similar trend on the national scene at the time. The effect was beneficial in every respect—in strengthening the Society, in enhancing its effectiveness in public affairs, and in affording public identification to the profession. This movement might have been delayed by the tightly reined administrative control from New York for the first half century, but once it gained momentum, ASCE lost its mantle of strict conservatism and became increasingly extroverted in outlook. By 1930 the Society had acquired an aura of restrained progressivism, along with the agressiveness to assume a leadership role in serving the public interest at both the national and local levels.

The Fiscal Resource

Owing largely to the astuteness of Secretary-Treasurer James O. Morse, ASCE actually gained financial strength during the comatose years of 1855-67. The sum of $1,592 was at hand when the Society resumed active status in October 1867. Through further careful management the net worth of

OPERATIONS — 1853 TO 1973

EXPENDITURES

General Services	*Public-ations*	*Technical Activities*	*Prof. Activities*	*Total*	*Year*
$115	—	—	—	$115	1853
6,216	$3,112	—	—	9,328	1875
27,476	17,863	—	—	45,339	1900
169,203	55,265	10,498*	7,841*	242,807	1925
373,942	313,188	20,987	31,624	739,736	1950
1,867,027	2,009,890	303,363	258,316	4,438,596	1973

the Society reached $30,554 in 1875, more than a third of which was invested.

Laxity in collection of dues led to difficulty in 1877-78, and payment of some bills was so delayed that the credit of the Society was jeopardized. A rigid policy on the dropping of members in arrears was adopted in 1879, and has prevailed ever since. In 1880 the Finance Committee noted "that this is the most prosperous year of the existence of this organization, and that we have evidently entered upon an era of increasing prosperity."

The progressive change in major income and expense operations of the Society is summarized in Fig. 1. The quarter-century reporting interval does not disclose the several periods of financial stress, as in 1877, 1920, 1932-39, 1947-51, 1970 and 1973.

By far the major source of revenue has been membership dues, and the several dues increases through the years (Fig. 8, page 113) have invariably been prefaced by strained or deficit budgets. Rapidly rising publication costs brought problems in 1920, when a $200,000 mortgage on the 57th St. property was outstanding.

The Great Depression of the 1930's severely tested the Society's fiscal resources. Major assets were the 57th St. property (a $30,000 mortgage still outstanding) and the library. Liquid assets were less than $10,000 when the membership peaked at 15,190 in 1931 and then began to decline. Even though more than a quarter-million dollars in dues were remitted for unemployed members, it was not until 1939 that the membership loss was recovered. Retrenchment of expenditures together with the $30,000 annual rental income from the 57th St. building saved the day.

Rising publication costs and increased activities prompted a proposal to increase dues in 1947 that was rejected by the membership, thus requiring strict economy and curtailment of services until 1951, when dues were raised. Greatly expanded programs, particularly in the professional area, were mainly responsible for the substantial dues increases in 1971 and 1973.

Until 1946 all publications were furnished to members without extra charge. In that year *Transactions* was made available by subscription; in 1962 this was done for the Directory, and in 1966 a subscription schedule was established for *Proceedings*. Except for *Transactions* the subscription rates for members did not cover costs; the Directory and *Proceedings* were still heavily subsidized from other sources of income in 1974.

When the regular publication of *Proceedings* was authorized in 1873 it was provided that "select advertisements, to be approved by the Committee on Library, be received and published therewith." Income from this source was nominal, however, approximating only about $2,000 annually when advertising in *Proceedings* was discontinued after the May 1903 issue.

When insufficient funds were available for publications in 1920 the Publications Committee "was authorized to increase the page size of *Proceedings* to 9 by 12 in. and to insert proper advertising matter at such time as in its opinion this change would be advisable." This authority, however, was never exercised.

Advertising income was extremely important to the economics of *Civil Engineering*, which was begun in 1930. It was not until 1955, however, when sale of advertising space was aggressively promoted, that such revenues were developed sufficiently to cover all production costs of the periodical. A move in 1971 by the Internal Revenue Service to tax income from *Civil Engineering* was successfully resisted.

After administration expense, publications have always taken the major share of ASCE funds. By 1925, however, a significant part of the fiscal resource was being devoted to professional development, and marked growth in this area continued thereafter.

From 1881 to about 1936 the reserve funds of the Society

were largely committed to real estate. Reserves were adequate during both World Wars to permit waiver of dues for members in military service, as was done for unemployed members during the depression. The surplus figures of $734,317 for 1950 and $2,285,829 for 1974 included sufficient liquid assets to meet any reasonable emergency.

The Membership as a Resource

As the secretariat of ASCE was essentially an administrative mechanism until 1930, practically all the real work of the Society before that time—and much since then—was performed by the membership. At first ad hoc committees were almost entirely relied upon to assume specific assignments, but gradually a more and more complex framework of interlocked standing committees was conceived. The trend was continuing in 1974, with about 400 technical committees and more than 100 administrative, professional and awards committees operating in addition to several task forces.

For many years members serving the Society gave not only of their time and energy but also received no reimbursement for their travel expenses. In 1899, the Board of Direction was requested "to consider the propriety of providing for the payment of the expenses of the members forming the Nominating Committee . . .," which was required to meet in New York. The Board responded negatively a year later referring to:

> ". . . the possible opening of the door to payment of other travelling expenses from the funds of the Society, for the attendance of Directors at monthly meetings and for the attendance of members of special and standing committees."

The breakthrough came in October 1909, with a resolution authorizing a mileage rate of 3 cents for attendance of Board members at any meeting of the Board other than those at the Annual Meeting or Annual Convention. The mileage cost that year came to $715. In January 1914, the mileage allowance was extended to include the Annual Meeting.

A few months later the Board decided to use "surplus funds . . . to enlarge the work of the Society by the encouragement of the work of the Special Committees, including research work." Thus, mileage reimbursement was opened to the Nominating Committee and others, effective January 1, 1914. The mileage rate was increased and supplemented by a per diem allowance in 1920, and adjusted upward several times later. In 1974, official travel for the Society was reimbursed by payment of actual transportation expense plus $35 per diem.

The progression of expenditures of the Society for official travel of elected officers and committee personnel is shown herewith:

OFFICIAL TRAVEL REIMBURSEMENT

Year	*Officers*	*Committees*	*Total*
1910	$ 1,040	—	$ 1,040
1920	12,340	$ 6,814	19,154
1940	13,600	18,400	32,000
1960	21,953	121,960	143,913
1973	31,422	377,961	409,383

It is doubtful that any other engineering society in the world has matched the generosity of ASCE with regard to travel reimbursement of committee personnel. While the policy made it possible for any member to participate in the work of the Society, there were also some significant disadvantages. Conduct of committee business by correspondence was discouraged. Also, the difficulty of prescribing rules for justification and location of committee meetings resulted in considerable preoccupation of both committee members and staff with travel reimbursement procedures.

In 1930, it was estimated that about 3,750 members were actively serving the Society as officers, committee members or contributors otherwise at the national and local levels. This figure had grown to more than 10,000 by 1970.

Estimates have been made of the demands on the time of those serving the Society in various capacities. A conscientious Local Section president might spend the equivalent of at least a month on ASCE duties during his year of office, a Director three months per year, and a President of the Society half time or more. These contributions are not reimbursed by the Society; they are proffered voluntarily, as a professional duty. The total value of such service by the 10,000 members at work in ASCE in 1974 would dwarf the annual budget.

The demands upon the time and energy of the President had become so great by 1973 that the desirability of compensating him for full-time service was given study. All travel expenses of the President and his wife were being reimbursed at this time. The proposal to provide an annual salary for the presidency was rejected by the Board of Direction in 1974.

Whether motivated by altruism or personal aggrandizement, or both, it is the working member of the organization who justifies its raison d'être. ASCE has been blessed by the caliber, generosity and loyalty of the talent available to perform its work.

It is not possible to document the dedication of the membership other than by reference to the results of their work. The stature of ASCE, its growth, its immense production of publications and its service to society are all measures of membership input. These are set forth in part in following chapters. Immeasurable benefits, though no less real, are the professional guidance, encouragement and inspiration that have been imparted to thousands of civil engineers as they have practiced their profession through the past century.

Standards for members, growth of the Society and other characteristics of ASCE membership are detailed in Chapter III.

The Society Badge

An appropriate badge to identify members of the Society was considered first in 1882 when Director George W. Dresser obtained some designs but apparently did not submit them. Two years later Captain O. E. Michaelis, M. ASCE, proposed at the Annual Meeting that a suitable badge be prepared "to be worn by members at meetings, and which may be worn by them at other times." His motion was prefaced by narration of the following incident while attending a Society excursion:

> "We were on a special train . . . running down to Niagara Falls, and the Secretary was furnished with a number of invitations to a collation there for distribution to the members of the Society. He called upon me to aid him in making the distribution, and we went through the train distributing the invitations. The membership of the Society is now so large that even our able Secretary is not always sure whom he is addressing, and it is very awkward in handing these tickets out to have a gentleman say: "Give me one of those tickets," and to have to say to him "Are you a member of the Society?" . . . It occurred to me that it might be appropriate on such occasions to have some modest badge which could be worn to distinguish members of the Society . . ."

The committee (appointed in January 1884 to design the badge), submitted a design comprising a shield with the name of the Society imprinted above the replica of a wye level, which was forthwith adopted by the Board. Production of the badge, illustrated in Fig. 2, was authorized for immediate sale to the membership.

Negative membership reaction came immediately, requiring issuance of a circular letter of explanation. While more than 300 of the new badges were purchased, 93 members requested that the question of the badge design be reopened and placed before the membership. The Board declined to act, however, for parliamentary reasons.

Dissatisfaction with the original badge surfaced again in 1892 in the form of an Annual Convention resolution asserting that the badge was "not expressive of the objects of the Society," and requesting the Board to consider new designs. A membership opinion survey in 1893 disclosed that a majority of respondents (303 to 214) favored a new design. One member expressed the opinion:

> "Our present Bowery pin . . . its supreme merit lies in its perfect and satisfactory ugliness. It is also redolent with the prolonged perspiration of hard work in the woods, and it seems to lift our noble profession only high enough to declare us competent to survey an old farm."

At the 1894 Annual Meeting the task committee reported that it had studied many designs, and now recommended adpotion of the simple shield carrying the name of the Society and its founding date, illustrated in Figs. 2 and 3. Thus, the present badge design was adopted by the membership in a general business meeting. A few months later the Board decreed:

> ". . . That each member receiving a badge he required to sign an agreement to return it to the Society, receiving a credit equal to its intrinsic value, should his connection with the Society cease from any cause other than death, and waiving his right to resignation until such return."

Only Members, Associates, and Fellows were eligible to wear the badge in 1894. In 1909 the blue badge was restricted to Members and Associate Members, and the same design in maroon was provided for Associates and Fellows.

Juniors were given a badge in 1923, in the form of a disk upon which the official emblem in blue was superimposed. In 1925 this badge, in maroon, was authorized for Student Chapter members.

Various further modifications were made in the use of the official emblem to distinguish Corporate Members from other grades. In 1974 the original blue shield was used by Fellows, Members and Affiliates; the Associate Members, badge had a white border surrounding a blue shield and the student badge the same design in maroon.

A miniature lapel pin, carrying the initials of the name of the Society in a monogram, was authorized in 1949 to accommodate changing clothing styles. This was replaced in 1963 by a reduction of the official badge to lapel pin size.

FIG. 2. — ORIGINAL AND PRESENT OFFICIAL ASCE BADGES

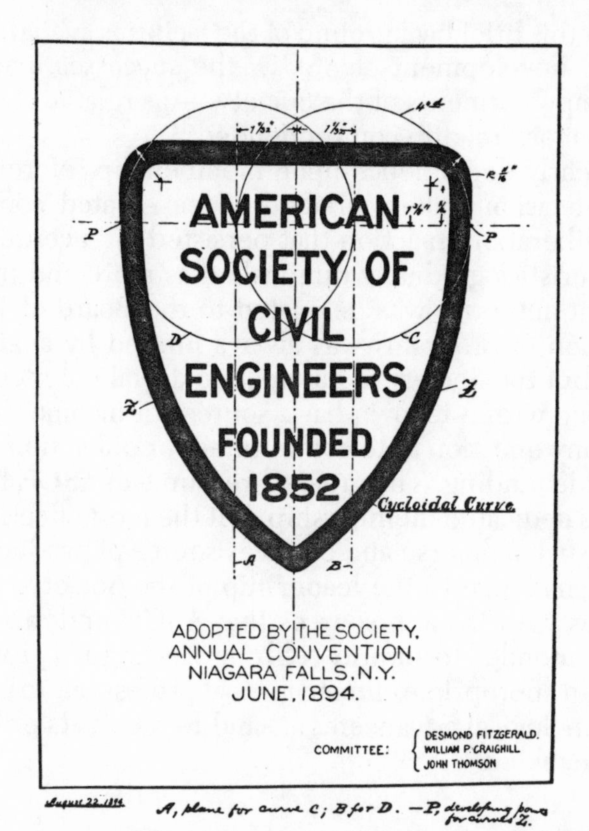

FIG. 3. — GEOMETRY OF PRESENT OFFICIAL BADGE

In the late 1960's special adaptations of the official shield were made in badges to identify the Past-Presidents and Honorary Members of the Society.

For many years the badges were numbered and inscribed on the back with the holder's name. This led to many unusual experiences in the recovery of badges that were lost. Outstanding among these was the finding of a badge still attached to the remnants of a vest that was taken from the stomach of a shark caught near Catalina Island in 1912. Subsequently, it was found that the badge had been overlooked when the member's wife gave a suit to a tramp.

* * *

With this brief background of the facilities available and the internal development of ASCE, the successes, frustrations, and disappointments of the Society — as related hereafter — will be more readily comprehended.

The early dependence upon membership referenda for all major decision-making created a deep-seated conservatism and deliberation in action that persisted for a century. These characteristics gradually diminished as more and more management authority was delegated to the Board of Direction. Execution of programs was always limited by availability of funds, but the Society maintained financial independence by operating within its own basic sources of income. There was never any question as to commitment or obligation by reason of outside funding. The greatest resource of ASCE has always been its dedicated membership, but the most effective utilization of such a diverse and complex source of productivity was ever a challenge to the leadership of the Society.

It was with these resources that ASCE undertook the activities intended to identify civil engineering as a profession, to create an appropriate climate for that profession, to provide for its technological advancement, and to satisfy its obligation to serve mankind.

CHAPTER III

DEVELOPMENT OF THE NEW PROFESSION

Any true profession must fulfill three responsibilities in setting itself apart from other vocations: The prescription and maintenance of appropriate standards for:

- the education of its members,
- the competence (qualifications) of its members, and
- the manner in which its members shall practice their profession.

The educational standards must primarily be effectively applicable to the institutions that prepare the members of the profession for their careers, but they must also impose an obligation upon the individual practitioner to remain apace of technological progress, through self-study and advanced education. The standards of competence are applicable through the membership requirements of the organizations representing the profession and also through the legal requirements of registration and certification boards. The standards of professional practice are implemented through codes of ethics or other rules delineating the principles under which the practitioner shall conduct himself in relationship to his clients, employers, professional associates, and the public.

Fulfillment of these responsibilities merely identifies a profession as such. Equally important obligations of a profession are the provision of a mechanism for collective service in the public interest, and the advancement of the welfare of its members so that they shall not be inhibited by economic, political, or social constraints in the "pursuit of their learned art."

These principles of professional obligation were not clearly identified in ASCE for many years. The constitutional purpose of the Society was generally interpreted, in the words of President-elect William J. McAlpine in 1868, as the advancement of "knowledge, science and practical skill among its members, by an interchange of thoughts, studies and experiences." As time passed, bringing proposals for action to meet the professional problems that arose, there was invariably extensive debate as to the propriety of non-technical functions within the terms of reference of the Society's objectives. Many noteworthy professional accomplishments were to be forthcoming, but they did not come easily.

Before recounting these developments in detail, it will be beneficial first to review the evolution of policy and structuring of the Society for non-technical activities.

The public interest as a professional responsibility of the civil engineer and of ASCE was first respected inadvertently in the guise of certain technical problems. Examples are the activities of ad hoc committees to study "inter-oceanic communication between the waters of the Atlantic and Pacific" (1870), the "means of averting bridge accidents" (1873), and the tragic failures of several dams (1874), as described in Chapter VI.

One such study of the need for rapid-transit facilities in New York City in 1874 evoked much controversy, and resulted in the first appraisal of ASCE policy on the role of the Society with regard to public issues of engineering import. The report "On Policy of the Society," adopted November 3, 1875, did bring about provision for the appointment of "special committees on engineering subjects." It was intended, however, that such committees would address themselves to the general aspect of an issue, without the delineation of specific engineering plans, as was the case with the New York City rapid-transit report.

It is unfortunate that this effort to clarify ASCE policy did not go farther in elucidating acceptance of responsibility for the stewardship of the standards of education, competence, professional practice, and public service that denote a profession. Had this been done, it would not have taken from 1873 to 1907 to initiate an education program, from 1893 to 1914 to adopt a code of ethics, and from 1897 to 1910 to assume a leadership role in engineering registration, as is documented later in this chapter. All these developments occurred as a result of pressures from the membership.

The voice of the membership swelled in volume and authority as the number of local associations of members, begun at St. Louis in 1888, grew to seventeen by 1915. The local groups sought a broadening of the objectives and more democratic administration of the Society, as well as official status and financial support for themselves.

A union-like activity in Chicago evolved into the American Association of Engineers (AAE) in 1915. Composed of a majority of civil engineers (many not in ASCE), the AAE aimed to improve the status and economic welfare of engineers, and its emergence was observed with some skepticism by the leadership of ASCE.

When the Cleveland Association of Members found considerable support for its proposal of a Joint Association Constitutional Conference, a movement toward progressive reform in the Society took shape in the authorization of a Committee on Development in June 1918. The preamble of the resolution creating the committee included these significant words:

> ". . .Sociological and economic conditions are in a state of flux and are leading to new alinements of the elements of society.
>
> "These conditions are affecting deeply the profession of engineering in its services to society, in its varied relationships to communities and nations, and in its internal organization.
>
> "A broad survey of the functions and purposes of the American Society of Civil Engineers is needed in order that an intelligent and effective readjustment may be accomplished so that the Society may take its proper place in the larger sphere of influence and usefulness now opening to the profession."

The committee was chaired by Onward Bates, M. ASCE, whose given name was most appropriate to this assignment.

The membership included six other members-at-large in addition to Chairman Bates plus one delegate from each of the 21 local associations then extant. A lucid and positively oriented precept by President Arthur N. Talbot left no uncertainties as to the committee's directive.

What may well be the most far-reaching committee report in the long life of ASCE was presented to the Board of Direction in October 1919 (*Proceedings*, 1919, page 889). Under the premise that ". . . the time has now come when this Society should adopt the principle of becoming an active national force in economic, industrial and civic affairs," the recommendations included the revolutionary proposals that:

- Government of the Society be centered in District organizations of the local associations.
- Membership grades be redefined.
- Activities of students and young engineers be encouraged.
- Greater emphasis be given to technical activities, research and standards.
- New emphasis be given to such functions in the "public affairs" sector as ethics; engineering education; licensing; arbitration and expert testimony; publicity; legislation; natural resource development; service to the community, state, and nation; patent law; coordination of government activities; and industrial affairs.
- Implementation of these aims to be accomplished by a comprehensive engineering unity organization encompassing the local, state and national levels.

A Joint Conference Committee, including ASCE, AIME, ASME and AIEE, studied an integrated approach to the resolution of similar objectives under study in all those societies.

A motion to submit the major elements of the report in a questionnaire to the membership was seconded by J. C. Ralston, M. ASCE, who revealed his impatience with parliamentary protocol with the comment:

". . . We see scores of young engineers trooping by the doors of our respective chapter, with sneers on their faces, rushing into the meritricious arms of the American Association of Engineers . . . We take the opportunity of urging that the American Society of Civil Engineers also respond to the vision of utility, and rise, if you please, above the inertia of pre-war inactivity to the mountain tops of post-war accomplishment."

The results of the mail ballot strongly favored the principles of broadened concern with "economic, industrial and civic affairs" and intersociety cooperation, but rejected the form of unity organization proposed. Also approved were the recommendations dealing with technical activities, local sections, and nomination and election of officers, but a suggested $5 to $10 increase in dues—if such should be necessary to finance the proposed expansion of internal and external activities —was narrowly unpopular.

The report of the Committee on Development was not implemented at once, but it was strongly manifest in the new ASCE Constitution adopted in 1921. The provisions reconstituting the Board of Direction, authorizing Local Sections and Student Chapters, revising nominating procedures, and expediting the appointment of special committees were all related to the report. Another important spinoff was the 1925 Committee on Aims and Activities, which provided the transition to the Functional Expansion Program of 1930. The progressiveness of the Committee on Development was evident throughout this chain reaction.

Designated as ". . . a systematized program designed to enhance the status of the civil engineer in the minds of his public," the Functional Expansion Plan brought into being a well-conceived operational facility for administration of an expanded range of Society activity.

Three new departments—Technical, Administrative and Professional—were created.

The Technical Department included the Technical Divisions and Research Committee; the Administrative Department encompassed the Membership Qualifications, Local

Sections, Student Chapters and Junior Member Committees; and the Professional Department comprised new committees dealing with engineering education, public relations, legislation, registration of engineers, fees, and salaries. Highly significant here is the Legislation Committee charged:

> ". . . to recommend to the Board of Direction such action as is deemed advisable with respect to legislation contemplated or in process, an engineering analysis of which will be calculated to be of help in the determination of a solution beneficial to the public . . ."

Finally, 78 years after its organization, ASCE had defined an acceptable professional development policy, and had set up the machinery to carry it out. The educational work begun in 1907, the registration activity initiated in 1910, and the ethical responsibilities assumed in 1914 were now to have systematic direction. Objectives and facilities at last were at hand to administer standards of professional education, professional competence, and professional practice, and to serve the public interest in a realistic way.

Timing was unfortunate, however, because of the onset in 1930 of the Great Depression, which produced complex economic and manpower stresses in ASCE as well as in the United States. Funds for new activities were limited, and major emphasis had to be given to the economic welfare of the membership. Nevertheless, the ensuing record will show that the 1930's were the turning point in ASCE professional development functions. In the very depths of the depression, 1934, a Committee on Aims and Activities was charged to study current programs" . . . with respect to human relations, welfare and public relations . . ."

The activity incident to creation of the National Society of Professional Engineers (NSPE) in 1934 was no doubt a stimulus to the emphasis given professional affairs in ASCE during this period. The AAE which had been a goading influence in 1919, was by this time declining as a significant factor in the profession.

Membership demand in the form of a petition, in January 1937, called for the creation of a Professional Activities Division to deal "... with subjects incident to professional practice and ethics, promoting understanding among engineers and between the profession and the public, and increasing the usefulness of the profession." Particular reference was made to "the humanities of the profession," apparently in recognition of social concerns.

The petition resulted in establishment of the Committee on Professional Objectives, with broad powers to coordinate the work of the whole spectrum of professional committees. It was expressly stated that the new committee would arrange its procedures "to strengthen and amplify the activities of the existing committees of the Society, and not to weaken or hamper such committees in any respect." This proved to be a wise admonition.

The committee undertook its charge in unusual fashion by staging a special program for the Annual Meeting in 1939. A panel of high-ranking spokesmen for the American Institute of Architects, the American Medical Association, the American Bar Association, and ASCE compared the progress being made to improve the social, economic, and professional status of the respective members of these organizations. In opening the program, Committee Chairman Enoch R. Needles (President, 1957) observed:

> "For 87 years, we have functioned almost exclusively as a technical society . . . This historic fact requires no explanation or apology . . . but if we look carefully into our Constitution, we note that one of the objectives is 'the professional advancement of its members.' And so today we are meeting the requirements of our Constitution in thinking of our professional improvement.
>
> "One of the most significant things about this meeting may lie in this start toward joint thinking about problems of the professional classes. Whether we are architects, physicians, lawyers, engineers, educators, or clergymen, we must acknowledge certain common fundamentals. A high degree of education for the professional man is taken for granted. His actions must comply with high ethical standards. His en-

deavors are largely devoted to service to his fellow man. He seeks respectability; and aspires to be deserving of respect. Finally, he desires a reasonably adequate compensation for his labors."

The Committee on Professional Objectives was continued until January 1947 and its tenure represents an extraordinarily productive era of professional development. Action programs were especially effective in the domain of economic welfare of the engineer, covering civil engineering salaries, collective bargaining, unionization, and consulting fees.

Although the depression was past, there was still some frustration with the difficulty of implementing the new professional programs with a limited budget and staff. As a result, there was considerable proccupation with "studies" and with bureaucratic manipulation of a growing number of committees. This tendency is shown by the following summary of special bodies dealing with professional policies and procedures from 1918 to 1974:

Year	*Activity*
1918-19	Committee on Development
1925-27	Committee on Aims and Activities
1929-30	Functional Expansion Program Committee
1934-38	Committee on Aims and Activities (ad hoc 1934-35)
1938-47	Committee on Professional Objectives
1947-51	Committee on Coordination of Professional Objectives
1952-53	Committee on Expanding Society Services
1955-57	Task Committee on Economic Advancement Objectives
1958-59	Task Committee on Administrative Procedure
1961	Legal Audit
1962-63	Task Committee on Society Objectives and Tax Status
1964-	Committee on Society Objectives, Planning and Organization (still extant in 1974)
1966	Committee Structure Reorganized
1968-71	Professional Activities Study Committee
1969	Task Committee on Social Action

The 1947 coordination effort reconstituted the Committee on Professional Objectives to comprise the Board Contact Members of the seven professional committees. Then, in 1953, this group of functions was designated as the Department of Conditions of Practice, to set it apart from the Department of Technical Activities. The study in 1955 was directed toward clarification of Society policy on the nature of programs intended to enhance salaries and employment conditions. The two-year study reaffirmed the collection and dissemination of guideline salary data and the recommendation of salary schedules in a "professional way" (without resort to trade-union tactics).

By 1958 the review, reappraisal, and reorganization of professional functions had become a continuous process, at times reaching such levels that the programs themselves were impeded by the preoccupation with bureaucratic detail. Committee operations and administration had become so involved that the Board created a Task Committee on Administrative Procedure ". . .to study and review the responsibilities and assignments of members of the Board . . . with a view toward a reduction of the routine functions that are now imposed on officers and directors." The seventeen recommendations urged elimination of some committees, reduction in size of others, and general simplification of procedures. Some of the recommendations were implemented, but the overall effect of this exercise was temporary, at best.

ASCE has always enjoyed the highest order of exemption from income taxes under the section of the Internal Revenue Code that applies to religious, scientific, charitable, educational, and literary organizations. A further requirement under this section, however, is that "No substantial part of. . . activities may be devoted to the carrying on of propaganda or otherwise attempting to influence legislation." Unlike the other Founder Societies, ASCE had never viewed this regulation as a constraint on its professional or public service activities, even though the Society had been reclassified for a short time in 1948 because of a misunderstanding on the part of the IRS. The 1961 "Legal Audit" was an in-depth analysis of all current programs of the Society by a leading Washington law firm, with respect to interpretations of the IRS Code.

The investigation produced an opinion that several current and planned policies and programs were subject to legal questions. After careful deliberation, the Board of Direction concluded that the Society would continue to act as necessary to serve the public interest and meet its responsibilities as a first-order professional organization. Should this course raise questions on the part of IRS under the terms of Section 501 (c) (3) of the Internal Revenue Code, a decision would then be made as to acceptance of the less favorable but less confining tax-exemption status under Section 501 (c) (6) as a "Business League."

Another issue from the Legal Audit was the standing Committee on Society Objectives, Planning and Organization, established in 1964 to provide continuous policy planning and guidance.

A structural reorganization of Society committees in 1966 replaced the Department of Conditions of Practice with a Department of Professional Activities, without significant change in constituent committee functions.

The 1968 ad hoc Professional Activities Study Committee was appointed—despite the existence of the standing Committee on Society Objectives, Planning and Organization—to evaluate in depth the professional activities of the Society and to bring recommendations to the Board of Direction. This task force triggered another wholesale reorganization in 1970, but more importantly, it also brought about a greatly increased commitment of Society resources to professional-development services and activities. As reorganized, the Professional Activities Committee was delegated to administer four new professional divisions, each encompassing its own executive committee and the subcommittees here listed:

ADMINISTRATIVE DIVISION
- Budget Committee on Professional Activities
- Professional Prizes and Awards
- Professional Publications
- Programs for Professional Sessions
- Public Relations

EDUCATION DIVISION
- Continuing Education
- Curricula and Education
- Career Guidance
- Educator/Practitioner Interchange
- Student Chapters
- Technician and Technology Education

MEMBER ACTIVITIES DIVISION
- Legislative Involvement
- Local Sections
- Minority Programs
- Public Affairs
- Younger Members

PROFESSIONAL PRACTICE DIVISION
- Employment Conditions
  - Employer/Engineer Relationships
  - Pensions
  - Salaries
  - Unionization
  - National Salary Guidelines
- Engineering Management
- Registration of Engineers
- Standards of Practice
  - Contingent Fees
  - Turnkey Contracts
  - Engagement of Professional Services

This imposing array of professional-development functions represents great progress from the halting efforts of the early years. The administrative process provided in this reorganization, however, was complex in structure. A proposal concerning education policy, for example, would be formulated in the Curriculum and Education Committee, after which review and clearance were required successively by the Education Division's Executive Committee and the Professional Activities Committee before presentation to the Board of Direction for action.

At the Annual Meeting in October 1969, a new dimension in professional concern was accented by a young black member, Ralph E. Spencer, who challenged the Society for its lack of involvement in social issues. He referred specifically to poverty and racial inequality as areas for action by ASCE. A Task Committee on Social Action was urged, to devise programs that would promote better living conditions, education, and employment opportunities for the socially underprivileged.

The Board of Direction responded to the admonition with a thorough evaluation of the prevailing attitudes, policies, and activities of the Society in relation to contemporary societal concerns. Considerable positive action was inventoried, both at the national and local levels, but a need for stronger and continuing recognition of social impact and humanitarian service in all facets of civil engineering practice was acknowledged. A direct issue from this evaluation was the Committee on Minority Programs, created in the 1970 reorganization to encourage and assist members of racial minorities to find successful careers in civil engineering.

While the recommendations of the Professional Activities Study Committee were being implemented, the Committee on Society Objectives, Planning and Organization was completing (1973) a three-year appraisal of "Goals of ASCE," as follows:

Goal 1—To Serve the Public: To provide a corps of civil engineers whose foremost dedication is that of rendering a public service.

Goal 2—To Advance the Profession: To improve the technical capability and professional dedication of civil engineers.

Goal 3—To Improve the Status of Civil Engineers: To continue to improve the professional stature and economic status of civil engineers.

Goal 4—To Improve ASCE Operations: To devote resources, organization, personnel, and operation in the efficient pursuit of these goals of serving the public, the profession, and the status of civil engineers.

FIG. 4. — MAJOR AND SUMMARY GOALS AND PRIORITIES OF ASCE — 1974

	MAJOR GOAL				
SUMMARY GOAL	*To serve the public*	*To advance the profession*	*To improve status*	*To improve ASCE*	*PRIORITY**
A) Civil Engineering Education	X	X	X		High
B) Continuing Education and Development of Civil Engineers	X	X	X		High
C) Public Information	X	X	X		High
D) Legislative Involvement	X	X	X		High
E) ASCE Service to Members		X	X	X	High
F) Practice of Civil Engineering	X	X	X		High
G) Public Relations	X	X	X		High
H) Meetings, Conferences and Publications		X	X	X	High
I) Economic Status and Employment Conditions		X	X		High
J) Local Sections and Branches		X	X	X	High
K) Communications		X	X	X	
L) Involvement in Professional Activities		X	X		
M) Student Chapters and Councils		X	X	X	
N) Unionization or Other Collective Endeavors		X	X		
O) Negotiating for Engineering Services	X	X	X		
P) Professional Ethics	X	X	X		
Q) Membership Growth		X	X	X	
R) Research	X	X	X		
S) Cooperation with Other Organizations		X	X	X	
T) Support Mechanisms for ASCE Activities		X	X	X	
U) Registration	X	X	X		
V) Areas of Civil Engineering Practice		X	X		
W) Unity of Engineers	X	X	X		
X) Standards	X	X			
Y) Role of Technologists and Technicians		X	X		
Z) ASCE Membership Requirements		X	X	X	
AA) Requalification	X	X	X		
BB) Award Programs		X	X		
CC) Certification			X	X	
DD) Liability of Engineers			X	X	

*High priorities are preliminary judgments of COSOPO. These high priority summary goals are listed in descending order of priority, and indicate a need for *increased* activity rather than the order of their basic importance. No order of priority is implied for the other summary goals at this time.

These goals are not new; they were expressed by predecessor evaluation and planning task forces in very similar words. But this committee did perform an important and unique service in providing a sequence of priority of resource commitment to 32 specific sub-goals. The findings of the committee, after input from all elements of the Society and approval by the Board of Direction, are summarized in Fig. 4. This represents the official policy of the Society in 1974.

Thus did the Society struggle for almost 80 years to define the context of its professional-development policy and then devote the next 40 years to devising the mechanism for implementing that policy. In the meantime, however, ASCE was by one means or another initiating and developing standards of professional education, competence, and practice—as well as serving the public interest in sundry ways—as befits a professional organization. The evolution of these functions is described in the remaining pages of this chapter and in the chapters to follow.

Let there be no doubt—in 1974—of the determination of the American Society of Civil Engineers to invest its every resource to the limit in serving the best interests of the public, the profession, and its membership in the professional domain as well as technologically. Never before in its history has the Society been better prepared to meet these responsibilities.

Education for the Civil Engineering Profession

"A profession is the pursuit of a learned art in the spirit of public service." This succinct definition by Dean Roscoe Pound became a prime guideline to ASCE leadership after it was quoted in a meeting of the Metropolitan Section in the late 1950's.

That engineering is a learned art was recognized in America as early as 1815, when Captain Sylvanus Thayer began his one-man crusade to develop a sound basic curriculum and able faculty at West Point. Because of the practical orientation of civil engineering in 1852, however, ASCE did not at once accept responsibility for the direction of civil engineering education. The 1870 requirement for admission to membership, equating graduation from a "school of recognized standing" with two years of practical engineering experience, was at best a grudging acknowledgment of the value of academic training.

The educational backgrounds of the first members of ASCE varied greatly. Available biographical data on 27 of the 55 members listed in the first annual report reveal that eleven of them attended West Point, of whom nine graduated; ten attended other colleges of whom six earned at least one degree (several earned more than one); and six had no college education. Of the latter, modest formal education was augmented by self-study, and technical education and training were acquired through apprenticeship on engineering projects of the day. Included in this group were James P. Kirkwood, James Laurie, and William J. McAlpine, all of whom later served ASCE in the office of President.

John B. Jervis, Hon. M. ASCE, who built a distinguished career upon a common-school education and 15 years of practical experience ranging from axman to Superintending Engineer on construction of the Erie Canal, spoke about engineering education in an address* at the first Annual Convention of ASCE in 1868; he was then 74 years of age.

"After a fair education in the ordinary elements," said Mr. Jervis, "the young man that designs to prepare for the profession of an engineer should study mathematics so far as to qualify him to make any computation of quantities, and to carry forward any investigations that he may find it necessary to make in pursuing the science of mechanical philosophy." Study in mechanical philosophy, "in which special attention should be paid to the character of all the materials required in the various structures . . . and the form and position of materials best adapted to the end it is sought to serve," would be followed by courses in hydraulics, surveying and the study of various structures under the tutelage of experienced en-

*ASCE *Transactions,* Vol. I, 1872, p. 137.

gineers. At this point the student would "enter on the field of practical duties" under professional direction, to round out his capacity to apply his knowledge to practical situations.

But there was much more to Mr. Jervis' curriculum. He gave great emphasis to the point:

> "No skill in forming lines and levels, and in devising structures, will complete the education of an engineer without an intelligent capacity for conducting business. This is an important item in his education, and indispensable to a successful practice . . . No profession needs more thorough business qualification."

It might be said that Mr. Jervis was describing the 1860 counterpart of the engineer-business administration graduate of the 1960's—except for the practical internship of the former.

Fig. 5 (page 100) indicates that the engineering educators of the latter part of the 19th century were providing a much more sophisticated technical training than that prescribed by Mr. Jervis. There was considerable difference of opinion among them concerning curriculum format and content, however, just as there was a hundred years later.

A most informative exchange of views was initiated by a paper by Thomas C. Clarke, presented in June 1874.* He noted the differences between the European and American approaches:

> "The methods of this education used in this country and in England differ materially from those on the continent of Europe. Here we begin by a course of study at some school, college or technological institution, and complete our education by serving as assistants on some class of public works, for which we receive more or less payment. In England, the course is the same except that there boys begin younger and with less perfect theoretical education; then they serve their time under some practicing engineer from three to five years,

*"The Education of Civil Engineers," by Thomas C. Clarke, ASCE *Transactions*, Vol. III, 1874, p.255.

> and pay him a premium of £300 to £500 for the privilege. This they do, because their chance of future employment depends on being personally known to some engineer in large practice. On the continent of Europe, practical training by serving on public works as an apprentice to some engineer is unknown. Education begins at the other end, by the compulsory acquirement of a high degree of theoretical knowledge. Partly with this and partly afterwards, a certain amount of practical information is given, but the leading idea is to make a man a thorough engineer theoretically before he begins or is even allowed to practice.
>
> "In this country, the tendency, of late years has been towards the continental methods. We have attempted with a wise eclecticism, to combine the advantages of both systems, and educate our engineers in the theoretical principles of the science first and then let them acquire practical knowledge by practice itself."

Mr. Clarke favored a classical education, strong in both natural science and the humanities, with mathematics limited "to the ordinary analysis." Only the student possessing unusual capability in mathematics would be encouraged to study "the higher calculus." Study of ancient and modern languages was urged, to improve communication capability.

After such general education, Mr. Clarke recommended that the student go into the field and office to learn the actual practice of his profession before returning to the technological school for training in his chosen field of specialty.

The paper elicited spirited discussion. Professor Estavan A. Fuertes favored a six-year program with strict entrance requirements in classical education and mathematics. After three years of general engineering study, the student would select a specialty area (railroad, bridge, water-works and canal, hydrographic, mining, etc.), which he would pursue under a "special" professor for one year to earn his bachelor's degree. An advanced degree would be conferred after two additional years of special study and practice.

Francis Collingwood supported a curriculum strong in mathematics and natural science, and agreed that the technological schools should offer training in the special branches of engineering. Professor De Volson Wood opined that inserting an internship between the general and the technical parts of the academic experience was not feasible. He favored a broader technological course, with specialization to be undertaken after entrance into practice. He also expressed the view that the technical courses should be taught " . . . by practical men — those who have worked in the field or shop and know what is wanted there."

These views are noteworthy in several respects. All participants agreed upon the need for practical experience in connection with the academic training. The divergence of opinion concerning curriculum balance between mathematics, natural sciences, and the humanities still prevails a century later. Likewise, the present bifurcation between engineering and technology appears to have taken root from seeds that were planted long ago.

Early Educational Activities

A feature of the sixth Annual Convention of the Society in New York City in 1875—and an early manifestation of interest in engineering education — was a visit to Stevens Institute at Hoboken, N. J. Having been founded only four years before, in 1871, for the specialized training of mechanical engineers, Stevens was outstanding at the time for the quality of its faculty, its facilities, its curricula, and its research programs.

The first recorded formal action of ASCE in the domain of engineering education resulted from a communication early in 1874 from Professor Estavan A. Fuertes, M. ASCE, of Cornell, who urged that the Society declare what should be the course of instruction in schools and colleges for students of engineering. The request was referred to a special committee comprising Professor De Volson Wood of Stevens Institute, Professor George W. Plympton of Polytechnic Institute of Brooklyn, and Charles MacDonald, all Members of the Society.

The committee's report, adopted on May 6, 1874, opens with the view that:

"... we are of the opinion that the Society is not an advisory body in such matters... The nearest practical approach to such a position would be the securing of a discussion upon the subject, in which each member would be free to express an opinion, and leave the educator free to draw inferences therefrom."

After observing that it is impossible for any body of men to outline a course of instruction that would generally be considered "best," because any course of study should keep apace of constantly changing circumstances, it was opined that "the institutions of learning should be left free to construct their courses, and no attempt should be made to mould them after a fixed pattern."

This cautious but liberal statement of opinion is followed in the report by a number of observations that are remarkably perceptive in view of the changing philosophy of engineering education through the next century:

"Without disparaging the functions of this Society, it is our opinion that schools of engineering have done more to elevate the standard of the engineering profession than any other single agency. All that pertains to theory and all the classified, practical sciences can and should be taught in the schools. There will always be a practical difficulty in determining the extent to which the details of construction should be taught. It is evidently absurd for the schools to attempt to make experts in any branch of the profession; but this is not a sufficient reason for not teaching, so far as possible, all the principles which pertain to practical operations; such as field-work, shop practice, office-work, etc.

"The subjects which are intended more especially to promote 'broad culture' as popularly understood, are not necessary parts of a professional course; they naturally precede technical courses. The system of technical education in this country — if it can be called a system — has peculiarities which have grown out of the circumstances surrounding it. Nearly all the technical schools here devote one or two years to general science and literature — which are no more technical than the old and more purely classical courses. These years are properly preparatory to those which follow. Had we a general system of education, as thoroughly graded as that in Germany, the technical schools would not necessarily be burdened with this preparatory work; but all things considered, we doubt not but more thorough work has been done by the system which generally prevails here than would have been by confining the instruction to professional subjects. We ought not to import a foreign system, but seek to build one here especially adapted to our times and circumstances. Our schools ought not to graduate men with a mere 'smattering' of the sciences they are to use, but the instruction should be thorough and the standard of graduation high. If an error is made in either direction, the schools should be too theoretical rather than too practical. The former gives a solid basis upon which to build the professional structure.

"It makes but little difference what degree is conferred at graduation; but if that of 'Bachelor of Engineering' instead of 'Civil Engineering' on the ground that the latter implies a certain amount of practical experience, then we respectfully submit that the latter should be conferred only by a body of practical engineers; such for instance as this Society. In the sense, however, that the candidate has acquired a thorough knowledge of the 'Science of Engineering,' we see no impropriety in conferring the latter at graduation. The student can become a 'Master of Engineering' only by long and varied experience."

Polarization of proponents of the "practical" and the "scholastic" elements of the profession brought about the Joint Committee of ASCE and AIME on Technical Education, which may have been the first intersociety activity in engineering. A program arranged by the joint committee at the Franklin Institute in Philadelphia during the 1876 Centennial was addressed to the questions: (1) Should a course of practical instruction precede, accompany, or follow that in technical schools? and (2) Is it practicable to organize practical schools under the direction and discipline of experts in engineering works? Although these questions were not answered, the airing of conflicting views was salutary. Chairman Alexander Holley, M. ASCE, closed the session on this positive note:

". . . Having recognized the grave and comprehensive character of the evil, the next step should be, not . . . to attempt any

> violent alteration in the existing conduct of engineering by the men who are now in active service, but to change the environments of the young men who are soon to take their places, in order that their development may be larger, higher and in better balance."

A resolution adopted at the July 1887 Convention requested the Board of Direction to consider appointment of a committee on technical education to examine and report annually on methods and progress of engineering education in America and Europe. An ad hoc Committee on Professional Training and Technical Education appointed by the Board to review the proposal made its report in January 1888.

The report stated ". . . Whatever the ultimate scope of the Society's work may be, it must always be borne in mind that the object of its formation and the aim of its present existence is not primary education in science, but the higher education of skilled scientists." Concluding that future resources may afford opportunity for study of primary education, the task committee recommended that no action be taken at this time to expend Society funds ". . . in discussing the early education of engineers so long prior to the time when they can, under any circumstances, become members of the Society."

When the Board adopted this recommendation and so reported to the membership, Oberlin Smith, M. ASCE, recorded his disagreement, saying:

> ". . . We members here have no idea of the immense importance of the education business; we think these students don't amount to much. Fifteen or twenty years hence, when a good many of us are dead, these young men are going to be in the places that we are in now; if we can improve their education by anything that we can do now, it is going to help their education forward and undoubtedly is going to make a better class of engineers than there otherwise would be."

Mr. Smith was an excellent prophet, for it was exactly twenty years later that the next significant action occurred in the field of engineering education.

Education — A Professional Responsibility

A development of great importance in engineering education took place as an issue from the Engineering Congress staged by ASCE with other national engineering societies as a feature of the 1893 Columbian Exposition in Chicago. The sessions conducted on engineering education generated so much interest, that organization of the Society for the Promotion of Engineering Education (SPEE) followed almost immediately thereafter; SPEE later became the American Society for Engineering Education (ASEE). Professor DeVolson Wood of Stevens Institute of Technology, the first president of SPEE, was a member of ASCE from 1872 to 1887. One of the first two vice presidents of SPEE, Professor George F. Swain of MIT, was also a member. Professor J.B. Johnson, M. ASCE, of Washington University (St Louis) was the first secretary and later president of SPEE.

A few months after a special Committee on Engineering Education was appointed in 1907, the SPEE invited ASCE to participate in a joint committee with ASME, AIEE, AIME, and SPEE that would study engineering education and report upon its "proper scope and proper direction for advance." After some reluctance, the Society's committee was directed to cooperate in the joint study, and to designate two of its members to serve in the joint committee.

Except for a routine progress report in 1909, the ASCE Committee on Engineering Education was next heard from in 1910, when a request was granted for an appropriation of $200 to cover the cost of compiling certain statistics that had been gathered by the Carnegie Foundation. In the meantime, the Board of Direction rather bruskly declined a proposal from SPEE that the latter organization hold its annual convention at the same time and place as ASCE. This would seem to have been a fine opportunity for interaction and collaboration, as was the case twenty years later when ASCE contributed its cooperation, support, and $2,500 to a Summer School for Engineering Teachers sponsored by SPEE at Yale University.

In 1912 the Committee on Engineering Education reported

upon a study "of the work of instruction as carried on by twenty of the leading technical schools and colleges in the United States." When the data were compiled and collated an evaluation meeting was held, ". . . attended by representatives of the schools and by those who feel that the present methods of instruction are not in the right direction. Different views of Engineering Education were unfolded, and reforms advocated and dissected."

The committee report went on to state:

> "As a result of the work, your Committee takes pleasure in reporting that the whole matter of Engineering Education will be taken up in a scientific manner by the Carnegie Foundation, which recently did the same thing for Medical Education, in which they expended $40,000, with grand results — rendering an inestimable benefit to that branch of scientific instruction."

The Carnegie Foundation study did come about, under the auspices of a joint commission encompassing representatives of the four Founder Societies, SPEE, and the American Chemical Society. The results of the study, however, did not become available until 1919.

A surprising difference in opinion between educators and practitioners was revealed by the Carnegie appraisal, which was made under the direction of Dr. C. R. Mann. A questionnaire circulated to members of the engineering societies brought 1,500 replies, from which the following essential characteristics for engineering were given the indicated weights:

Characteristic	*Weight*
Character (integrity, responsibility, resourcefulness, initiative)	41%
Judgment (common sense, scientific attitude, perspective of life)	17.5%
Efficiency (thoroughness, accuracy, industry) Understanding of Men (executive ability)	28.5%
Knowledge of Engineering Science Fundamentals	7%
Technique of Practice and of Business	6%

Although the validity of such data might be questioned, Dr. Mann drew an interesting conclusion from the survey when reporting on it at the 1916 Annual Meeting:

> "If this is the correct definition of an engineer, the schools must reorganize very fundamentally. Very little conscious attention is there being paid to those qualities on which 87% of the success of the engineer depends. These personal qualities are usually regarded as a sort of by-product. I do not say that the school does nothing to develop character or resourcefulness, nor that it pays no attention to them. I do say that the development of these important personal qualities is not in the focus of attention of the school. Their attention is focused on the last two items, which have a weight of 13 per cent.
>
> "If you engineers will give the schools a clear definition of what the engineer is, the schoolmen will know how to use it to strengthen the school . . . If you will define the product for the schools, the schoolmen will prove competent to produce it."

The final report ("A Study of Engineering Education," Carnegie Foundation Bulletin No. 11) dealt with the prevailing status of education in twenty schools, each of which had been visited by Dr. Mann. Among the problems he found were that only 40% of all engineering students completed their course, that "specialization and subdivision of curricula has gone too far," and that there was wide inconsistency in course content, testing, and grading, and in the utilization of shopwork. His recommendations included the conduct of research on aptitude tests and student performance, reduction in the number of courses to four or five each term, and recognition of professional experience as a qualification for teaching. Noting that "the spirit of research is part of university life," he urged that teaching and research be conducted in such balance as to realize the best values of both. The inculcation of professional attitude and discipline in the student was held to be a responsibility of the school and industry, and experimentation was needed on how this might best be done.

With the acceptance of this report by the Board of Direction the special Committee on Engineering Education was discharged, having been under the chairmanship of Past-President Desmond FitzGerald for its entire twelve years of existence.

For the next several years ASCE did not undertake independent educational activity, but was closely associated with the operations of the SPEE. Under another Carnegie Foundation grant to SPEE, a significant "Report of the Investigation of Engineering Education (1923-29)" was produced with Dr. W. E. Wickenden as director. ASCE was well represented in both the advisory and managing committees for this study.

As the number of engineering schools increased, questions began to arise as to the interpretation of the term "recognized standing," as used in the membership requirements of the Society. Since 1928 graduation from "a school of recognized standing" had been equated with four years of active practice. In 1930 the Board of Direction found it expedient to appoint from its own membership a Committee on Accredited Schools to resolve these questions. Through this medium ASCE developed and maintained its own list of accredited civil engineering schools until 1938.

A giant step forward was taken by the engineering profession on October 3, 1932, when the Engineers' Council for Professional Development (ECPD) was formed for the purposes of accrediting engineering schools and enhancing the professional recognition of engineers. Included in the original membership were ASCE, AIME, ASME, AIEE, SPEE, the American Institute of Chemical Engineers (AIChE), and the National Council of State Boards of Engineering Examiners (NCSBEE). Although the leadership in organizing ECPD came from ASME, particularly in the person of Clarence E. Davies, ASME Secretary, ASCE was a strong supporter of the movement from its beginning.

Several years were required for ECPD to develop its accreditation standards and procedures. In 1938, the ASCE Committee on Accredited Schools recommended that the Society adopt the accrediting service of ECPD, with suitable provision for the schools heretofore accepted by ASCE but not by ECPD. The ASCE committee was then discharged.

A notable contribution to ECPD accrediting procedures was made in 1965 by the ASCE Committee on Engineering Education. This was the "Guide for Civil Engineering Visitors on ECPD Accreditation Teams," which has since served as a model for other ECPD societies.

World War II brought unusual problems to the colleges and universities of the United States. In January 1943, the Board of Direction addressed itself to these conditions by the adoption of two significant resolutions. The first of these dealt with "Emergency Conditions in Engineering Education," and urged that the federal military agencies and War Manpower Commission maintain a strong basic engineering education and student body to insure successful prosecution of the war as well as postwar reconstruction and rehabilitation. Attempts to convert liberal arts and non-scientific curricula into "so-called engineering courses" were opposed.

The second resolution, on "Long-Range Problems and Objectives of Engineering Education," affirmed the conviction:

> ". . . that active participation in defining and promoting the basic educational standards of the profession of civil engineering is a necessary and proper function of the Society, and that such participation is hereby adopted as policy of the Society."

This position is almost diametrically opposite the attitude of the leadership of the Society in 1874! The resolution went on to authorize a joint study of civil engineering education with the Civil Engineering Division of SPEE, as basis for:

> ". . . recommended standards of professional education and achievement requisite for qualification for admission to the practice of civil engineering as a profession."

The joint study developed in two phases, both highly successful. The first part, reported in 1944, was based upon a detailed study of civil engineering curricula in 114 schools. The conclusions were most perceptive, and their soundness has been proved by subsequent experience. The following curriculum outline recommended for the four-year bachelor's-degree program is a milestone in that it incorporates parallel development of the scientific-technological and the humanistic-social sequences in engineering education:

Subject	*Percent*
Humanistic-social	20
Physical sciences, including geology	15
Drawing	4
Mathematics, not including trigonometry	10
Mechanics, hydraulics, strength of materials	11
Engineering subjects other than civil	10
Civil engineering	30

While the 1944 report was under review by the engineering school administrators, a survey of 2,700 members of the Society elicited a 38% response to a detailed questionnaire relating to education needs in civil engineering. This provided documentation for a second report in 1945. In summary, graduates of the 1935-45 decade were judged to be lacking in oral and written communication skills, and to have little interest in public affairs. They were rated favorably, however, in ability to get along with people, leadership capacity, grasp of fundamentals, and in the personal qualities of accuracy, diligence, and dependability. It was especially noteworthy that the survey endorsed the curriculum structure recommended in the 1944 report.

The two reports were published, (*Proceedings*, March 1946, p. 375) and given wide circulation. They were undoubtedly of timely value to many engineering schools with low enrollments at the close of World War II, and thus were in good position to make adjustments in their programs.

An Exploratory Conference on Engineering Education in 1947 brought together the chairmen of all ASCE committees related to the subject, together with representatives from

ECPD, ASEE, and the staff. The objective was greater interaction among committees and with other organizations to further career guidance, the educational process, and postgraduate professional development. Recommendations from the conference urged that ECPD expand its guidance program and also implement the 1945 ASCE curriculum recommendations; that local sections promote part-time employment opportunities for engineering teachers; that salary studies and guidelines be provided for engineering educators; and that ASCE professional activities be coordinated toward these goals. The Committee on Salaries promptly carried out the survey of qualifications, responsibilities, and compensation of civil engineering teachers, reporting to the Board of Direction in 1948.

The Committee on Engineering Education found another opportunity for service in 1948, when it noted that many young graduates with advanced degrees were becoming teachers without having acquired practical experience. The committee utilized the local sections of the Society in its efforts to find employment for the young teachers, and to encourage the university administrators to release them for professional summer work.

New Dimensions in Education

In 1953, recognizing a need for studying civil engineering education independently from the system of engineering education in general, Dean F. M. Dawson, of the University of Iowa, convinced the Board of Direction that a Task Committee on Civil Engineering Curricula should be formed. As the task committee approached its charge, it foresaw the desirability of broader inquiry, which was authorized by the Board in 1954 with a change of name to Task Committee on Professional Education.

When the committee requested funds for an opinion research survey of the membership as a source of data, an appropriation of $15,000 was provided under the condition that the survey would cover areas other than education. A concurrent Task Committee on Economic Advancement Objectives was thus able to utilize the survey for its purposes. This was a major undertaking for the Society at that time.

The task committee report, including the highly interesting results of the membership opinion survey, was published in *Civil Engineering*, (February 1958, pp. 111-123). Twelve very broad conclusions and recommendations were set forth, of which seven pertained to such matters as public relations, economic welfare, unionism, and local section activity. The education-related findings urged higher salaries for teachers, attention to the needs of the construction industry, more emphasis on career guidance, and stronger post-baccalaureate programs. Most important, however, was the following recommendation, which was no doubt responsible for an important conference on civil engineering education that was staged by ASCE and other sponsors several years later:

> "*Recommendation*: Although certain undergraduate courses are commonly required for all students in engineering, civil engineering curricula must maintain their separate identity and objectives and the civil engineering departments of our colleges should receive support consistent with the civil engineer's great contribution to society."

While the Task Committee on Professional Education was at work, a significant general study was being completed by ASEE under the direction of Dr. L. E. Grinter, Hon. M. ASCE. This "Evaluation of Engineering Education (1952-55)*" had a profound effect toward the sophistication of all engineering education by its encouragement of stronger training in mathematics and the sciences, and in the application of scientific fundamentals to engineering problems. This was a natural result of the postwar acceleration of technological progress.

**Journal of Engineering Education*, ASEE; September 1955, pp. 26-63.

Through the initiative of Professor Felix A. Wallace, F. ASCE, of The Cooper Union, a major Conference on Civil Engineering Education was held at Ann Arbor, Mich., July 6-8, 1960. Joint sponsorship was provided by The Cooper Union, ASCE, and ASEE, with funding assistance from the National Science Foundation. About 250 delegates and participants attended, with the objective to "redefine and reorient the educational processes to meet the needs of the present and the challenges of the future," in the words of the 1958 Committee on Professional Education.

At the conclusion of the prepared program, eleven resolutions were placed before the conference for appropriate action. Extensive debate ensued as to whether such resolutions should merely be accepted or rejected for publication in the Conference Proceedings, without implied approval or disapproval, or alternatively, each resolution should be voted upon for specific approval or disapproval. When the oratory and parliamentary jousting ended, it was agreed that nine of the eleven resolutions would be published. The thrust of these resolutions was as follows:

(1) That the Conference favors the growth of a pre-engineering, undergraduate, degree-eligible program for all engineers, emphasizing humanistic-social studies, mathematics, basic and engineering sciences, with at least three-quarters of the program interchangeable among the various engineering curricula; to be followed by a professional or graduate civil engineering curriculum based on the pre-engineering program and leading to the first engineering degree, with a civil engineering degree awarded only at the completion of the professional or graduate curriculum. Further, that increasing opportunities be provided for qualified students to earn graduate degrees at the master's and doctor's levels.

(2) That the four-year undergraduate program terminating

in the B.S. degree in specific fields of engineering be retained.

(3) That abandonment be recommended of the degree "Civil Engineer" as a non-resident, non-academic degree.

(4) That the Conference favors establishment of graduate professional schools offering the degrees of Master of Engineering and Doctor of Engineering in the several engineering specialties, and also favors continuation of current programs leading to the Master of Science and Doctor of Philosophy degrees.

(5) That the Conference favors development in Colleges of Arts and Sciences of a three-year undergraduate pre-engineering program for all engineers, emphasizing humanistic-social studies, mathematics, basic science and communications, followed by a three-year engineering program of engineering science and professional courses taught in professional schools of engineering, extending about as far as present Master-Degree programs and leading to a professional degree in engineering.

(6) That it is the responsibility of the practicing engineers to achieve truly professional standards of performance and ethics, and to create an atmosphere through their individual efforts and engineering societies to sustain such standards.

(7) That the civil engineering educator take an articulate position in the profession in order to maintain an appreciation among engineers and the public of the significance of basic scientific and general cultural education, to the end that the engineering student shall acquire an awareness of these areas of study and a willingness to apply himself to them.

(8) That the Conference favors the growth of graduate programs in schools able to sustain them; that each institution respond on a school-by-school basis to the

needs of students, parents and employers so that the broad spectrum of needs in the civil engineering profession may continue to be met; and that experimentation be encouraged so that special programs may be perfected for the more gifted students.

(9) That it is the sense of the Conference that the time has come to increase the length of the curriculum for the first degree in Civil Engineering from four years to five years.

These far-reaching proposals were incorporated into specific guidelines in 1962 by the Committee on Engineering Education, following a survey of 112 ECPD-accredited civil engineering programs. The policy had the following major elements:

(1) A five-year first degree combining a pre-engineering core with a professional program, with a broad base of engineering science.

(2) Establishment of professional schools of engineering having responsibility for professional-type postgraduate programs.

(3) The need for at least two types of postgraduate programs — one to serve the needs of the scholarly bent, including those pursuing careers in research and development; the other to serve the needs of those desiring to become practitioners of high competence in a specialty field.

(4) Encouragement for all schools to adopt these recommendations in the manner that will best serve the objectives and particular conditions of their own institutions.

These guidelines lasted for four years, and were applied at least in part in a number of schools. In 1966 a major appraisal of all engineering education was made by ASEE with National Science Foundation funding to the extent of $307,000. Designated as a study of "Goals of Engineering Education," it involved circulation of some 4,000 questionnaires to engineering practitioners and others.

The most important issue from the Goals Study was a recommendation that "The first professional degree in engineering should be the Master's Degree, awarded upon completion of an integrated program of at least five years' duration." Such Master of Engineering degree would not identify any field of specialization. The ASCE Board of Direction promptly voiced its strong disagreement to this proposal to shape all engineers by the same template, as there was already a growing opinion that civil engineers required different training from those engaged in the manufacturing industry. This pronouncement was followed, in 1968, by replacement of the 1962 ASCE policy with a new one pointing up the need for education of the civil engineer to function with other design professionals in solving complex environmental and specialized problems. Four years of study leading to a Bachelor's Degree from an accredited civil engineering curriculum was held to be minimal education for those entering the civil engineering profession. A fifth year was considered minimal, however, for one aspiring to be a civil engineering specialist.

Costly as it was, the Goals Study failed to recognize an emerging development that was to render its findings obsolete almost as soon as they were published. The strong scientific orientation of engineering education in the post-Sputnik years had created a shortage of engineers capable of handling the practical planning, design, and construction operations sought by their employers. In consequence, a new member of the technology spectrum was introduced by a number of schools in the person of the four-year Bachelor of Engineering Technology. By the mid-1970's several thousand BET graduates were being produced annually. Obviously, adjustments were found to be necessary in career-guidance procedures, in the education system, and in the registration process, to accommodate the BET.

In 1972 the ASCE Committee on Engineering Education was replaced by an Education Division, encompassing six committees: Continuing Education, Curricula and Accreditation, Guidance, Interchange Between Education and Practice, Student Chapters, and Technician and Technology Education.

The new Education Division promptly decided that it was time for another appraisal of civil engineering education. So came about the ASCE Conference on Civil Engineering Education held at Ohio State University in February 1974, which may prove to be the Society's most important contribution in this area of activity. The theme of the conference was "Civil Engineering Education Related to Engineering Practice and to the Nation's Needs." The specific purposes were to evaluate the need for modification of educational policies and patterns to accommodate such current developments as the Bachelor of Engineering Technology, the environmental assessment of engineering functions, and the growing demand for multidisciplinary interaction.

ASCE sponsored the conference without outside funding, and seventeen civil engineering-related organizations participated. Of the 350 conferees, a third were non-educators — an unprecedented ratio for such gatherings. The program emphasized education for planning and construction as well as design. There was much discussion of social responsibility, broadening of curricula, and "professional schools" of civil engineering. Assessment of the results of the conference was not immediately possible, but it is certain to have a beneficial long-range impact upon the civil engineering profession.

Continuing Education, Guidance, and Special Activities

In the 1960's, the education functions of ASCE were expanded considerably beyond a concern with education policy and curriculum development. To improve communication among the educators themselves, the Society began in 1961 to sponsor regular regional meetings of the heads of civil engineering departments. These were immediately successful, and were followed in 1966 by the scheduling of breakfast gatherings of department heads and ECPD accreditation-team visitors at all national ASCE meetings. Committee and staff participation in these discussions kept the Society closely in touch with current affairs.

The addition to the staff in 1966 of an Assistant Secretary for Education began a new era. Society policy in continuing education favored collaboration with existing programs in schools, engineering societies, and industrial organizations rather than the creation of new competing programs. First emphasis was given to the offering of seminars in the local sections and at national meetings of the Society, usually under joint sponsorship with local institutions.

Programmed learning courses for self-study were also made available to the membership, beginning in 1970. The first of these was developed by ASCE under the title "Engineering Economy." Off-the-shelf courses entitled "Management Games Seminar" and other correspondence courses also found excellent response. Further self-study assistance and encouragement were provided in the form of bibliographies of civil engineering books in various categories including both technical subjects and professional orientation.

A set of guidelines to encourage and assist the local sections in continuing education functions was compiled in 1970. A particularly successful effort through the sections made possible the presentation in fifty cities of a seminar assembled by the American Institute of Steel Construction on "Plastic Design of Braced Multi-Story Frames."

To keep the membership informed of opportunities in continuing education available to them, a listing of current seminars, conferences, and short courses was regularly carried in *Civil Engineering.*

Addition of a professional continuing education specialist to the staff in 1973 presaged substantial further development of these important services.

Career guidance services were of modest order prior to 1958, comprising some printed pamphlets and a few high school visitation activities in local sections. Because of static civil engineering enrollments in the late 1950's, a unique brochure, *Your Career in Civil Engineering,* was designed to encourage junior high school students to take the mathematics and science courses that are prerequisite to engineering school enrollment. About $24,000 was invested in the production and distribution of five editions aggregating 500,000 copies of this piece.

An effective paperback, entitled *Your Future in Civil Engineering*, was authored by Alfred R. Golze, F. ASCE, for guidance use. An expenditure of $30,000 made possible the free distribution of 200,000 copies of this publication to parents and students.

In 1967 a colorful illustrated brochure, titled *Is Civil Engineering for You*?, became the primary item of career guidance literature. About 250,000 copies of it were distributed at a cost of $50,000; a new edition was produced in 1974.

The 1969 color moving picture film, "A Certain Tuesday," proved to be a guidance medium of extraordinary value. Designed for showing in high schools, with narration preferably by a civil engineering student who was an alumnus of the high school, the film found wide usage through the local sections.

In 1971 ASCE became an enthusiastic participant in a guidance program called "The World of Construction." It was a junior high school level industrial arts course intended to involve the student in a construction-related education experience that might arouse career interest. More than 1,500 schools were offering the course in 1973.

A 1965 Task Committee on Civil Engineering Management Education recommended that ASCE provide partial funding for a number of pilot courses in civil engineering management in selected universities. The aim was to encourage greater emphasis toward such training at the graduate level. The program was implemented by the establishment of fellowships at Carnegie-Mellon, Drexel, Purdue, Stanford and the University of Illinois, each in the amount of $2,000 per year for five years. The purpose of the pilot program was successfully accomplished when it was terminated in 1972.

The heavy commitment of financial resources to guidance and other educational functions has been possible only because of the "Voluntary Fund," a special reserve receiving contributions and annual dues from Life Members who continue such payments even though they are not obliged to do so. The fund was initiated in 1947, and has made a total of some $300,000 available for special needs of the Society.

The educational functions of ASCE acquired an international dimension in 1958, when it assumed responsibility for the secretariat of the Committee on Engineering Education and Training of the Conference of Engineering Societies of Western Europe and the United States, known as EUSEC. The Executive Secretary of ASCE at the time served as General Secretary of EUSEC as well as secretary of its education committee. Input was provided in the form of fund-raising services, administrative services, and professional experience.

About $60,000 was raised to produce a classic comparative study of engineering education in nineteen countries. A three-volume report* — published in English, French, and Spanish — provided detailed information about the general education systems in each country, the university level systems, practical training requirements, graduate education, and criteria for professional recognition. The project was funded by the Ford Foundation, the Organization for European Economic Cooperation, and the various EUSEC societies, and the report has provided guidance to many international studies on engineering education since its publication in 1961.

---

*_Report on the Education and Training of Professional Engineers_ (1960), Conference of Engineering Societies of Western Europe and the United States of America.

FIG. 5. — RENSSELAER POLYTECHNIC INSTITUTE: COURSE OF STUDIES AND ENGINEERING EXERCISES FOR THE DEGREE OF CIVIL ENGINEER (1852)

FIRST YEAR.

Division	Departments of Instruction.	Subjects of Study and Practical Exercises.
DIVISION C—THIRD CLASS.	MATHEMATICS.	Algebra.—Elementary Geometry.—Nature and use of Logarithmic and Trigonometric Tables.—Trigonometry.—Mensuration.
	PHYSICS.	General Properties of MATTER.—Nature of the Physical FORCES.—Phenomena and Laws of GRAVITY.—Phenomena and Laws of HEAT.
	GRAPHICS.	Use of Drawing Instruments.—Graphical Constructions of Chain and Compass Surveys.—Copying of Mechanical Drawings.
	CHEMISTRY.	Principles of Chemical Philosophy.—Study of the Non-Metallic Elements.—Laboratory Practice.
	GEODETICAL ENGINEERING.	Operations in the Field,—Measuring of Lines; Chain and Compass Surveys of Fields and Farming Estates; Dividing Land.—Computations of Areas, etc.—Mapping of Surveys.
	PHYSICAL GEOGRAPHY.	Structural and Systematic Botany.—Vegetable Physiology.—Geographical Distribution of Plants.
	ENGLISH COMPOSITION AND CRITICISM.	Section Lecture Exercises with Criticisms.—Keeping of Lecture Books.—Writing of Special Theses, etc.
	FRENCH LANGUAGE.	Elements of French Grammar.—French Exercises and Translations.

SECOND YEAR.

Division	Departments of Instruction.	Subjects of Study and Practical Exercises.
DIVISION B—SECOND CLASS.	MATHEMATICS.	Analytical Trigonometry.—Analytical Geometry.—Differential Calculus.—Integral Calculus.
	PHYSICS.	Terrestrial Magnetism.—Statical and Dynamical Electricity.—Electro-Magnetism.—Magneto-Electricity.—Acoustics.—Optics.
	GRAPHICS.	Descriptive Geometry.—Measuring and Sketching Engineering and Architectural Structures, and Construction of Working Drawings from these data.—Topographical Drawing.
	CHEMISTRY.	Chemical Study of the Metals.—Exercises in Qualitative Analyses.
	GEODETICAL ENGINEERING.	Theory and Adjustments of Field Instruments.—Trigonometrical Determination of Heights and Distances.—Field Exercises in General Geodetic Operations.—Field Practice in Hydrographical Surveying.
	GEOLOGY.	Descriptive Mineralogy.—Systematic and Descriptive Geology.—Economic Geology.
	ENGLISH COMPOSITION AND CRITICISM.	Section Lecture Exercises with Criticisms.—Writing out of General Lectures.—Writing of Theses on Scientific and Practical Subjects.
	FRENCH LANGUAGE.	Double Translations, (French and English.) Reading of French Scientific Authors.

THIRD YEAR.

Division	Departments of Instruction.	Subjects of Study and Practical Exercises.
DIVISION A—FIRST CLASS.	RATIONAL MECHANICS.	General Statics and Dynamics of Solids, Liquids, and Gases.
	PRACTICAL ASTRONOMY.	Investigation of Astronomical Principles and Data for the solution of the Practical Problems of the *Meridian, Time, Latitude,* and *Longitude* of a place.—Sextant and Transit Observations, including Computations for, and Reductions, made by students.
	CONSTRUCTIVE ENGINEERING.	Equilibrium and Stability of Architectural and Engineering Structures.—Materials for Construction.—Theory of Machines.—Road Engineering.—Hydraulic Engineering.—Steam Engine and Hydraulic Motors.
	GEODETICAL ENGINEERING.	Higher Geodetic Surveying of Extensive Areas by Trigonometrical and Astronomical Methods.—Topographical Surveying, with Field Practice, Reductions, and Construction of Maps.—Surveys, Location, etc., of Engineering Works.
	PRACTICAL CHEMISTRY AND PHYSICS.	Chemical Study of the Principal *Economic Minerals*:—Blow-pipe Examinations.—Practical Exercises in Determining the Specific Gravties of Solids, Liquids, and Gases.
	PHYSICAL GEOGRAPHY.	Meteorology, General Hydrology, and Topography of the Earth's Surface.—Distribution of Plants and Animals.—Relations of Physical Geography to Engineering Works of Inter-Communication.
	GRAPHICS.	Descriptive Geometry, embracing Projections of Shades and Shadows, and the Principles and Practice of Natural and Isometrical Perspective.—Sketching of Machines, and Construction of Working Drawings of same.—Topographical Drawing.—Drawing Maps and Sections of Surveys for Lines of Transit.
	ENGLISH COMPOSITION AND CRITICISM.	SECTION LECTURES,—*Extemporaneous* and *Written*,—with full Criticisms.—*Written Reviews* of Machines, Structures, and Processes in the vicinity of Troy.—*Scientific* and *Practical Theses.*

FIG. 6. — RENSSELAER POLYTECHNIC INSTITUTE: BACCALAUREATE CIVIL ENGINEERING CURRICULUM (1973)

FIRST YEAR

First Semester		Second Semester	
Mathematics I	4	Mathematics II	4
Chemistry I	4	Physics II	4
Physics I	4	Hum. or Soc. Sc. Elective	3
Elementary Engineering	3	Engineering Electives (2)	6
Hum. or Soc. Sc. Elective	3	Phys. Ed. or ROTC	—
Phys. Ed. or ROTC	—		

SECOND YEAR

Mathematics III	3	Differential Equations	3
Physics III	4	Engineering I	3
Materials	3	Parameter Systems	3
Mechanics	4	Engineering Elective	3
Hum. or Soc. Sc. Elective	3	Hum. or Soc. Sc. Elective	3
Phys. Ed. or ROTC	—	Phys. Ed. or ROTC	—
		Engineering Seminar	—

THIRD YEAR

Engineering Lab. I.	2	Engineering Lab II	2
Thermodynamics I	3	Hum. or Soc. Sc. Elective	3
Mechanics	3-4	Engineering Electives (4)	12
Hum. or Soc. Sc. Elective	3		
Engineering Electives (2)	6		
Engineering Seminar	—		

FOURTH YEAR

One course each in structures, transportation, and soil mechanics and foundation engineering	9
One course from either construction, professional practice, environmental engineering or water resources	3
Four additional civil engineering electives	12
Humanities or social science electives	6

121 Years of Change

The cumulative change that has taken place in engineering education since the organization of ASCE may be observed by comparison of the 1852 and 1973 civil engineering curricula at Rensselaer Polytechnic Institute (Figures 5 and 6). The difference is even more pronounced when it is noted that the 1973 Professional Program at Rensselaer requires five years for completion — the same three-year Pre-Engineering Curriculum plus two years of civil engineering and humanistic electives leading to the Master of Engineering degree. The Pre-Engineering Curriculum is common for all branches of engineering, and the engineering electives are mainly in the areas of engineering science, economics, basic science and management.

On the other hand, the 1852 Course of Studies must be considered a rigorous one for its time. Surprising in its breadth and in its coverage of mathematics and the sciences, it clearly represents an education of professional level.

While ASCE was conservative and reactionary in many of its attitudes, policies, and endeavors, that certainly was not the case with its concerns for education. As early as 1874 it realized that the schools should not be restricted to inflexible patterns, and that innovation was to be encouraged. Trends in civil engineering education were examined in depth no less than a dozen times between 1874 and 1974. This frequent evaluation curbed swings toward extremes of practical or scientific emphasis, while accommodating the increasing sophistication of technology.

Since 1940 there was a growing awareness that civil engineering education must acquire a humanistic, social, and political dimension if practitioners of the profession were to be prepared properly. After 1960 the need for civil engineers to be able to assess both the positive and negative impact of their works upon the quality of the environment added another complication to their system of education. It was by this time quite clear that the kind of engineer sought by "industry" was not necessarily prepared to plan, design, build, and manage public works to satisfy 1974 standards.

It is obvious from the record that continuing review is essential to the administration of professional engineering education. Only through such periodic self-appraisal can a profession maintain an educational system that will advance its "learned art" to keep apace of a fast-changing world.

STANDARDS OF COMPETENCE: MEMBERSHIP REQUIREMENTS AND REGISTRATION

Because the Institution of Civil Engineers of Great Britain (ICE) was so influential in the organization concept of ASCE, it is not surprising that the membership standards of the Society have always limited membership in the higher grades to those having education and experience identifying them without doubt as professional level engineers. The ICE was legally chartered to serve this function; membership in the Institution automatically provides a legal right to practice under the terms of the Royal Charter.

Although ASCE was not so ordained, it has always been intended that membership in the Society in itself constitutes adequate proof that an individual is fully qualified to practice civil engineering at the level represented by his grade of membership. Thus "professional identification" was a tangible value to the early member, recognized by the courts as well as the public.

The professional identification dimension of ASCE membership has been jealously guarded by the stringency of the requirements and by their strict and conscientious administration. When legal registration was first proposed in 1897, there was great reluctance toward acceptance of the principle at the national level, as it was held that nothing more than ASCE membership was necessary to set the qualified civil engineer apart from the incompetent practitioner. It is paradoxical that engineering registration found its origin in local sections of the Society despite the resistance of national leadership.

Membership Standards as a Measure of Competence

Although membership admission procedures were quite informal in the beginning, there was still a high order of selectivity exercised. The initial "grandfather" privilege was extended to an invited list of engineers of national reputation in the following terms:

> ". . . Such gentlemen only as receive this circular are eligible as members of the Society by a bare notice of their desire to become such and a compliance with the accompanying forms (on or before December 1st, 1852). All others will be elected by ballot in conformity with the requirements of the Constitution."

The constitutional requirements for membership were not at all definitive:

> "Civil, geological, mining and mechanical engineers, architects, and other persons who, by profession, are interested in the advancement of science, shall be eligible as members."

Processing of membership applications apparently involved only review and acceptance by those in attendance at any meeting. At the first Annual Meeting, on October 10, 1853, a total of 55 members was reported: 6 Honorary, 1 Corresponding, and 48 in the grade of Member. Most, if not all, of these are believed to have been on the original invitation list.

Membership actually declined by one in the year 1853-54, and a most unusual action was taken when the dues of Resident Members were reduced from $10 to $5 and for Non-Residents from $5 to $3. At this time only 10 of the 47 in the Member grade were Residents. The Board action was taken in order to "render membership less onerous, and with the hope that by so doing new members might be induced to seek a connection with the Society." It was also recorded that "the labors of the members more than their money is wanted to make the Society useful."

The rejuvenation of the Society on October 2, 1867, was accompanied by a healthy membership growth, despite the restoration of the original dues of $10 for Resident Members and $5 for Non-Residents. Fifty-four candidates were elected at the December 4, 1867, meeting, of whom 32 eventually qualified by payment of dues. By 1870 total membership was approaching 200.

The national character of the Society was well established by 1873. An analysis in September of that year showed that about 70% of the total was Non-Resident (beyond 50 miles of New York City).

A detailed account of the evolution of membership standards from 1852 to 1974 is not feasible here. Instead, the requirements are summarized in Fig. 7. This matrix affords convenient comparison of the qualifications for the various grades at six significant stages in their 122-year metamorphosis.

The overall increasing rigor of the requirements is substantial, but it will be noted that the greatest changes occurred in the periods 1870-91 and 1950-72. The frequency of bridge and dam failures as a result of faulty design may have induced the emphasis on professional competency in the earlier period. Since 1950, postwar sophistication of engineering practice is reflected in the membership requirements.

The "General Eligibility Requirements" were essentially the same throughout the full 120 years, although the language was varied. The most important adjustment here was the 1930 change in equivalency of the engineering degree from two to four years of professional practice.

Very little change occurred in the requirements for the "Eminence Grade" of Honorary Member, but nomination and election procedures were substantially modified over the years. The sanctity of this grade as the Society's "highest honor" has been carefully preserved by the selection and election process. A maximum of twenty Honorary Members was established in 1868; in 1972 there was no prescribed maximum, but no more than one Honorary Member for each

FIG. 7. — EVOLUTION OF ASCE

	1852	*1870*
General Eligibility Requirements	Civil, geological, mining and mechanical engineers, architects and other professionals interested in advancement of science.	Same as 1852
Eminence Grade	*Honorary Member:* Unanimous selection.	Same as 1852
Advanced Professional Grade	*Member:* General eligibility requirements above.	*Member:* Five years engineering practice, including responsible charge; college diploma equivalent to two years practice.
Professional Grade	None	*Associate:* General eligibility requirements above.
Entrance Grade	None	None
Non-Engineer Professional Grade	None	None
Financial Contributor	*Fellow:* Contributor of funds to the Society.	*Fellow:* Same as 1852.
Foreign	*Corresponding Member:* Non-resident of the U.S.A.	*Corresponding Member:* Same as 1852.
Unrestricted Grade	None	None

MEMBERSHIP REQUIREMENTS

	1874	*1891*
General Eligibility Requirements	Civil, military, geological, mining and mechanical engineers, architects and other professionals interested in advancement of science.	None
Eminence Grade	*Honorary Member:* Eminence in engineering with 30 years practice.	*Honorary Member:* Eminence in engineering or related science.
Advanced Professional Grade	*Member:* Civil, military, mining or mechanical engineer with 7 years of practice, or 5 years with C.E. degree, including one year in responsible charge.	*Member:* Civil, military, naval, mining, mechanical, electrical engineer or architect 30 years of age; 10 years practice, 5 years responsible charge; qualified to design; engineering degree equivalent to 2 years practice.
Professional Grade	None	*Associate Member:* Professional engineer or architect 25 years of age; 6 years practice, one year in responsible charge; engineering degree equivalent to 2 years practice.
Entrance Grade	*Junior:* Two years engineering practice or college degree and one year of practice.	*Junior:* Age 18, with 2 years practice or engineering degree. Must transfer to higher grade by age 30.
Non-Engineer Professional Grade	*Associate:* Public works manager, scientist, industrialist or other non-engineer professional qualified to cooperate with engineers.	*Associate:* Non-engineer professional qualified to cooperate with engineers in advancement of professional knowledge.
Financial Contributor	*Fellow:* Same as 1852.	*Fellow:* Same as 1852.
Foreign	*Corresponding Member:* Same as 1852.	None
Unrestricted Grade	None	*Subscribers:* Persons age 21 not eligible to other grades. (To 1895)

FIG. 7. — EVOLUTION OF ASCE

	1930	*1950*
General Eligibility Requirements	Corporate Member: Civil, military, naval, mining, mechanical, electrical or other professional engineer, architect or marine architect. Engineering degree equivalent to 4 years practice.	Civil engineer or person qualified in engineering or allied profession. Engineering degree equivalent to 4 years practice.
Eminence Grade	*Honorary Member:* Same as 1891.	*Honorary Member:* Same as 1891.
Advanced Professional Grade	*Member:* Age 35, with 12 years practice, 5 years responsible charge; qualified to direct, conceive and design engineering works.	*Member:* Same as 1930.
Professional Grade	*Associate Member:* Age 27, with 8 years practice, one year responsible charge.	*Associate Member:* Age 25, with 8 years practice, one year responsible charge.
Entrance Grade	*Junior:* Age 20, with 4 years practice or engineering degree. Must transfer to higher grade by age 33.	*Junior Member:* Age 20 with engineering degree or 4 years practice. Must transfer to higher grade by age 32.
Non-Engineer Professional Grade	*Affiliate:* Non-engineer professional age 35 qualified to cooperate with engineers; 12 years practice, 5 years responsible charge.	*Affiliate:* Same as 1930.
Financial Contributor	*Fellow:* Same as 1852.	None
Foreign	None	None
Unrestricted Grade	None	None

MEMBERSHIP REQUIREMENTS (Continued)

	1959	*1972*
General Eligibility Requirements	Same as 1950.	Same as 1950.
Eminence Grade	*Honorary Member:* Same as 1891.	*Honorary Member:* Eminence in engineering or related arts and science.
Advanced Professional Grade	*Fellow:* Transfer from Member only. Age 40; legally registered engineer with 5 years responsible charge as Member in engineering design or management.	*Fellow:* Transfer from Member only. Legal registration as engineer or land surveyor, with 10 years responsible charge as Member in engineering design, surveying or management.
Professional Grade	*Member:* Age 27; engineering degree with 10 years practice or 12 years practice; 3 years responsible charge.	*Member:* ECPD-accredited degree, or approved degree with legal registration, plus 5 years practice in responsible charge.
Entrance Grade	*Associate Member:* Age 20, with engineering degree or registration as Engineer-in-Training. Must transfer to higher grade within 12 years.	*Associate Member:* ECPD-accredited degree, or approved degree with registration as Engineer-in-Training, or Masters degree. Must transfer to higher grade in 12 years or become Affiliate.
Non-Engineer Professional Grade	*Affiliate:* Engineer or non-engineer professional qualified to cooperate with engineers, age 27, with degree and 10 years practice or 12 years practice, 3 years responsible charge.	*Affiliate:* Graduate architect, scientist, lawyer or other licensed or certified professional, or graduate in engineering technology with 5 years acceptable experience.
Financial Contributor	None	None
Foreign	None	None
Unrestricted Grade	None	None

7,500 members was authorized to be elected in any one year. Only 259 Honorary Members were named during the period 1852-1974, an average of but two per year. A roster of Honorary Members is given in Appendix IV.

The "Advanced Professional Grade" shows the greatest development, reflecting the progressive sophistication of standards of competence. The recognition given to legal registration since 1959 is important. Only registered engineers were made eligible for the new Fellow grade in 1959, and the Board rarely exempted a candidate from this requirement unless he was a non-resident of the United States. This was the first recognition given to registration in the membership standards of any of the Founder Societies.

The "Professional Grade" of Associate Member/Member reflects increasing emphasis on education and legal registration, as is the case with the "Entrance Grade" of Junior/Junior Member/Associate Member. The "Non-Engineer Professional Grade" or Associate/Affiliate was remarkably consistent until 1972. The substantial amendments to membership requirements at that time are noteworthy, first, for the elimination of the minimum-age limits and, second, for the automatic transfer to the Affiliate grade of Associate Members who fail to advance to a professional grade in the allotted time. Actually, in 1972 the Affiliate grade for the first time accepted both engineers and non-engineers.

The Fellow grade, which existed from 1870 to 1950, was available to anyone who contributed to the "Fellowship Fund," originally established to finance the publication of papers read before the Society. The fund was abolished in 1891, after which Fellowship fees were credited to "the permanent funds of the Society." The one-time Fellowship fee began at $100, but was increased to $250 in 1873. The number of Fellows in this grade ranged from 43 in 1870 to 40 in 1900 to only one when the classification was abolished in 1950.

The Fellow grade established in 1959 was decidedly different, in that it represented a truly advanced professional engineering category of members. The grade was attained by application, however, and was not conferred.

The grade of Corresponding Member was available from 1852 to 1877 to eminent foreign engineers who were willing to communicate with the Society at least once each year. Few Corresponding Members were enrolled, however, and after 1877 qualified foreign engineers were admitted as Non-Resident Members.

The Society maintained an "open" membership classification for those unable to qualify for other grades of membership for only four years, from 1891-95. This "Subscriber" classification also found little demand.

The evaluation and processing of membership applications have never been taken lightly in ASCE; the Society may well be the most "exclusive" organization to which many of its members belong. Since 1885 the candidate has been required to furnish a statement of his qualifications, and to name endorsers who can verify his good character and reputation as well as his education, technical training, and professional experience. An official application form, requiring signature by the applicant and certified endorsement by his proposers, was adopted in 1872. Lists of candidates were first posted in advance of election; later they were mailed to all members, and from 1930 until mid-1968 they were published in *Civil Engineering.*

Originally all membership admissions and transfers were voted upon by the entire body of members. In 1879 the election of Honorary Members was delegated to the Board of Direction; in 1890 the Board was authorized to elect applicants to all grades other than the Corporate (voting) grades of Member and Associate Member, and in 1909 the Board took over all membership functions.

By 1930 the volume of applications required further administrative structure in the Board, and the Committee on Membership Qualifications was created. In 1931 provision was made for Local Membership Committees to furnish "reports with respect to the admission or transfer of all applicants residing in the territory assigned to each local committee." These committees, renamed "Local Qualifications Committees" in 1946, have interviewed and reported upon thousands of membership candidates. They are believed to be unique in their function, at least in the engineering profession.

Since 1950 a membership application upon receipt was verified by the staff with regard to educational data and referred to the Committee on Application Classification. Meeting for a full day each month, this committee cleared those applications on which there was no question and these were submitted to the Board for election by mail ballot after a favorable consensus was received from the references. Where there was a question as to qualifications, or if the consensus of references was in doubt, the case was referred to the Local Qualifications Committee for investigation and interview. Doubtful cases were considered by the Committee on Membership Qualifications for specific recommendations to the Board.

This general procedure resulted in a declination rate of 6% to 10%. The most frequent cause of declination was a deficiency in qualifications, but some applications were declined where the candidate's professional attitude or reputation was held to be not in accord with the objectives of the Society.

Fig. 8 records the progression of annual dues in ASCE for the period 1852-1974. The differential in annual dues between Residents and Non-Residents prevailed until 1954, more than a century. The amount of fees and dues was determined by membership referendum until 1970, when such authority was delegated to the Board of Direction.

An entrance fee has always been a requirement for ASCE membership, in the amounts shown in Fig. 9. Since 1959 the $10 entrance fee has been waived for Student Chapter members who apply for Associate Member grade in the Society within a few months of their graduation. This inducement replaced the waiver of the first year's dues for "graduates of a school of recognized standing," which prevailed from 1926 to 1959.

Responsive to the economic pressures suffered by many

FIG. 8 — ANNUAL MEMBERSHIP DUES IN ASCE IN DOLLARS

Membership Grade	*1852*	*1853*	*1867*	*1868*	*1870*	*1872*	*1874*	*1891*	*1921*	*1930*	*1931*	*1950*	*1954*	*1971*	*1973*
M/F (R)	10	10	10	10	20	20	25	25	25	25	25	25	25	35	45
(N-R)	5	3	5	5	10	10	15	15	20	20	20	20			
AM/M (R)	—	—	—	—	15	15	—	25	25	25	25	25	25	35	45
(N-R)	—	—	—	—	10	10	—	15	20	20	20	20			35[3]
J/AM (R)	—	—	—	—	—	—	15	15	15	15	15	15	15	20	25
(N-R)	—	—	—	—	—	—	10	10	10	10	10	10			
Assoc./Aff. (R)	—	—	—	—	—	—	15	15	15	20	25	25	25	35	45
(N-R)	—	—	—	—	—	—	10	10	10	15	20	20			35[3]
Fellowship	—	—	—	100[1]	100[1]	250[1]	150[1]	250[1]	250[1]	250[1]	250[1]	[2]	—	—	—

(R) denotes Resident; (N-R) denotes Non-Resident.
[1]Lump sum single payment. [2]Fellowship abolished 1950. [3]Reduced dues paid by members outside U.S.A.
Grade of Corresponding Member (1891-95) required no payment of dues.
Grade of Subscribing Member (1891-94) required annual dues of $10 for Residents. $5 for Non-Residents.

FIG. 9. — ASCE ENTRANCE FEES

Membership Grade	*1852*	*1870*	*1872*	*1874*	*1891*	*1931*	*1959*	*1973*
M/F	$10	$10	$30	$30	$30	$30	None[1]	None[1]
AM/M	—	10	10	—	25	25	$25	Same
J/AM	—	—	—	20	10	10	10	Same
Assoc./Affil.	—	—	—	20	20	30	25	Same

[1]Admission to Fellow grade only by advancement from Member grade.

engineers during the Great Depression of the 1930's, ASCE acted in 1932 to waive the annual dues of some 2,033 members.

Only the Advanced Professional Grade of membership has carried the Corporate privileges of voting and of holding office throughout the life of the Society. The Regular Professional Grade, however, has also carried these privileges from the time it was instituted in 1891. Honorary Members not elected from the Corporate Member grades were given full privileges in 1950. The Entrance Grade was granted the right to vote in 1947, and the right to hold office followed in 1950. Also, in 1950, the non-engineer Associate/Affiliate Grade was given all membership privileges, but the rights to vote and to hold office were withdrawn in 1959.

An unusual feature of the ASCE membership requirements since 1891 has been the provision that a member in the entrance grade must transfer to a higher professional grade by a certain age or within a specified time. Several hundred members have been dropped each year for failure to meet this requirement. Many of these reinstate later. The effect of this requirement, that a young member achieve full professional status in order to become a permanent member of the Society, was studied in 1957. It was found that the loss of entrance-grade members was largely offset by lower annual loss rates in the higher grades of membership than are experienced by other engineering societies. Apparently there was some weeding out of individuals whose professional orientation was marginal.

The membership growth of ASCE is briefed in the following summary; a detailed record is furnished in Appendix IX:

Year	*Total Membership*
1852	55
1875	408
1900	2,221
1925	11,275
1950	28,105
1974	69,671

Only the Institute of Electrical and Electronic Engineers (IEEE) exceeds ASCE in membership among all the engineering organizations in the United States serving a single major engineering field.

ASCE has become one of the largest engineering societies in the world, despite its precarious beginning and its high admission standards. These requirements and the painstaking manner in which they are administered have limited solicitation of new members because of the embarrassment inherent to an invitation to a candidate who might later be denied admission. Hence, until 1974, ASCE did not conduct membership campaigns, and the distribution of application forms had been carefully controlled.

A change in policy in 1974 with regard to membership development resulted for the first time in the publication of advertisements in the technical press inviting membership applications. More aggressive promotional effort was also encouraged in the local sections. The effect of this new policy on the stature and prestige of Society membership must await future assessment.

It will be evident from the foregoing that ASCE has always given primacy to the quality of its membership over mere numbers. A heavy investment in budgetary resources, staff effort, and thousands of man-hours annually in voluntary committee service by members of the Society is required. This is the only way to provide the meaningful "professional identification" that is compatible with the responsibilities of the civil engineering profession.

Role of ASCE in Engineering Registration

Toward the end of the 19th century it was becoming apparent that ASCE membership was not sufficiently recognized in itself as a mark of professional competence to protect the public from the consequences of incompetent engineering. The numerous and usually tragic failures of railway bridges and dams in the period 1870-95 were certainly in many cases the result of the inadequacies of unqualified persons holding themselves to be civil engineers (see Chapter VI).

The movement to bring about the enactment of legislation that would enable only those qualified by education and experience to offer their services as civil engineers originated within the profession. The reaction against this proposal was so strong, however, that almost 40 years were to elapse before the regulation of engineering practice by registration, or licensure, was to be wholly endorsed by ASCE. Nevertheless, the Society played a dominant role in the development of the registration process.

At the 1897 Annual Meeting of ASCE the following resolution was proposed for adoption by S. C. Thompson, M. ASCE:

> "In consideration of the fact that the engineering profession and the people of this country have no legal protection against any incompetent or unscrupulous person who chooses to advertise or sign himself with the title of Civil Engineer, thereby reflecting upon and injuring the entire profession, and
>
> "In consideration of the fact that it has been found advantageous in most states to grant legal protection to members of the legal and medical professions:
>
> "Therefore Be It Resolved: That the American Society of Civil Engineers places itself on record as favoring judicious legal restrictions against the unauthorized and improper use of the title of Civil Engineer."

The resolution was promptly tabled, as it was when presented by Mr. Thompson again at the 1898 Annual Meeting and still again at the 1899 Annual Meeting. But Mr. Thompson's initiative and persistence had brought the issue into focus. At the 1901 Convention similar questions were posed and were discussed to great length in light of the fact that at this time any person who chose to do so could practice civil engineering without regard for his education, training or experience. These questions were:

> "1. Do the interest of the profession, and the duty of its members to the public, require that only those who are competent be allowed to practice as Civil Engineers?
>
> "2. Under what authority, through what agency, and upon what evidence of competency should applicants be admitted to the practice of Civil Engineering?"

Discussion of these cogent points elicited a proposal that licensure be administered at the state level by a Board of Examiners, with three classes of license: The first class to identify engineers of acknowledged eminence, the second class to encompass a lower echelon of engineers of proved qualification, and the third class to include recent graduates of acceptable engineering colleges or non-graduates in their early years of practice. Advance in class of license would be by examination.

In the closure of the discussion it was observed:

> ". . . There can be no doubt that National legislation would be preferable (to many state laws) if it were possible, but it must be remembered that legislation of the character contemplated is, by the Constitution of the United States, distinctly reserved to the several states, and the Congress has no authority to enact such laws."

The 1901 exercise resulted in the appointment of a Committee on Regulating the Practice of Engineering, chaired by Samuel Whinery, M. ASCE. (Mr. Whinery was the first signer of the 1893 request from the Cincinnati Association of Engineers that formualtion of a code of ethics be undertaken by the Society.) The committee submitted an exhaustive report in May 1902 (*Proceedings*, August 1902, p. 188), in which it described laws regulating the practice of civil engineering in Canada, of engineering in Mexico, and of medicine, law, pharmacy and dentistry in the United States. It was also reported that laws regulating the practice of architecture had been enacted in California, New Jersey, and Illinois, and the successful administration of the Illinois law was noted.

The committee recommended, however, "that no further action be taken in the matter by the Society at this time," giving as reasons:

- The difficulty of obtaining consistent laws in all the states.
- The desirability that any laws cover all branches of engineering, and the lack of interest in the subject outside of ASCE.

- The fact that, unlike physicians and lawyers, only "engineers engaged in public service" actually required regulation by law in the public interest.

The Board accepted the recommendation.

The Birth of Engineering Registration

The issue of licensure or registration was avoided at the national level of the Society during the next eight years, but this did not deter significant developments by ASCE members in their own states. Most important was the action taken in Wyoming in 1907, which resulted in enactment of the first state registration law for engineers in the United States.

Clarence T. Johnston, M. ASCE, State Engineer of Wyoming, became greatly concerned about the widespread practice of engineering by incompetents. In his own words*:

> ". . . Those proposing to use the waters of the state were obliged to file an application for permit in the office of the State Engineer on prescribed forms and accompanying the same with a map showing streams, canals and reservoirs and lands to be irrigated. Under the law a survey was required. However, I discovered that lawyers, notaries and others were making the maps and signing them as engineers and surveyors."

Mr. Johnston was one of eleven engineers in Wyoming enrolled in the Member grade of ASCE in 1907. In consultation with two of them—William Newbrough and Edward Gillette—a bill was drafted in his office and submitted to the state legislature. Enactment followed after nominal delay brought about by the self-styled engineers and surveyors who created the problem. With Mr. Johnston as chairman of the new Board of Engineering Examiners, examinations were promptly held. In the next two years 134 engineers were registered in the categories of Land Surveyor, Topographic Engineer, Hydraulic and Topographic Engineer, Construction and Designing Engineer, and Administrative Irrigation Engineer.

Proceedings, National Council of Engineering Examiners *(1970)*.

The Wyoming law set off a chain reaction. Louisiana followed with a licensure act in 1908, and by 1910 the legal regulation of professional engineering practice was a live issue in many states across the land. By 1920 twelve states had enacted laws.

When engineering licensure legislation was introduced into the New York State Assembly in 1910, the author of the bill, the Hon. J. L. Raldires, requested that it be supported by ASCE. No stated position was taken on the bill by the Society, but it was withdrawn after a delegation headed by President John A. Bensel conferred with the New York State Commission on Public Education in Albany. (Seven years later, in 1917, the minutes of an Executive Committee meeting stated that the Society ". . . had taken action in defeating vicious legislation in New York some years ago.")

In anticipation of the consideration of further licensing legislation by the New York State Legislature, the Board of Direction adopted this resolution in April 1910:

> "Resolved, that it is the sense of the Board that it is the duty of the American Society of Civil Engineers to use its influence in the proper formulation of all legislation by the general government or by any of the States of the Union, which affects the practice of Engineers and the Board recommends the appointment by the Society of a Committee whose duty it shall be to formulate the general lines on which such legislation shall be based, and that said Committee be requested to report at the next Annual Convention."

The First Model Law

This resolution was implemented with great expediency, for in January 1911 ASCE adopted the first "model law" entitled an "Act to Provide for the Licensing of Civil Engineers." To insure a uniformly high standard of competency in the profession the guide prescribed fifteen sections embodying "proper requirements" to be imposed in licensing the practitioner of civil engineering. The following minimum qualifications were recommended for the licensed Civil Engineer:

1. That he shall be not less than 25 years of age.
2. That he shall be of good moral character.
3. That he shall have been actively engaged in Civil Engineering work, as an assistant to a licensed practitioner, for at least six years, and as such shall have had responsible charge of engineering work for at least one year.
4. For a graduate of a school of engineering of recognized reputation, the terms of actual engagement as an assistant shall be four years.

The committee further recommended that licensing laws be administered by a State Board of Examiners comprising five qualified Civil Engineers, each with at least 10 years of practical experience, and that the annual registration fee not exceed $25. In order to accomplish nation-wide uniformity in the enactment and administration of laws three Corporate Members of the Society from each state of the Union were appointed to serve in the Standing Committee on Registration of Engineers.

The resolution adopting these recommendations is particularly significant:

"*Whereas:* There are National Societies of Engineers in the United States, membership in which can only be secured after rigid examination of the fitness of applicants to practice as Engineers; and

"*Whereas:* The public has ample protection if they will employ only those who have thus demonstrated their ability; be it

"*Resolved:* That the Board of Direction of the American Society of Civil Engineers does not deem it necessary or desirable that Civil Engineers should be licensed in any State; and be it further

"*Resolved:* That if notwithstanding this, the Legislature of any State deems the passage of a statute covering the practice of Civil Engineering desirable for the protection of the public, the accompanying draft of such a statute, which has been prepared by the Board as embodying proper requirements for that purpose, is recommended."

Thus, the Society became involved in the licensing process unwillingly, only because of the urgency imposed by unilateral action in the States toward the enactment of licensure legislation.

It is noteworthy here that the American Institute of Architects (AIA) also opposed the advent of licensing in that profession some years earlier, at the turn of the century. AIA also was influential in shaping the pattern of registration, while at the same time openly disfavoring the movement toward such legislation in the states.

This position was steadfastly maintained, although a special Committee on Licensing of Engineers was formed in 1911 to keep the Board of Direction informed as state after state adopted registration laws. The local sections were encouraged to become involved with licensing legislation only when it became a live issue in the state.

A 1913 proposal by C. H. Higgins, M. ASCE, to put the issue of statutory regulation of engineering practice to a ballot of the membership, was summarily quashed by a tabling action. The refusal even to discuss the public-interest aspects of the matter casts a shadow upon the image of the Society at this time.

In 1923 the Committee on Licensing of Engineers presented the following resolution, which was adopted by the Board:

"*Whereas,* Laws requiring registration or licensing of engineers and (or) land surveyors are in force in twenty-five states; and,
"*Whereas,* This Board of Direction, without endorsing the registration of professional engineers, considers that any State laws adopted should be uniform.
"*Be It Resolved,* That the President appoint a Standing Committee on Registration of Engineers, at least one of whom shall be a member of the Board of Direction; that the duties of this Committee shall be:

"(1) To keep this Society informed as to the progress of registration or licensing of engineers.

"(2) To draft a registration law for professional engineers and land surveyors for approval by the Board of Direction.

"(3) To keep in touch with the Local Sections of this Society, lending assistance in securing the adoption of the approved draft of a new law where registration of this character is proposed by a State Legislature."

The new permanent Committee on Registration of Engineers undertook immediately its difficult task of drafting a model bill that would provide the guidance so sorely needed in states planning licensure legislation. To be truly effective, of course, the model required the endorsement and support of as many of the engineering societies as possible. For the next several years the committee labored diligently on its assignment, making many amendments after wide circulation and joint committee discussion on the several successive drafts. Finally, in 1929, a Model Law was issued under the title "Recommended Uniform Registration Law for Professional Engineers and Land Surveyors." The primary aim was to encourage as much consistency and uniformity as possible in the various state laws.

The ASCE draft law was then presented to other engineering societies in a series of special conferences in an effort to obtain their endorsement. Three years more of intersociety committee discussion and revision took place before the following resolution emerged:

"*Resolved:* To recommend the adoption of the Model Law for the Registration of Professional Engineers and Land Surveyors by all National, State and local organizations of engineers as a model to be followed in the framing of all new registration laws and the amending of existing laws, with a view to attaining a uniform high standard throughout the United States."

Only three national organizations were sufficiently interested to join ASCE initially in adopting the Model Law: the American Society of Heating and Ventilating Engineers, the AAE, and the NCSBEE. The ASME gave approval "in principle." Possibly there was some difficulty here in reconciling the public-works orientation of the civil engineer with the industrial orientation of the other branches.

Continuing joint negotiation and amendment brought about endorsement by eight national organizations by 1937. Intensive efforts in 1946 brought the list to thirteen societies.

A policy adopted in 1935 finally delineated sharply and clearly the attitude of the Society toward engineering registration. The enactment of registration laws was endorsed in all states not having such legislation, and the improvement of all inadequate laws was encouraged. A more comprehensive policy was given effect in 1957, including these elements:

(1) Support and endorsement of the registration of engineers and land surveyors as being in the best interests of the public.
(2) Support and endorsement of the Model Law.
(3) Continuing support and cooperation with NCSBEE.
(4) Recommendation that Engineers Joint Council cooperate with NCSBEE.
(5) Recommendation of specific conditions under which corporations might be registered or licensed.

Since 1960 periodic review and maintenance of the Model Law have been managed by the National Council of Engineering Examiners (NCEE), formerly the NCSBEE. Although ASCE has continued to take an active part in the operation, through its participation in NCEE, certain conflicts with ASCE registration policy have precluded endorsement of the Model Law by ASCE since 1946. Difference of viewpoint over the provisions dealing with practice in engineering corporations has been mainly responsible for this paradox. The ASCE dissension arises from its representation of the interests of the large segment of engineers in private practice in its membership. The situation illustrates the difficulty of achieving absolute unity of opinion throughout such a heterogeneous profession as is engineering.

In March 1965, the Board of Direction adopted a strong position paper in which it questioned the propriety of NCEE activity to influence the leglislation that its member state boards are required to administer. Specifically, objection was expressed concerning:

(1) Efforts to integrate registration boards more closely with the state administrative structure.
(2) Extension of the registration requirement to all engineers, regardless of the basic public protection function of registration.
(3) Efforts to register corporations.

The paper states:

> "As the professional society whose members are perhaps more deeply concerned with registration than those of any other, and the society more than any other responsible for the development of registration law, ASCE endorsement [of the Model Law] is essential. To earn that endorsement, the law must square with ASCE policy . . . ASCE must resume its traditional leadership in this field; it must do it quickly, and it must commit its strength fully to the struggle."

Civil engineers far outnumber any other branch of engineering in identifying themselves legally as engineers through the registration process. Even before registration was made a requirement for the new Fellow grade in 1959, surveys showed that about two-thirds of all ASCE members were registered in at least one state, and that an additional one-sixth were engineers-in-training. Taking into account the substantial number of foreign members not subject to USA registration, the overall percentage of domestic ASCE members registered and in the EIT phase is on the order of 90%.

By 1970, there were 55 engineering registration boards in operation, including the 50 states, the District of Columbia, the Commonwealth of Puerto Rico, the Panama Canal Zone, Guam, and the Virgin Islands.

Engineering registration appeared to be acquiring a new

facet in 1974, in the proposals in some states to make renewal of the license contingent upon a showing by the licensee that he was keeping apace with advancing technology in his field. The "requalification" requirement, if widely adopted, would create new demands upon the professional societies. In ASCE the impact upon continuing-education activities would be especially great because of the substantial number (80-85%) of members who are registered.

Other Proposals for Professional Recognition

It is appropriate here to mention the two proposals that have been made toward the identification of engineering qualification by a system of certification, even though they did not originate in ASCE.

The first of these moves was made by ECPD in 1934, when it proposed to publish an annual roster of all engineers who could qualify under the following definition:

> "Graduation from an approved course in engineering of four years or more in an approved school or college; a specific record of an additional four years or more of active practise in engineering work of a character satisfactory to the examining body (the examining body, in its discretion, may give credit for graduate study in counting years of active practise); and the successful passing of a written and oral examination covering technical, economic, and cultural subjects, and designed to establish the applicant's ability to be placed in responsible charge of engineering work and to render him a valuable member of society; or alternatively,
>
> "Eight years or more of active practise in engineering work of a character satisfactory to the examining body, and the passing of written and oral examinations designed to show knowledge and skill approximating that attained through graduation from an approved engineering course, and also examinations written and oral covering technical, economic, and cultural subjects designed to establish the applicant's ability to be placed in responsible charge of engineering work and to render him a valuable member of society."

The certification process would have been administered by

ECPD, and a roster of 40,000 to 100,000 names was anticipated.

The ASCE Board of Direction approved the plan in April 1935 after it had previously been approved by ASME and NCSBEE. Apparently there was resistance by other ECPD societies, however, for in 1936 the ECPD Committee on Professional Recognition referred only to professional education, professional registration, and engineering society membership in a professional grade as the means for formal identification of the engineer. The annual report of that committee went on to state:

> "The proposal of any additional procedure of certification or recognition would only be adding a fourth method to the three methods of progressive recognition already established. It would introduce new competition or conflict and new difficulties of correlation, and would therefore not be a solution of the problem of harmonious coordination."

In 1970—noting that less than half the graduate engineers in the United States were registered—NCEE proposed that a system of "accreditation" be adopted as an adjunct to licensure as a means of identifying the engineering profession more completely. Under the plan an engineer could attain professional identification either through the conventional registration process or he could become an "Accredited Engineer" through a "coordinating agency" to be established by the engineering societies.

Except for passing interest on the part of NSPE, the proposal aroused little enthusiasm. Through the medium of Engineers Joint Council it was still under review and consideration by ASCE and other engineering societies in 1974.

The early lack of enthusiasm in ASCE for legal registration can be understood and respected, as this meant the relinquishment of an important professional responsibility to a quasi-political function. Once it was realized, however, that licensure by state law was a permanent reality, ASCE provided the leadership and direction that were sorely needed at the national level.

The progression of model laws advanced by the Society all centered upon administration by state boards composed of professional engineers of acknowledged competence and reputation. This insured management of the operation within the profession. Moreover, the local sections of the Society maintained continuing awareness of appointments to the state boards, as well as amendment of the laws, and were quick to react to any threat of undue political influence.

Had it not been for the conservatism of ASCE, it is possible that the registration movement might have gone too fast and too far in "closing" the profession as a self-serving measure. While it is evident from the early discussions that the resistance to registration before 1930 was anything but altruistic in its motivation, it is equally clear that the public interest was given primacy in ASCE deliberations on the subject since that time. When the concept of licensure was finally accepted as a professional responsibility, it was accorded the highest order of professional treatment.

Standards of Ethics and Professional Practice

ASCE concern with ethical matters was plagued by the same ultra-conservatism and intransigence that characterized its approach to other professional responsibilities in the 19th century. Yet, again according to pattern, its recalcitrance turned into leadership for the engineering profession. Both AIEE (1912) and ASME (1912) were to antedate ASCE in adopting a code of ethics. Enactment of the ASCE code in 1914, however, was accompanied by a conscientious enforcement policy, the implementation of which was to give the Society stature in this area that was unmatched anywhere else in the engineering world.

Informal though it might have been in the mid-1850's, the practice of civil engineering was strongly imbued with the professional ethic of public service. This appeared to be tacitly accepted as a necessity on the part of those responsible for the direction of large and expensive public-works enterprises. However, professional ethics were considered to be a strictly personal responsibility—a matter of honor. The practitioner was expected to police his own actions through his conscience.

John B. Jervis, having just been elected an Honorary Member, gave considerable emphasis to the ethical relationships of the engineer in a classic address presented at the first Annual Convention, in June 1869. Speaking generally on the subject, he said:

> ". . . the engineer eminently depends on character. The interests of others, in various ways, are committed to him. On his capacity for his profession, and his integrity as a man, reliance must be placed. He will meet many difficulties of a physical, and not a few of moral nature."

More specifically, on the relationship between the engineer and his client, Mr. Jervis observed:

> "He [the engineer] has to deal with men who, as a class, are proverbially sharp in the conduct of their affairs, with whom many questions arise that are not to be determined by a simple computation, and even computations will be questioned. With these men the engineer holds delicate relation, as the umpire between the contracting parties; and will often be placed between plausible claims on one hand, and a sense of duty on the other. In these circumstances, he will have great opportunity to obtain a business knowledge of men. Some he will find upright though they may have mistaken views of their rights; and others he will find under much pretension to seek what they should not have. Under various conflicts the engineer must aim to do justice between the parties. They have committed to him the duty of adjusting all questions, and in this he must examine the bearing of all claims, and though he may be annoyed at what he thinks an unjust demand, he is in duty bound to render equity, according to the terms of the contract. The engineer being in the employ of one of the parties, it is indispensable that he maintain the reputation of an upright man, for on this the contractor must depend. If he shows a disposition to take undue advantage, not warranted by the

terms and spirit of the contract, the contractor will lose confidence if this is against him, and, if in his favor, the other party will be dissatisfied. In all such cases, committed to the judgment of the engineer, he will need the best experience as a business man, and especially to cultivate the golden rule of doing as he would be done by."

and further to this same point:

". . . There is another point in this connection that will often be more trying than the above. The managers of such enterprises have been known to conduct their affairs with a view to make them subservient to their private interest or ambition. They may do this more or less, without interference with the duties of the engineer, in which case he may know nothing of it, or, if he does know, he may not be bound as a duty to take notice of it. But it is very likely to come in conflict, and he will be expected to shape his professional duties in a way that will promote private interests at the expense of the institution. Great skill and adroitness will be practiced, and if the engineer has any weak side it will surely be found. These things will always be done under profession of serving the institution. To avoid wrongdoing and, at the same time, give no just offense under the circumstances, will surely try the business capacity of an engineer. The matter in issue may bring a crisis that will compel resignation."

The first proposal of ethical nature to come before the Society was the following resolution submitted in a meeting on October 18, 1877, by Secretary Gabriel Leverich:

"*Whereas:* A Civil Engineer, in the practice of his profession, is sometimes restrained or overruled by his employers, in matters involving serious risk to property and life which he *only*, as the engineer should determine; whence he must either discharge his duties in a manner contrary to his best judgment or resign his position;
"*Resolved:* That in the opinion . . . of the American Society of Civil Engineers, . . . it is unprofessional for a civil engineer to continue the discharge of his duties when so restrained or overruled; or to accept an engagement which is generally known to have been vacated because of such interference . . .

until the judgment of his predecessor has been formally disapproved by other disinterested civil engineers . . ."

This resolution, and an alternately worded one of similar import offered by Past-President William J. McAlpine, were considered at the 1877 Annual Meeting and were referred to the Board of Direction. In so doing, however, it was:

"*Resolved:* That it is inexpedient for this Society to instruct its members as to their duties in private professional matters."

With this mandate, it is not surprising that the Board took no further action upon the Leverich and McAlpine motions.

Some fifteen years later, in 1893, there was persistent reference in *Engineering News* to the need for a code of ethics in engineering. The issue came before the Board of Direction in a communication from the Cincinnati Association of ASCE Members, which stated that "the time has come when it is both possible and desirable that a Code of Ethics for the guidance of the members of the profession of Civil Engineering should be formulated and adopted." It was urged further that a special committee be appointed to draft a code. The letter was signed by fourteen members of the Cincinnati group, the first being Samuel Whinery, M. ASCE, who some years later was to head an important committee dealing with licensing.

The proposal was presented for membership consideration at the 1893 Annual Meeting. Speaking in favor of it, Clemens Herschel (President, 1916) cited the need for ethical standards if civil engineering were to take its proper place with medicine and law, as follows:

"Is a civil engineer a member of a profession or is he merely following some calling? . . . . What are the relations of civil engineers to each other? . . . It is by reason of the absence of a code of ethics that civil engineers presumably have not been considered professional men, and have not the respect that is given to professional men by the general public.

". . . The Society from the beginning has endeavored to say to the public that whoever is a member of the Society may undoubtedly be considered a civil engineer. That also is one of the reasons for having the engineering profession considered a profession, namely, that there shall be some society which shall pass upon the qualifications of the candidate and say whether or not he shall be considered a civil engineer, and it is all right that within such a society the professional standing of its members shall be subjected to definite rules—a court of ethics and a code of ethics . . ."

Extensive discussion produced an action referring the appointment of a special committee to the Board of Direction, but the Board—following its conservative tradition—did not favor such a move, and so recommended to the membership at the July 1893 Convention. Further general discussion elicited a proposal that the code-of-ethics question be presented to the Corporate Membership for a referendum. The motion was lost, and the issue was not to emerge again for almost a decade.

The topic, "The Regulation of Engineering Practice by a Code of Ethics," was a program feature of the May 1902 Convention. The 1893 experience was reviewed, but the membership sentiment exposed by the discussion was still opposed to the formulation of a code. The consensus was expressed in an address by President Robert Moore:

"And it is, I think, safe to say that for the kindling of professional enthusiasm, and the establishment of high professional standards, the Society and its members will continue to rely, as they have done in the past, upon these vital and moral forces, and not upon the enactment of codes or upon any form of legislation."

Another decade passed, during which the only documented reference to professional ethics was made in several papers published in *Transactions*.

A letter addressed to the Secretary of the Society by Percival M. Churchill was presented to the membership at the 1913 Annual Meeting in which the following proposal was set forth:

"*Moved:* That this Society shall consider the adoption of the Code of Ethics for Engineers recently proposed by a committee of the American Society of Mechanical Engineers and published in *Engineering News,* January 2, 1913.

"That the proposed Code be printed as a letter ballot and submitted to the members of this Society with the request that each member vote to accept or reject the Code article by article; that where a member so desires he shall—after voting against a certain article—submit a substitute in the form he desires the article to take. In the same manner additions may be presented.

"This ballot to be closed on April 1st, 1913.

"The Board of Direction shall then send out a second letter ballot giving the Code as first proposed and also any suggested changes and additions. Members shall again vote article by article. The ballots to be opened at the next Annual Convention, and any article having a majority of the votes cast shall be declared adopted.

"The Code as thus adopted to be then printed in the *Proceedings* and the *Transactions* of the Society."

The communication was referred to the Board of Direction, which acted in February 1913 to authorize appointment of a committee "to report on the whole question as to whether it is desirable for the Society to adopt a Code of Ethics, and if they think favorably of it, to present to the Board a short Code of Ethics."

It will be noted that the committee's instruction contained no reference to the ASME code mentioned by Mr. Churchill. That instrument was similar to the one adopted by AIEE in 1912. There is also marked similarity both in format and content to the present Canons of Ethics of Engineers, as promulgated many years later by ECPD. Incidentally, two members of the five-man committee that drafted the ASME code, Charles T. Main and Spencer Miller, were members of ASCE.

The special committee, consisting of Mordecai T. Endicott, John F. Wallace and Henry W. Hodge, transmitted its draft of a "short" code, as requested, to the December 1913 meeting of the Board. The draft was reviewed in the next two meetings, amended by change of language in two articles and by addition of one new article, and then approved for favorable recommendation to the membership at the June 1914 Convention. After further general oral discussion and presentation of several letters on the subject, the draft code was adopted by the meeting and ordered submitted to the Corporate Members for mail ballot.

Code of Ethics Adopted

On September 2, 1914, the result of the ballot was canvassed and the Code of Ethics was ordered adopted by vote of 1,997 ballots in favor from a total of 2,162 ballots cast. The Code, in the language as adopted, is reproduced in Fig. 10.

The Code of Ethics has undergone amendment eight times since its original adoption in 1914 (1934, 1941, 1949, 1950, 1956, 1961, 1962 and 1971). To detail all of these amendments is not appropriate here. Instead, the 1914 and the 1974 Codes are reproduced in Fig. 10 so that they may readily be compared. It will be noted that not a single one of the original articles has escaped revision; even the preamble has been modified.

The story of the Code of Ethics is not complete, however, without recounting the fate of one article that does not even appear in the 1974 instrument.

The fourth article of the original 1914 Code read as follows:

> "4. To compete with another Engineer for employment on the basis of professional charges, by reducing his usual charges and in this manner attempting to underbid after being informed of the charges named by another."

In 1949 a new companion article was added:

> "4. To participate in competitive bidding on a price basis to secure a professional engagement.
>
> "5. To compete with another Engineer for employment on the basis of professional charges, by reducing his usual charges and in this manner attempting to underbid after being informed of the charges named by another."

both of which read as follows after the 1956 amendments:

FIG. 10 — COMPARISON OF 1914

Adopted by the Society by Letter Ballot, September 2, 1914

It shall be considered unprofessional and inconsistent with honorable and dignified bearing for any member of the American Society of Civil Engineers:

1. *To act for his clients in professional matters otherwise than as a faithful agent or trustee, or to accept any remuneration other than his stated charges for services rendered his clients.*

2. *To attempt to injure falsely or maliciously, directly or indirectly, the professional reputation, prospects, or business of another Engineer.*

3. *To attempt to supplant another Engineer after definite steps have been taken toward his employment.*

4. *To compete with another Engineer for employment on the basis of professional charges, by reducing his usual charges and in this manner attempting to underbid after being informed of the charges named by another.*

5. *To review the work of another Engineer for the same client, except with the knowledge or consent of such Engineer, or unless the connection of such Engineer with the work has been terminated.*

6. *To advertise in self-laudatory language, or in any other manner derogatory to the dignity of the Profession.*

AND 1974 CODES OF ETHICS

As Amended to October 19, 1971

It shall be considered unprofessional and inconsistent with honorable and dignified conduct and contrary to the public interest for any member of the American Society of Civil Engineers:

1. *To act for his client or for his employer otherwise than as a faithful agent or trustee.*
2. *To accept remuneration for services rendered other than from his client or his employer.*
3. *To attempt to supplant another engineer in a particular engagement after definite steps have been taken toward his employment.*
4. *To attempt to injure, falsely or maliciously, the professional reputation, business, or employment position of another engineer.*
5. *To review the work of another engineer for the same client, except with the knowledge of such engineer, unless such engineer's engagement on the work which is subject to review has been terminated.*
6. *To advertise engineering services in self-laudatory language, or in any other manner derogatory to the dignity of the profession.*
7. *To use the advantages of a salaried position to compete unfairly with other engineers.*
8. *To exert undue influence or to offer, solicit or accept compensation for the purpose of affecting negotiations for an engineering engagement.*
9. *To act in any manner derogatory to the honor, integrity or dignity of the engineering profession.*

Under the Code of Ethics of the American Society of Civil Engineers, the submission of fee quotations for engineering services is not an unethical practice. ASCE is constrained from prohibiting or limiting this practice and such prohibition or limitation has been removed from the Code of Ethics. However, the procurement of engineering services involves consideration of factors in addition to fee, and these factors should be evaluated carefully in securing professional services (Added July 1972).

The Society has also endorsed the *Fundamental Principles of Professional Engineering Ethics* of the *Canons of Ethics* as adopted by Engineers' Council for Professional Development on September 30, 1963, by Board action on May 11-12, 1964.

†On foreign engineering work, for which only United States engineering firms are to be considered, a member shall order his practice in accordance with the ASCE *Code of Ethics*. On other engineering works in a foreign country he may adapt his conduct according to the professional standards and customs of that country, but shall adhere as closely as practicable to the principles of this Code. (Adopted by ASCE Board of Direction October 7-8, 1963.)

"4. To invite proposals for the performance of engineering services or to state a price for such services in response to any such invitation when there are reasonable grounds for belief that price will be the prime consideration in the selection of the engineer.

"5. To compete with another engineer for employment on the basis of professional charges, by reducing his usual charges and in this manner attempting to underbid after being informed of the charges named by another."

and were combined into this single concise provision in 1961:

"3. To invite or submit priced proposals under conditions that constitute price competition for professional services."

The competitive bidding problem was of major concern as early as 1925. It assumed extraordinary importance in 1953 when the Chief Highway Commissioner of South Carolina published a formal "Notice to Bridge Engineers," stating that sealed bids to be submitted on official proposal forms would be received for engineering services in connection with a bridge project. A number of consulting firms responded to the invitation, and the award was made to the lowest bidder. As a result, the Highway Commissioner and principals in thirteen of the firms that filed bids were charged with violation of the Code of Ethics by their participation "in competitive bidding on a price basis to secure a professional engagement." After thorough investigation and deliberation in the Committee on Professional Conduct and the Board of Direction, with careful observation of "due process" requirements, the fourteen members of the Society were found guilty of the charges. The Highway Commissioner was expelled from membership; one consulting engineer was suspended for five years and the remaining twelve bidders were suspended for one year.

This landmark case began an era of great activity on the part of ASCE — later to be joined by the NSPE and the Consulting Engineers Council — in encouraging prospective clients to engage professional engineering services by the process of "professional negotiation." In this procedure the quality of the service to be rendered is given primacy, and competition on a price basis is precluded. The campaign was waged by the distribution of published manuals and pamphlets, hundreds of letters, telegrams, and telephone communications, and by many personal staff contacts with public- and private-enterprise agencies involved with the employment of consulting engineers. The total effort was primarily educational in character, but the point was emphasized that members of ASCE who engaged in competitive bidding were acting in an unethical and unprofessional manner. There were occasional professional conduct proceedings involving various degrees of disciplinary action for violations of the competitive bidding provisions, but none attained the importance of the 1953-54 "South Carolina Case."

The total effect of this effort from 1954 to 1971 was to bring about almost universal acceptance of "professional negotiation," as defined by ASCE, among the users of professional engineering services. Early in 1971, however, as a result of insistence on competitive bidding on the part of certain federal bureau officials, the Society was charged by the U.S. Department of Justice with violation of the anti-trust provisions of the Sherman Act. The charges centered on the "Article 3" competitive bidding clause in the Code of Ethics, and its administration.

The action of the Department of Justice was not altogether unexpected. In 1956, ASCE headquarters was visited by an agent of the Department who investigated in great detail the ethical standards and activities of the society relating to engagement of professional services. There was never any official communication from the Department after that visit. In 1961, a "Legal Audit" of Society policies and programs elicited a warning from counsel that administration of the competitive bidding policy as an ethical matter was vulnerable to Sherman law interpretation. This was followed soon after by inquiries on the part of the Justice Department concerning similar competitive bidding provisions in the professional practice codes of the American Institute of Certified Public Accountants and the American Institute of Architects.

Thus in 1971 the Board of Direction, under the sage leadership of President Oscar S. Bray, was confronted with a grave and far-reaching decision. There was exhaustive deliberation of *(a)* the advice of eminent legal counsel as to the chances for successful defense of the suit, *(b)* the dire inhibitive and possible fiscal consequences of an unsuccessful defense, and *(c)* the possibilities for a negotiated settlement that would permit the Society to continue to advance its philosophy of professional negotiation by education, persuasion, and legislation, if not as a matter of ethics.

The Board acted with considerable courage and conviction in October 1971 when it moved voluntarily to delete Article 3—"To invite or submit priced proposals under conditions that constitute price competition for professional services"—from the Code of Ethics. About 500 members (of 67,000) reacted with letters, telegrams, and telephone calls protesting the deletion, but most of these had missed the earlier reports concerning the Department of Justice action that had been published in *Civil Engineering*. They were largely mollified when the circumstances were explained.

The voluntary elimination of the ethical standard in question opened the door to a Consent Decree settlement of the Department of Justice suit in mid-1972. The settlement left essentially intact the vitally important Manual of Practice No. 45, *Consulting Engineering — A Guide for the Engagement of Engineering Services*, which sets forth the traditonal ASCE professional negotiation procedure that had been recommended for so many years. Ironically, only a few months after settlement of the suit, Congress enacted legislation — with vigorous ASCE support — prescribing a professional selection and negotiation procedure for engagement of architect-engineer services (Public Law 92-582, 92nd Congress).

Particular attention is drawn to the footnotes appended to the 1974 Code of Ethics in Fig. 10. The note added in July 1972 concerning quotation of fees for professional services is in compliance with one of the terms of the Consent Decree resulting from the Department of Justice suit. The footnote added in October 1963 is a practical expedient to enable American engineering firms engaged in foreign work to compete with firms from other countries in which there are no ethical constraints. The provision, promptly dubbed the "When in Rome Clause," has been highly controversial; several efforts to invalidate it have been narrowly averted.

Professional Practice Guidance

Apparently the "short" Code of Ethics ordered by the Board of Direction in 1913 permitted sufficient latitude of interpretation to bring about some differences of opinion. About 1925 a committee of the Northeastern Section of the Society presented to the Board of Direction a "Code of Practice" primarily "for those engaged in construction," and providing practical guidance to the individual engineer in ruling his professional conduct. The document was approved by the Board of Direction after membership reaction was solicited and some revision by the Committee on Professional Conduct. It was published in 1927 as Manual of Engineering Practice No. 1, to initiate a distinguished series of ASCE publications.

One of the foremost authorities on ethics in the history of engineering was Prof. Daniel W. Mead, Hon. M. ASCE, of the University of Wisconsin. During his tenure as President of the Society in 1936 he observed that "most engineering ethical codes seemed to apply almost exclusively to the engineer in general practice and not to the more than 90% of the profession who are public or private employees." When he took this view to the Committee on Professional Conduct and the Board of Direction it was pointed out that the word "client" where appearing in the Code of Ethics was to be inclusive of the term "employer." At the same time, Professor Mead was invited to prepare "an inclusive paper on this subject as a basis for a wide and full discussion, and that such paper after discussion be made readily available to all engineers." The result of this action was Manual of Engineering Practice No. 21, *Standards of Professional Relations and Conduct,* adopted for publication in October 1940. Many thousands of copies of this classic monograph have been distributed in the years since.

It was not until 1961, however, that an official "Guide to Professional Practice Under the Code of Ethics" was to be promulgated. The product of several years of diligent effort by the Committee on Professional Practice and rigorous review by the Committee on Professional Conduct, the interpretive Guide has been published since its adoption as an adjunct to the Code of Ethics. The Guide as amended to 1974 is reproduced in Appendix X.

Although its own Code of Ethics has always been the basis for formal administration of professional conduct matters, ASCE has supported other ethical standards in joint sponsorship with other professional engineering organizations. The ECPD adopted its Canons of Ethics in 1947; ASCE became a signator to the Canons in January 1950. This status continued until 1964, when ECPD made certain amendments. When called upon to accept the revised instrument ASCE endorsed the three "Fundamental Principles of Professional Engineering Ethics," as follows, but withheld formal action on the amended Canons because of the possibility that there might be some ambiguity in the interpretation of the Society's own Code:

ECPD FUNDAMENTAL PRINCIPLES OF
ENGINEERING ETHICS

The engineer, to uphold and advance the honor and dignity of the engineering profession and in keeping with high standards of ethical conduct:

I. Will be honest and impartial, and will serve with devotion his employer, his clients, and in keeping with high standards of ethical conduct:

II. Will strive to increase the competence and prestige of the engineering profession;

III. Will use his knowledge and skill for the advancement of human welfare.

Through the years there has been hope that a single code of engineering ethics could be devised that might be acceptable to all engineering societies. The unsuccessful effort in 1913 to obtain acceptance of a code drafted by ASME was the first attempt. In January 1921 the four Founder Societies, together with the American Society of Heating and Ventilating Engineers, formed a Joint Committee on Proposed Universal Code of Ethics, which made a valiant try to achieve its aim. The committee reported in 1922, recommending a short form of code for universal adoption, and also that each society appoint a Committee on Professional Conduct. When these recommendations came to the ASCE Board of Direction the proposed universal code was rejected forthwith, but there was strong sentiment favoring the idea of a Committee on Professional Conduct, which would afford more than a guilty or not-guilty judgment of unprofessional actions. As a result the Board of Direction created the standing Committee on Professional Conduct that has been maintained in ASCE ever since.

In recent years there have been efforts to bring compatibility to the ethical standards of ASCE and the NSPE. They have met with only limited success. The problem appears to center upon the difference in ethical emphasis for those engaged primarily in manufacturing industry, such as mechanical, electrical, and mining engineering, and those engaged primarily in the public works sector, as are civil engineers. The ASCE membership includes a much greater representation of engineers in private practice than is the case with other societies, and this is markedly reflected in ASCE ethical attitudes.

ASCE leadership in the field of professional ethics has had a profound effect on similar activities in other engineering organizations. The Code of Ethics has served as a pattern for several other domestic societies and registration boards, and there has also been significant impact upon foreign ethical standards. The presence of the ASCE Code is unmistakably manifest in the codes adopted by the Pan American Federation of Engineering Societies (UPADI) in 1961, the Conference of Engineering Societies of Western Europe and the USA (EUSEC) in 1963, and the World Federation of Engineering Organizations (WFEO) in 1969.

There is much more to the creation of an appropriate environment for professional practice than the formulation of ethical standards alone. Such standards have meaning only if they are administered and enforced with determination, fairness, and consistency. It is believed that the record of ASCE in thus implementing its Code of Ethics is unequalled by any other engineering organization in the world!

The 1852 Constitution and those thereafter have required a member of the Society at the time of his admission to certify by signature that he would subscribe to the Constitution and Bylaws. Since 1878 the Constitution has provided for the expulsion of members for cause as established by due process. The first recorded cases were two in 1899; in one the charges were dropped, in the other a member was requested to resign.

The Board of Direction dealt directly with all disciplinary matters until 1923, when the Committee on Professional Conduct was authorized to advise the Board on all matters of ethical nature referred to it. Until 1968 the Committee on Professional Conduct was a committee of the Board; since that time it has comprised past Board members so that no current members of the Board would be required to function both as prosecutor and as judge in any proceeding. The duties of the committee include the investigation and judicial review of all charges of unprofessional conduct as basis for possible disciplinary action by the Board.

Fig. 11 summarizes disciplinary activities in the Society through the years. It is a record that is unique in the engineering profession both for the number of cases and the manner of their disposition.

Mention has already been made of the "South Carolina Case," which was of particular significance. Three other cases merit brief comment here because of their unusual nature.

In the late 1920's a flood control district in a Western state undertook construction of a dam, which ultimately was abandoned before completion because of foundation problems. A number of irregularities were alleged to have taken place in the course of the project, and a member of the Society was

FIG. 11 — ASCE DISCIPLINARY ACTIONS, 1852-1974

Period	Cases Filed	Charges Dropped	Members Admonished	Members Suspended	Resignation Requested*	Members Expelled
1852-75	—	—	—	—	—	—
1876-1900	2	1	—	—	1	—
1901-25	40	31	4	—	4	1
1926-50	104	73	17	—	5	10
1951-74	185	130	26	25	4	12
Totals	331	235	47	25	14	23

*Includes resignations requested and those accepted while charges pending.

highly critical in his public expressions on the matter. Some of his comments were alleged to reflect upon the professional reputation of the chief engineer of the flood control district, who had been exonerated from any connection with the alleged irregularities. After an investigation and hearing, all in strict accordance with the rules of the Society, the critical individual was expelled from membership, with considerable attention on the part of the public press. The unusual aspect of the case was the repeated effort to induce the Board of Direction to rescind the expulsion. No less than four requests for reconsideration were filed, none of them presenting any new or additional evidence beyond that submitted in the original hearing. The last request was received following the death of the expelled member 35 years after the expulsion occurred. Obviously, it would have been inappropriate for a subsequent Board to reverse the decision of the Board that heard the case unless there had been new evidence to consider.

Another case, in the 1960's, involved a young member employed by a consulting firm, who was charged with unprofessional conduct incidental to the commencement of his own private practice before terminating his employment. When the matter came to the point of a hearing before the Board, the defendant in a dramatic proceeding secured an injunction against such procedure. At considerable expense the Society succeeded in having the injunction set aside, and the hearing went forward some months later in conventional fashion. The defendant was found guilty of the charges against him, and he was suspended from membership for five years. The cost of his case to the Society for special legal services and related expense was in excess of $20,000.

The most spectacular ASCE professional conduct case resulted from charges against three members of the Society who were involved in the political influence-peddling activities in Maryland which resulted in the resignation in 1973 of Spiro T. Agnew from the Vice Presidency of the United States. Two of the defendants represented consulting firms who made contributions in return for favored treatment in the award of engineering contracts; the third, as a public employee, was a go-between in arrangement for the payments. All were charged with violation of the ethical provision against undue influence or the offering, solicitation, or acceptance of compensation "for the purpose of affecting negotiations for an engineering engagement."

Even though the defendants had gained immunity from criminal action by providing evidence of the Maryland corruption, they were found guilty and expelled from membership in the Society in 1974 after being given full opportunity to answer the charges in writing and in a hearing before the Board of Direction. It appeared likely at the time that other members of the Society might also face charges in connection with the Maryland situation.

* * *

When Past-President and Honorary Member Daniel W. Mead criticized the ASCE ethical standards in 1936 for their strong emphasis toward the professional relationships of the engineer in private practice, he sounded a note that was to echo long thereafter. Other critics had observed that the Code was overbalanced toward "business ethics" at the expense of "personal ethics." The Environmental Age of the 1960's brought further allegations that all engineers were subordinating the "social ethic," under which they were obliged to accept responsibility for any negative impacts of their works upon man and his surroundings.

It must be admitted that these criticisms were valid. Certainly, the ASCE Code of Ethics does not highlight a consciousness for human relationships and social responsibility. This concession is made, however, without acknowledgment that the engineering profession alone is at fault for all of the excesses and crises that have befallen society with the progress of technology. The civil engineer — specialist in those works that determine the quality of the environment — had applied his art with a concern for the public welfare that was implicit in the general provisions of his ethical standards. He was ready and willing at all times to meet any level for environmental quality that was in accord with the political, economic, and public support that he received.

In this context, the ASCE role in establishing and implementing norms of professional practice was on the one hand late and lacking in breadth, and on the other hand more specific and rigorously administered than anywhere else in engineering. From all indications in 1974, the objectives of the Society were ranging more and more into the social concerns of civil engineering practice. This trend will undoubtedly be reflected in the future interpretation of the Code of Ethics and in its further development.

CHAPTER IV

CREATING THE PROFESSIONAL ENVIRONMENT

A profession cannot limit its concern to the education, competence, and mode of practice of its members, despite the essentiality of these elements. The professional person must operate in a favorable milieu if he is to be effective. Adequate compensation, appropriate recognition in the employing entity, provision for professional development, authority to exercise judgment, and freedom from undue political, bureaucratic, or hierarchial constraints are some of the qualities of that milieu. Although the professional person accepts primary responsibility for these conditions as they affect him as an individual, he expects his professional society to provide the mechanism for any collective action in such matters.

ASCE has been aware of these expectations on the part of its members, and has served where it could, to the extent of its resources. At the same time the Society has carefully avoided the attitudes and methods of the trade unions.

This chapter is best introduced by a brief review of the various segments that make up the engineering profession as well as those constituting civil engineering. Problems have arisen from the stresses that have sometimes generated between these segments. Activities aimed toward the enhancement of the economic welfare of civil engineers are covered here. The place of the young engineer and of women in the Society is discussed, as are the provisions made to encourage personal professional development through recognition of accomplishment.

Categories of Engineering Employment

Civil engineering stands apart from the other major branches of engineering in the classification of its members with regard to public, private, and industrial practice. The following data (compiled in 1964) are indicative of this difference as represented by the membership of several national professional organizations:

Area of Practice	*ASCE*	*AIME*	*ASME*	*IEEE*	*AIA*	*ABA*	*AMA*
Private Practice (Principals and Employees)	29%	9%	10%	10%	68%	76%	63%
Government (Federal, State and Local)	40	4	9	9			
Industry (Including Construction)	23	76	73	68	32	24	37
All Other	8	11	8	13			

Of particular significance here is the strong orientation of civil engineering to practice in the public works sector, whereas the other branches of engineering are overwhelmingly engaged in the manufacturing industry. Only about one in eight of all civil engineers is employed in industry other than construction. Civil engineering also has strong representation in the private practice sector, although not to the same extent as in architecture, law, and medicine. The trend in civil engineering, at least since 1900, has been very slowly from government service toward private practice.

This analysis of the "composition" of the major branches of engineering is significant because it explains the marked differences in professional objectives and attitudes on the part of the civil engineer, as compared to his industry-oriented brothers. It is also noted that civil engineering does not have a majority component, and that the public practice and private practice groups are predominant.

Polarization between the practitioners in government service, as represented by the Corps of Engineers, and the non-military civil engineers was probably taking place at the time ASCE was being formed. As early as 1870 there were differences in judgment and opinion between the engineers of the Corps, usually graduates of West Point, and those of non-military origin.

The problem surfaced at the St. Louis Convention in 1880 when Charles MacDonald (President, 1908) bitterly protested the education:

> ". . . at public expense, of a privileged class of engineers in military service, to whom are entrusted the design and supervision of works of public improvements, to the exclusion and prejudice of engineers in the civil service, whose education has not been a tax on the public treasury, who have proved themselves perfectly competent to execute works of the greatest magnitude in the best and most efficient manner."

His pleas resulted in appointment of a Committee on the Engagement of Civil Engineers upon Government Works,

> ". . . to prepare a memorial to Congress asking that Civil Engineers may be placed in full charge of the works of public improvement carried on at Government expense."

With Mr. MacDonald serving as chairman, the committee brought its draft "memorial" before the 1881 Annual Meeting, but with the admonition that

> ". . . it may be inexpedient for the Society to place itself in the position of advocating before Congress the claims of a certain class of its membership, in seeming conflict with any other class whose interest may be in a different direction."

The memorial was submitted for the use of any civil engineers who might wish to endorse it as a personal communication to Congress, with the suggestion that ". . . the Society may with propriety decline to consider the subject further." This disposition of the matter was approved.

The issue arose again just five years later when the Civil Engineers Club of Cleveland sought ASCE cooperation in exploring the desirability ". . . of new ideas which shall provide for the better condition of Civil Engineers employed on Government works other than military." The Society declined to participate, citing the conclusions reached in 1880.

The dichotomy became an open controversy in 1885 in the paper, "Ten Years Practical Teachings in River and Harbor

Hydraulics," presented by Elmer L. Corthell (President, 1916) and published in *Transactions*, XV, p. 269. He maintained that civilian engineers were better qualified by education and experience than military engineers to direct major public works projects. Forthright and spirited discussion supported both sides. Arthur P. Boller (Secretary 1870-71 and Vice President 1911-12) offered a compromise view in his observation that there were well-trained and competent engineers both within and outside the military, and he urged that the specialist services of capable civilian engineers be utilized in the direction of government projects. Actually, by this time the influence of West Point as a source of planners and builders of major public works had already begun to decline.

This confrontation appears to have cooled the issue for some years, although it did not go away. In 1920, Charles MacDonald headed a group "with the object of urging suitable recognition of Civilian Engineers employed on River and Harbor Works, in the proposed legislation providing for an increase of the Engineer Corps of the U.S. Army." This effort was successful in 1914 in bringing about inclusion of a provision requiring that all civilian assistant engineers engaged in the Engineer Bureau of the War Department have at least five years experience, and that they be given the rank of Captain or Major in the Corps.

The controversy emerged again in 1921 in a somewhat different format. A resolution of the Board of Direction objected to the lack of consideration given by the War Department to civilian engineers who had been given responsibility over important river and harbor works during World War I. With the war ended, the civilians were replaced in their supervisory posts by commissioned officers, and demoted to their former duties. Congress was urged to enact legislation that would afford protective status to civilians under a new classification of "United States Engineer." The resolution drew fire from General Lansing Beach, M. ASCE, Chief of Engineers, who considered it a reflection on his office. There was, however, no other apparent result.

The last confrontation between the civilian and military segments came in 1924, as a result of the Society's energetic support of legislation that would have placed responsibility for all non-military federal public works in a new department, outside the War Department (see page 274). This time General Beach appeared before the Board of Direction to voice his objections. The legislation was not enacted.

Although the Corps of Engineers did not concur in the ASCE view on unification of federal construction services, it welcomed the aid of the Society in 1949 when Congress considered a drastic reorganization of the Army. The proposal would have abolished the existing technical services (Corps of Engineers, Signal Corps, Quartermaster Corps, Transportation Corps, and Ordnance Department), and would have assigned any or all duties of these branches to the general branches such as Infantry, Field Artillery, etc. ASCE moved, through its Committee on Military Affairs and through Engineers Joint Council, to support "the retention and further development of the technical and professional branches of the Army, and the maintenance of professional status to a high degree among the officers of the technical branches of the Army."

After World War II any schism between the military and civilian interests appeared to vanish (see page 281). The relationship between the Society and the civil engineering units of the Army, Navy, and Air Corps was most cordial and compatible. Mutual cooperation in professional matters was excellent.

The Great Depression brought a new order of contention between engineers in the private and public sectors—the growing bureaucracy in all government agencies involved with public works. This included such bodies as the Bureau of Reclamation, Bureau of Public Roads, the Geological Survey, the Coast and Geodetic Survey, the Corps of Engineers and, later, such new agencies as the Atomic Energy Commission and the National Aeronautics and Space Agency.

A 1938 resolution drafted by the Committee on Fees and

approved by the Board of Direction expressed concern for ". . . an increasing encroachment of organizations under civil service into fields which heretofore have been occupied by members of the engineering profession engaged in private practice . . ." The resolution had a double thrust: (1) disapproval of all proposals that would unduly restrict the field of activities of the engineers engaged in private practice, and particularly its disapproval of any laws that would serve to deprive the public of the benefits to be derived from the services of the private engineer and his organization, and (2) reaffirmation of safeguards for tenure of position under civil service for engineers in the service of the public.

The issue arose again in 1941 when the Board recorded its disapproval of "the accelerated trend toward socialization of the engineering profession," and urged the abandonment of efforts by federal agencies to perform "in-house" the services normally furnished by engineers in private practice.

In 1945 a Committee on Engineers in Private Practice was created to represent the interests of that segment of the membership. The committee promptly began discussions directly with the Secretary of the Interior, specifically with regard to the operations of the Bureau of Reclamation. The discussions were highly successful, leading to policy decisions in 1946 that were mutually satisfactory to the Interior Department and the Society.

In January 1948 the Board of Direction adopted a policy statement that was still providing effective guidance in 1974. The statement:

(1) Recorded the intent of the Society to cooperate fully with all government agencies employing engineers, pursuant to the best interests of the public and the profession.
(2) Discouraged unfair competition by engineers in public employ with those in private practice.
(3) Recommended that only administrators having suitable technical training be appointed to head agencies employing large numbers of engineers.

(4) Encouraged the use of private engineering consultants by government agencies whenever this procedure would best serve the public interest.

The policy was reaffirmed by the Board in 1959.

In 1963 a flurry of incidents brought on threat of legislation that would have barred a government agency from "hiring" another agency to perform engineering services. These included mapping services in Ethiopia by the Army Map Service, hydrographic studies by the Coast and Geodetic Survey for the Bureau of Reclamation, and efforts by the latter Bureau to extend its services to local irrigation districts. The legislation was not enacted, but since that time there have been accusations only of minor encroachment by government bureaus into the domain of private engineering practice. By 1974 the agencies of the federal, state, and local governments had become by far the greatest users of the professional services of engineering firms in private practice.

After eight years of productive existence the Committee on Engineers in Private Practice was merged into a more comprehensive Committee on Professional Practice when the Department of Conditions of Practice was created in 1953. In 1957 the engineers in government were given their day when a Committee on Engineers in Public Practice was created by the Board of Direction. For thirteen years it gave attention to the status of the engineer in federal Civil Service, as well as to grade classifications and salary schedules for engineering employees in state and local agencies. The organizational merry-go-round came full circle in 1970, however, when the responsibilities of the Committee on Public Practice were delegated to the new Professional Practice Division. Thus, the conditions of practice for all members of the society — regardless of the manner of their employment — were again to be serviced in the same group of committees.

Representation of the various segments of civil engineering practice in the Board of Direction has not followed the pattern of numerical apportionment, as shown in the table on pages 34 and 148. Although 40% of the ASCE membership was engaged in some level of government practice, this segment had never been fully represented in the Board. This reflects the constraints which have limited participation in professional activities by many engineers in public employment. Some public bureaus openly discouraged such activities; others charged the time required against annual leave allowances. By the early 1970's these considerations were becoming increasingly important in the recruitment of personnel for engineering positions in government.

Membership Welfare Services

Almost from its beginning ASCE has been confronted with a dilemma in deciding the extent to which it should go in advancing the economic self-interests of its members. Admittedly, a professional organization must be concerned with the economic circumstances under which its members will carry on their practice. The practitioner should be accorded suitable surroundings and facilities with which to work; he should be given the recognition appropriate to a professional; and he should be reimbursed at fee and salary levels commensurate with the true worth of his services to society.

Only a small segment of the membership has ever felt that ASCE should fulfill the same role as a trade union, with the power politics and profligate disregard for the public interest that has characterized much of the union movement in the United States. But how far should the Society go in serving the very real economic needs of its members, while retaining its identity as a professional body? That has been a long-standing question.

The founders and early members of ASCE were mature and well-established engineers, enjoying top incomes in industry and government. In 1850-75 civil engineers were very well compensated indeed, in comparison with other workers, professional and otherwise. Many of the leaders of the Society at this time were independently wealthy. Thus it was some years later, when a substantial number of younger members were enrolled, that there was concern for the economic welfare of the civil engineer employee.

Indicative of this attitude was the reaction to a proposal in 1875 that "the Civil Engineers' Insurance League" be set up as a voluntary agency within ASCE to aid and provide benefits to the families of deceased members. When the plan was submitted to the members, however, there was so little interest that the idea was abandoned.

A circular letter to the membership in 1872 invited "those seeking engagement and those requiring engineering service" to communicate their needs to the Secretary, so that "such may be brought together." Any such employment and professional exchange services were provided informally by the secretariat.

The practice of civil engineering and its economic climate underwent considerable change through the turn of the century, during which time the membership of ASCE changed from those mostly in executive-management positions to a heavy predominance in the employee category. The following proposal, presented by Percival M. Churchill, A.M. ASCE, at the Annual Meeting in 1912, indicates that a problem was at hand:

> "That the President appoint a committee of eight members to look into the conditions of employment of Civil Engineers throughout the country, the compensation they receive, the duration of employment, the expenses for which they are reimbursed by the employer, the expenses due to the work paid by the engineers themselves, the net yearly income, the prices charged for different classes of private work, and any other facts necessary to clearly set forth the problem. The report to set forth recommendations for action by the Society looking forward to improving existing conditions and to include a report on the feasibility of this Society operating an employment bureau for its members . . ."

The resolution was referred to the Board of Direction, which appointed a Special Committee to Investigate Conditions of Employment of, and Compensation of, Civil Engineers. In doing so, however, the Board observed "that it does not feel that it would be practicable or wise for the Society to operate an Employment Bureau for its members."

Slightly more than half the 6,805 total membership of the Society at this time were in the Associate Member and Junior grades.

Mr. Churchill persisted in his plea for an "Exchange for the Marketing of Engineering Service" in another resolution offered at the 1913 Annual Meeting. This time he urged that ASCE seek the cooperation of the American Institute of Consulting Engineers (AICE), ASME, AIEE and others in exploring such a venture. There was no action on the proposal.

The Special Committee made several questionnaire surveys of the membership and made its final report to the Board of Direction at the 1917 Annual Meeting (*Proceedings*, XLII, pp. 1603-10). Although the report was an excellent compilation of data and information on compensation and employment conditions, it offered no specifics in the way of Society action. The average yearly compensation of 6,358 civil engineers was $3,985. Members averaged $4,141 compared to $3,389 for non-members. Interestingly, graduate engineers averaged $3,982 against $3,993 for non-graduates.

The persevering Mr. Churchill was not content with the committee's report, and by letter he filed another resolution at the 1917 Annual Meeting requesting appointment of a committee "to formulate a plan for the systematic marketing of Engineering Services." The committee was to be given a budget of $5,000, and empowered to employ a "competent progressive Engineer" to work out details of such a plan in cooperation with other national engineering societies. His motion was supported by an eloquent statement in which he noted that:

> ". . . The time has therefore arrived when the energies and resources of the Society must largely be directed in other channels (than technical affairs) . . . along the business and human sides of engineering activity."

His arguments were persuasive, for a new committee was appointed to study the advisability of establishment by the Society of an employment bureau.

Through extensive inquiry this committee found "a widespread demand for action by the Societies in this matter," as well as "a general feeling that our Society should, in the future, give systematic study to the practical matter of employment." It was recommended that a cooperative engineering employment agency be formed, under separate management from the existing societies but supported and controlled by them. It was proposed that implementation be undertaken either by a joint committee comprising the secretaries of the four Founder Societies (ASCE, AIME, ASME, AIEE) or by the recently organized Engineering Council.

After an exchange of resolutions with the Engineering Council, the Engineering Societies Employment Bureau was finally established late in 1918, under the direction of the Engineering Council, but under the management of the Committee on Engineering Employment, comprising the four Founder Society secretaries. The precarious but useful life of this agency until its demise in 1965 is related in Chapter VII, which deals explicitly with intersociety operations. Mr. Churchill's ideas did not turn out exactly as he envisioned, but they were responsible for an important and productive joint professional enterprise.

A 1921 proposal to set up a facility for providing financial aid to needy members was explored by a Committee to Investigate the Desirability of a Benevolent Fund. After a year's study it was recommended that such a fund be established on a national basis and operated through a central organization of the local sections. Donations, bequests, and annual subscriptions were to be solicited to afford assistance to "necessitous members" and their families, and to the families of deceased members. When the plan was put before the membership it was received with no enthusiasm and some objection, and it passed into oblivion to join the "Insurance League" proposed in 1875.

The 1921 controversy concerning civilian engineers in the Corps of Engineers (page 148) gave origin to a Committee on the Status of the Civil Engineer in Government Work and His

Compensation. Intended to gather information and to advise the Board of Direction, the committee reported "progress" five times in the next eighteen months, but was discharged in 1923 without having made any recommendation of record.

After its 1917 study, the next ASCE action on engineering salaries was collaboration in and endorsement of the excellent 1920 report of the Engineering Council Committee on Classification and Compensation of Engineers (*Proceedings*, XLVII, pp. 9-11). This committee was chaired by Arthur S. Tuttle, Hon. M. ASCE (President, 1935).

Salary Studies Initiated

As a result of engineering salary problems in various government agencies across the country, the Committee on Engineering Employment in Public and Quasi-Public Offices was created in April 1927. Also in that year the Society authorized a delegation to call upon the Mayor of New York City on behalf of 3,600 engineering employees; in consequence salary increases aggregating $2,500,000 were granted.

The committee was renamed the Committee on Salaries in 1930, beginning a 27-year era of great productivity. In the beginning the committee assembled vast amounts of information on job classification and compensation, and disseminated it in useful reports for application to specific problems. The onset of the Great Depression in 1929 captured the full attention of the committee in the early 1930's when emphasis was given to emergency measures to cope with the severe unemployment and economic distress then prevalent. The commendable leadership of E. P. Goodrich, M. ASCE, who chaired the committee from 1927-39, merits acknowledgment here.

An outstanding service was performed by the committee in 1931 in a special report dealing with unemployment. Noting that at least 15% of civil engineers were unemployed, and that at least 10% of these were in straitened circumstances, it was recommended that the local sections of the Society set up committees to gather funds, determine those engineers in need of relief, locate jobs for them, and use the funds for wages or disbursement according to need. The committees were also to endeavor to generate useful work wherever possible to utilize the services of the unemployed.

This plan was implemented in large cities across the country with great success. Local sections of other societies joined in the good work, and interprofessional relations benefitted thereby. The Professional Engineers Committee on Unemployment in New York City collected more than $100,000 and generated more than $300,000 in wages, while placing 1,389 unemployed engineers in useful jobs. Herbert Hoover, Hon. M. ASCE, then President of United States, contributed $5,000 to the aid of engineers nationally. It was recorded in 1932 that 41 ASCE Local Sections were involved in this worthy movement.

Prior to 1934 the survey data gathered by the Committee on Salaries was freely disseminated to members and to engineering employers on an informal basis for guideline purposes. The first official issue of a recommended schedule of position classifications, with appropriate salary ranges for each of the six classifications, was endorsed by the Board of Direction and published in 1934 under the title, "Prevailing Salaries of Civil Engineers." Reprinted in 1935 as a separate report, this document found wide use, including adoption by several federal agencies employing large numbers of engineers. Particularly important was its acceptance by the federal Civil Works Agency as basis for the correction of inequities in the compensation of engineers engaged for service under the Emergency Relief Act.

In 1935 the Committee on Salaries was able to turn from the welfare stresses of the depression and to return its full attention to salary investigations. A noteworthy report was published in 1936, and intense activity in 1939 produced the highly significant "Grading Plan and Compensation Schedule for Civil Engineers," published in March 1940. This document put forward, under the full authority of the Society, recommended salary schedules:

"... intended to be used as standards against which to measure salaries now being paid, and to be used with judgment as to any need for differentials by reason of higher costs of living and higher general salary levels prevailing in any given city or other geographic area."

Issuance of the 1939 employment and salary guidelines was followed later that year by an intersociety "Joint Conference on Engineering Salaries," sponsored by ASCE.

Probably the most aggressive membership welfare services ever to be undertaken by the Society were performed during the years 1940-49. A near crisis existed in several state highway departments and some large city engineering departments as a result of low salary budgets, threatened unionization, and competition from non-government employers. When the assistance of the Society was sought a program was devised under which a staff specialist was made available for extended service in the field, to work in the agency, to study job classifications and salaries, and to develop specific recommendations for appropriate adjustments. The service was extended upon request of the local sections and with the consent of the employing agency, which also bore the subsistence expense of the staff expert.

This service was extended to the Arizona and Nevada State Highway Departments in 1940, and to the Nebraska State Highway Department in 1941. The Arizona and Nebraska studies were used by the North Dakota State Highway Department in 1941 to update salary schedules. These activities are reported in detail in ASCE Manual of Engineering Practice No. 24, *Surveys of Highway Engineering Positions and Salaries*, published in July 1941.

The year 1944 brought further positive action in creation of the standing Committee on Employment Conditions. An updated "Classification and Compensation Plan for Civil Engineering Positions" was published by the Committee on Salaries. More staff field consultations were made, in five municipal departments in Milwaukee, followed in 1945 by similar work for Los Angeles County, the Maryland State Employees Standard Salary Board, and the Louisiana Department of Highways. A special survey of qualifications, responsibilities, and salaries of civil engineering teachers was made in 1947.

A new edition of the 1944 classification and compensation plan was published in 1946, and the final staff field consultation was rendered in 1948 for the Office of Personnel, Commonwealth of Puerto Rico. The service was then discontinued because of complaints that the staff salary analyses and consultations transgressed the field of private practice of management consultants.

Beginning in 1951, the Committee produced a series of salary survey reports that differed from those produced from 1934-40 in that they carried only statistical data on prevailing salaries, with no recommendations concerning job classification and appropriate salary levels. The change in policy was accompanied by publication of Manual of Engineering Practice No. 30, *Job Evaluation and Salary Surveys,* providing guidance for the conduct of salary surveys at the local level. This manual was designed to enable a local entity to analyze problems in compensation of engineering personnel without having to call for outside assistance.

The national surveys have continued at two- or three-year intervals, although the functions of the Committee on Salaries were assumed in 1957 by a broader based Committee on Employment Conditions. A new service was initiated in 1956 in the "Engineering Salary Index," published quarterly in *Civil Engineering.* The index reports regional changes in salaries in a manner permitting trends to be observed and compared.

Year-to-year comparison of the results of the salary surveys is complicated by the wide variation in their format through the years. The following data are reasonably compatible, and are illustrative of the trends:

Year	*Median Annual Salaries*	
	Starting Rates for Graduates	*Professional Grades*
1930	$1,824	
1940*	1,856[1]	$3,721[2]-$5,503[3]
1951	3,150	
1961	5,550	8,223
1971	9,450	16,032

*Averages for Juniors[1], Associate Members[2] and Members[3], respectively.

From 1959 to 1971 the ASCE reports included data from the surveys of income of engineering educators as made by EJC. In 1968 the mean salary at the rank of professor was reported as $17,763; mean professional income from all sources, however, was $22,277.

Unionization and Collective Bargaining

The impetus given the American labor union movement by enactment of the National Labor Relations (Wagner) Act in 1935 was not without impact on the engineering profession. Wage rates for skilled and unskilled labor rose to levels that were incompatible with professional engineering salaries, especially in the case of the young engineer. The original Wagner Act made it readily possible for professionals to be drawn into unions whether or not they so chose, and the unions took full advantage of this loophole in the law. A few engineers favored the union route to immediate income benefits. Membership pressures called for ASCE action.

A Committee on Unionization of the Engineering Profession was formed in 1937, and a useful report was published in *Civil Engineering* in March 1938. The status of unionization of engineers was summarized, and it was concluded that employers of engineers would have to meet their responsibilities in providing equitable remuneration and working conditions for their engineering employees if collective action through the union mechanism was to be avoided. The recommendations of the committee are interesting:

(1) Trade union membership is a matter of personal economic determination, and should have no bearing on a member's status in the Society.
(2) ASCE should not endeavor to bring about amendment of the Wagner Act to exclude professionals, but should support amendments to clarify the position of professionals and subprofessionals under the Act.
(3) When necessary ASCE should cooperate with other professional bodies to establish temporary or permanent agencies affording a dignified means of collective representation for engineers.
(4) The Society should "seek actively" to encourage acceptance and implementation of its recommended classification of professional employment grades and salary schedule, and should cite any member for unethical conduct who does not comply with these provisions.

In 1944 a standing Committee on Employment Conditions assumed the charge of the Unionization Committee, giving top priority to a problem that was being given little or no attention by other engineering societies. A plan was evolved in 1943-44 under which the local sections might initiate the formation of professional collective bargaining groups that would eventually become independent of the Society. Guidance to members was provided in Manual of Engineering Practice No. 26, *The Engineer and Collective Bargaining*. It was obvious, however, that the only positive relief rested in amendment of the Wagner Act.

When the 80th Congress undertook overhauling the Wagner Act in 1946, ASCE was ready to press aggressively for incorporation of the following specific principles into appropriate amendments:

"(1) Any group of professional employees, who have a community of interest and who wish to bargain collectively, should be guaranteed the right to form and administer their own bargaining unit and be permitted free choice of their representatives to negotiate with their employer.
"(2) No professional employee, or group of employees, desiring

> to undertake collective bargaining with an employer, should be forced to affiliate with, or become members of, any bargaining group which includes nonprofessional employees, or to submit to representation by such a group or its designated agents.
>
> "(3) No professional employee should be forced, against his desires, to join any organization as a condition of his employment, or to sacrifice his right to individual personal relations with his employer in matters of employment conditions."

By this time other engineering societies were aware of the attention being given by ASCE to the unionization problem since 1937. EJC adopted the above three-point ASCE policy, and set up a Labor Legislation Panel. NSPE and the ASEE added their support, and a strong presentation was made to the Congress. It was this effort that brought about enactment of the "professional employee" provisions set forth in the Public Law 101—80th Congress, the "Labor Management Relations Act," better known as the "Taft-Hartley Act."

The new law recognized the differences in basic objectives and interests of professional and non-professional employees, and that it was not "appropriate" to include both types in a heterogeneous collective bargaining group. It provided the needed statutory definition of a professional employee. It afforded professionals the right to form collective bargaining units of their own, if desired. In essence, the professional employee was finally given the right to determine his own course of action in his quest for economic advancement.

The labor lobby resisted the professional employee amendments, and unsuccessful moves were made in the 81st Congress to effect their repeal. The Labor Relations Panel of EJC, with strong ASCE representation, was a major factor in defending against these attacks.

Under the Taft-Hartley Act, ASCE is free to inform its members concerning labor relations matters, and to adopt policies concerning them, but action by the Society toward formation of collective bargaining groups is open to challenge. Because the Society includes both employers and employees, it may not provide assistance of any kind toward the establishment of units under the Act that must be restricted in membership to employees alone.

ASCE has monitored statistically the progress of unionization in the civil engineering profession ever since the enactment of the Taft-Hartley Law. This is done by the Committee on Employment Conditions, through detailed membership surveys made at five-year intervals to 1973 and quadrennially thereafter. It is of interest to compare some of the data over the two decades 1953-73, as follows:

ASCE MEMBERSHIP ATTITUDES TOWARD COLLECTIVE BARGAINING

	1953	*1958*	*1973*
Member of CB Group	3.6%	2.3%	4.6%
Favor Collective Bargaining	36%	16%	27%
Would Join CB Group: Voluntarily	-	13%	24%
To Hold Present Job	-	33%	39%

The periodic surveys of employment conditions also afforded an opportunity for the questionnaire respondents to express themselves on matters of economic welfare. The 1968 survey, possibly as a result of then prevalent aggressive efforts of the building trades unions on behalf of their members, elicited an unusual number of letters and marginal comments chiding the Society for its lack of attention to the static nature of engineering salaries as compared to the soaring wage rates of skilled and unskilled labor. In reporting this groundswell of opinion editorially, the Executive Director observed that "Professionalism in engineering can survive only in a realistic and equitable economic environment."

Positive response by the Society was soon forthcoming. By 1973 a general manual, entitled *Guidelines to Professional Employment for Engineers and Scientists,* was in use with the endorsement of sixteen national organizations. This was supplemented by a more specific ASCE "Guideline: Employer-Engineer Relationship" for the use of civil engineers and their employers. A recommended salary guide—similar to those issued with Board of Direction sanction in the early

1940's—had been produced. The headquarters staff had been augmented by a specialist assigned to work with the local sections in the implementation of programs aimed toward improvement of employment conditions. Efforts were under way to develop a profession-wide pension plan that, hopefully, would afford portability of pension privileges and equities as well as other benefits.

Fees for Professional Services

As far back as 1874, Past-President William J. McAlpine affirmed in a communication to the Board of Direction ". . . the propriety of engineers making charges, the same as other professional men, for advice given." Consultation services at first were provided on an individual basis, but in time engineers were employed in private-practice firms headed by an individual or operated as partnerships or corporations. By 1974 almost a third of all civil engineers were engaged in consulting firms either as principals or employees.

The study of compensation of civil engineers reported in 1916 did not cover consulting fees, but it referred to them as a source of compensation problems in the profession:

> ". . . engineers in private practice sometimes employ men of extensive experience, and presumably of good ability, at salaries which young graduates with little or no experience are able to command, but which are less than those of an ordinary mechanic who has a labor organization behind him. It may be urged that the competition for work on the part of engineers who may employ a technical staff is so keen that it is necessary to take advantage of the needs of those seeking employment in order to secure professional work which is frequently let to the lowest bidder."

ASCE has long been concerned with procedures for professional-service contract negotiation and the manner in which such fees are determined and charged. Fees must be adequate to insure the highest quality of service to the client as well as a level of compensation to employed engineers that is commensurate with the degree of professional responsibility they are expected to assume.

One of the earliest official reviews of the cost of engineering services was made in 1920 by R. L. Parsons, M. ASCE, in a special investigation for the Engineering Council's Committee on Classification and Compensation of Engineers. He concluded that engineering costs would amount to about 5% for railroad work, and about 7.5% for water works and municipal engineering.

A special Committee on Charges and Methods of Making Charges for Professional Services was authorized in 1927. Its extraordinarily comprehensive report — published in *Proceedings*, September 1929, pp. 236-83—was to be the forerunner of an important series of ASCE Manuals of Professional Practice. The 1929 edition, however, covered such matters as engineer/architect relationships and engineer/contractor relationships, and it presented forthright recommendations for fee schedules. For general engineering services, exclusive of resident supervision, recommended fees ranged from 9.5% for a net project construction cost of $25,000 to 4.25% at the net construction cost level of $2,000,000.

Again, in 1938, a schedule of approved fees for structural and foundation engineers engaged on federal housing projects was issued. The schedule was developed through conferences with the AIA, the American Engineering Council, and ASME, primarily to apply to programs of the U.S. Housing Authority and the Federal Housing Administration.

The Committee on Fees served from 1930-45 in providing guidance on professional service charges, and was followed by the Committee on Private Engineering Practice until 1950. Since that time this function has been served by the Committee on Professional Practice. Manual of Engineering Practice No. 29, *Manual of Professional Practice for Civil Engineers*, was issued in 1952; Manual No. 38, *Private Practice of Civil Engineering*, in 1959, and Manual No. 45, *Consulting Engineering — A Guide for the Engagement of Engineering Services* appeared in 1968. The data in these publications were carefully presented as prevailing rates—not as recommended schedules. This was to avoid possibility of question by the Department of Justice with regard to price-fixing.

Manual No. 45 assumed particular importance in 1972 when the competitive bidding clause was removed from the Code of Ethics by agreement with the Department of Justice (see p. 137). Some minor revisions in language were required but the "Procedure for the Selection of the Engineer" remained intact. This section incorporates the professional negotiation process that has been fundamental to ASCE policy for many years.

On several occasions the Society has demonstrated concern for the economic circumstances of the membership by remission of dues during periods of national emergency. This was done first in 1918, when about 1,100 members (15% of the total membership) were exempted from payment of annual dues for the duration of their military or other war-related service. Again, during the Great Depression, action was taken in 1933 and 1934 to remit the dues of more than 2,000 members, who otherwise would have been dropped from the rolls for non-payment. Deferment of annual dues was again extended in 1942-45 to certain members in military service in World War II and in the subsequent involvements in Korea and Vietnam. More than 5,000 members were in the uniformed military services in 1943, about 26% of the total membership.

Group Insurance Programs

There was little membership interest when the "Civil Engineers Insurance League" was proposed in 1875 and the Benevolent Fund for "necessitous" members in 1921. However, a 1949 plan for the Society to sponsor group disability income insurance for its members, found a warm welcome. The program has operated continuously ever since under the direction of a professional administrator, with about 3,000 members enrolled in 1974.

The success of the Disability Income Plan led to the offering of other group insurance programs, as follows:

1955-56—Hospital and medical insurance
1960—Life insurance
1961—Accident insurance
1968—Supplemental hospital insurance
1972—Catastrophic health and accident insurance

It was the intent of the Society that, except for the administrator's fee, the entire group insurance premium should be used for the purchase of benefits to the policyholder. The programs were periodically modified and updated to meet changing needs and loss experience.

Almost 25,000 certificates were in force under all group insurance programs in 1974. Coverage was available to members of student chapters, and special plans were provided for members over 65 years of age.

In 1957 consideration was given to the sponsorship by the Society of a special program covering "errors and omissions insurance," for the particular benefit of members in private practice. The idea was dropped on the premise that liability insurance was already readily available to responsible engineers desiring it.

In 1974 possibilities for further ASCE group insurance service were being explored in such areas as supplemental pension annuities, automobile and homeowners insurance, and dental insurance. Full advantage was being taken of the preferred risk values that were represented in the ASCE membership.

Joint Economic Welfare Activities with Other Societies

A significant development in 1973 offered promise that most of the ASCE objectives for the economic welfare of the civil engineer might be furthered through joint action with other engineering and scientific societies. A set of "Guidelines to Professional Employment for Engineers and Scientists" had been promulgated through collaboration between EJC and the NSPE. The recommendations covered recruitment practices, terms of employment, professional development, and conditions of termination and transfer. ASCE, AIChE, ASME, IEEE and NSPE promptly endorsed the guidelines, and it was indicated that many other major engineering and scientific organizations would do likewise; as of mid-1974 a total of 25 organizations had endorsed the guidelines.

Leadership by the American Chemical Society early in the 1970's had produced an ambitious "portable" pension-plan proposal, which immediately captured the interest and support of a number of major engineering societies, including ASCE. This was under review by the Internal Revenue Service in 1973 and, if found feasible, would require some years to implement. In the meantime, ASCE initiated formation of a Joint Pension Committee as a mechanism for cooperative effort toward improvement of "the legislative and regulatory climate in which engineers/scientists pension plans exist."

Although unemployment was not a serious problem to civil engineers in 1974, it was sufficiently prevalent in other segments of engineering to merit attention by EJC. A proposal that the U.S. Department of Labor establish a cooperative engineering job-placement service under professional control was advanced by EJC, and was finding considerable support by ASCE and other organizations. The plan differed from previously unsuccessful engineering employment services in the intent that it would operate with government support rather than in competition with employment services offered by government agencies.

The Young Engineer in the Society

The founders of ASCE were mature, prestigious engineers, established in the top rank of their profession. Because of preoccupation with organizational problems there was no concern for the upcoming generations until 1873, when the Junior grade of membership was established. Even this move was not entirely a concession to the young engineer, as the term "Junior" was carefully defined:

"The term Junior is to be understood as not referring to the age of a person but to his classification in the Society for the time being; he is *junior* to Members in the sense that his professional experience has had a more limited scope than theirs . . ."

Nevertheless, the Junior grade was a breakthrough, even if it did not carry the voting privilege. Only two years of engineering practice were required, or a college degree with one year of practice. About 5% of the total membership were Juniors in 1875; ten years later the ratio had risen to 11%.

There was also some discussion of the desirability of a student grade of membership in the early 1870's, but no action resulted.

At the 1887 Annual Convention Robert E. McMath, M. ASCE, proposed a membership grade for students "For the purpose of taking the sense of this meeting, and perhaps eliciting some discussion . . " Not only were these aims fulfilled, but also a committee was directed to consider and report on ". . . the advisability of adding a new grade of membership to be called 'students' . . ." In the face of objection that the new grade would lower the membership standards of the Society, the committee recommended that a Student grade be created. It would have included those under 18 years of age with a minimum of one year of study in a technical school or two years of engineering study and practice, with advancement to higher grade required in seven years.

Most of the 1888 Annual Meeting was devoted to discussion of this proposal. Those opposing it claimed:

"This Society is not an institution for primary education, but is intended to be an association of skilled and experienced engineers for mutual improvement by the interchange of ideas and experiences of a character more advanced than can be fully understood by novices in professional work. Admission to any grade of membership should be contingent on experience had and work accomplished, and not on the mere desire to learn and the hope of future benefit by association."

This reasoning prevailed, in a resolution holding it ". . . undesirable at this time to amend the Constitution by adding a student class." Although the Society was later to authorize student chapters, there was never a student grade of membership to 1974.

Francis Collingwood (Secretary 1891-94) in 1894 presented a $1,000 bond to the Society to fund the Collingwood Award for Juniors. In his letter transmitting the gift he said:

> "It will doubtless be conceded, without argument, that the most rapid advancement of the Society will be promoted by interesting the young men of the profession in its affairs. The professional advantage accruing from friendly intercourse, and especially from the presentation of well-written papers at our meetings, is not fully appreciated by our older members and it is not strange that there should be among the Juniors a tendency to lose interest in the Society after a limited experience in its membership."

When the award failed to attract papers of high caliber in its early years Mr. Collingwood urged in 1899 that "steps be taken for holding meetings of Juniors of the Society for the reading and discussion of papers." Unfortunately, the ASCE members sharing Mr. Collingwood's concern for the young engineer were very much in the minority well into the 20th century.

Student Chapters Authorized

At the close of World War I a special Committee on Student Branches brought specific recommendations to the Board of Direction concerning the formation of student chapters. A strong written dissent was filed by M. E. Cooley, Hon. M. ASCE, for many years Dean of Engineering at the University of Michigan. Dean Cooley saw the student chapter as a device for narrowing the viewpoint of the student toward professional subjects, whereas he considered the real need to be for "the breadth that comes from the study of other than technical subjects" in order to produce:

". . . men of vision—generals, if you like—qualified to coordinate and direct the great special forces of individuals. No profession is overcrowded with men of vision. The world has never needed such men more than now."

Holding further that the student chapter would cheapen membership in the Society, Dean Cooley forcefully opposed their authorization.

The Board could not have disagreed altogether with Dean Cooley's views, but it acted nevertheless in January 1919 to adopt the recommended "Regulations for Student Chapters." The wisdom of the move was demonstrated very soon and in almost dramatic fashion.

The first student chapters were formed in 1920 at Stanford, University of Cincinnati, Rensselaer, Drexel, Iowa State, Penn State, University of Pennsylvania, and Washington University (St. Louis). By 1925 the number of chapters had increased to 71, with a total enrollment of 4,107 members.

More remarkable, however, was the effect on membership in the entrance grade of the Society, as showed in the following tabulation:

Year	*Student Chapters*	*Entrance Grade Members*	
		Number	*Percent of Total*
1875	None	73	4.8
1920	8	506	5.4
1925	71	771	6.8
1930	96	2,329	16.4
1935	110	3,046	20.4
1950	128	9,667	38.1
1974	183	25,530	37.3

Without question, the student chapter has proved to be vitally important in the transition from academic training to the professional practice of civil engineering. It has served to generate professional attitudes, to introduce principles of professional practice, and to instill career enthusiasm in the student.

Student chapters were authorized only at institutions in which the civil engineering program was approved by the Board of Direction, and accreditation by ECPD was always accepted as adequate qualification. Where the curriculum was not accredited by ECPD the Society had authorized student clubs, and nine of these were in operation in 1974. Total membership in all chapters and clubs at that time was somewhat in excess of 10,000. A listing of the currently active student chapters and clubs is shown in Appendix VII.

From time to time there had been discussion of the relative advantages of the student chapter and the student member grade directly in the national organization. A study made in 1956 (*Civil Engineering*; July 1957, p. 73) disclosed that the student chapter plan, as conducted in ASCE, compared most favorably with the student member system employed by the other four Founder Societies. Almost nine out of ten of the Junior Members at that time had entered the Society as graduating students.

Although only about half the student chapter members became members of the Society at the time of graduation, many applied for admission later. The vitality and level of member interest in the student chapter were a direct function of the professional orientation, enthusiasm, and leadership capability of the Faculty Adviser of the chapter. In an effective chapter, it was not uncommon for all graduating members to apply for Society membership.

The policies and operations of the student chapters have been administered by a Committee on Student Chapters since 1923. The committee has functioned since 1953 under the professional activities arm of the Society.

Younger Member Affairs

Once the young engineer became represented in the Society in appreciable numbers he began to gain in recognition. First designated as a Junior, in 1873, his grade was changed to Junior Member in 1949, and to Associate Member in 1959. More important, however, was his enfranchisement to voting status in 1947, and his privilege three years later to hold national office.

A 1947 *Civil Engineering* editorial on the constitutional amendment referendum that would give Juniors the right to vote and also authorize a dues increase highlighted the issue in forthright terms:

> "Here, then is the challenge of the Society's younger men—those more closely concerned with and affected by the professional activities of the Society—to the older members whose duty it is to perpetuate the profession. Here is ready recognition that the expanding responsibilities the Society has undertaken in the professional behalf of its members, together with the diminishing dollar purchasing power now being encountered on all sides, require an increase in revenue if those activities are to be continued. Here is the younger member's frankly stated view that it is not enough for his Society merely to provide him with good tools, in the form of outstanding technical training and engineering knowledge, but that he also expects it to strive jointly with him for a proper environment in which he can utilize those tools advantageously—an environment of professional recognition."

When the ballots were counted, the Juniors had been given the enfranchisement they sought, but the increase in dues did not receive the necessary majority.

In 1935, when Juniors comprised about 20% of the total membership, the Board of Direction decreed that they be brought more actively into the operations of the Society. Upon recommendation of the Committee on Aims and Activities, the Board approved the policy that in so far as possible a Junior be given a place in every Society committee, including the local sections and technical divisions, so that "the committees may have the benefit of the point of view of the younger men of the profession."

Another policy adopted in 1951 requested every local section to appoint at least one Junior Member to each of its committees, and to delegate complete responsibility to a Junior Member for at least one meeting of the section each year. Other actions provided for young members to be appointed as Junior Contact Members for student chapters, Local Section Conference delegates, and as members of other committees. A Committee on Juniors was appointed in 1930 to administer to the interests of this group; it became the Committee on Junior Members in 1950 and was renamed the Committee on Younger Members in 1959.

Several surveys have been made to determine the actual involvement of young members in activities of the Society. Following is a comparison of data gathered in 1956 and 1970:

	1956	*1970*
Entrance Grade Percentage of Total Membership	45	42
Participation by Entrance Grade Members (%):		
Authorship-Journals	10	38
Authorship-*Civil Engineering*	13	26
Technical Committes	7	16
Local Section Committees	29	—
Local Section Officers	24	35

In 1970 the Board of Direction requested annual reports on young members' participation in Society affairs, even to the extent of setting age 35 as the upper limit for a "Younger Member." A summary of 1971 and 1972 data follows:

	1971	*1972*
Members and A.M. under 35 (%)	32	33
Paying Local Section Dues	45	46
Committee Participation (%)		
National Technical	7	12
National Professional	13	17
Local Section Participation (%)		
Section Officers	33	38
Branch Officers	35	59
Technical Groups	47	57
A.M. Forums	73	99

Junior/Associate Member Forums

Thirty-one years after Francis Collingwood urged in 1899 that "steps be taken for holding meetings of Juniors of the

Society for the reading and discussion of papers" a progressive group of Juniors in the Los Angeles Section, inspired by A. W. Dennis, M.ASCE, formed the first Junior Forum. The Forum met an hour preceding each meeting of the Section. In addition to the presentation of papers, the Forum gave the Juniors and student chapter members opportunity to enjoy association and intercourse in the professional affairs of the Society with a view toward their greater ultimate participation and contribution.

The idea was well conceived and implemented, and was picked up in other local sections across the country. Following the Los Angeles Section Junior Forum in 1930 were similar organizations in the San Francisco Section (1932), the Metropolitan Section (1933), the Illinois Section (1937), and the Sacramento Section (1939). By 1950 there were ten local section subsidiaries for Juniors, and in 1973 there were fourteen of them, all designated by this time as Associate Member Forums. Junior branches in the Central Illinois, Cleveland, Illinois, and Northwestern Sections did not survive. The movement found its greatest success in the larger local sections.

Where young member interest and leadership were sustained the Junior/Associate Member Forums were generally effective as implementing mechanisms for certain local section programs. In addition to the conduct of technical seminars, the forums carried on career guidance work; continuing education activities, including review courses for candidates for registration examination; manpower programs directed toward encouragement of the disadvantaged into technical careers; and activities drawing attention to economic and employment problems in the profession. On a number of occasions the Forums were successful in the instigation of important actions at the local section and even the national levels.

The Forums in the Los Angeles, Sacramento, San Diego, and San Francisco Sections were particularly energetic and diligent in their efforts. A measure of unity was achieved by these four Forums through the conduct of annual regional joint conferences, usually on a theme of timely concern. In 1968 the topic was "The Professional Engineer and the Union," and it activated the Los Angeles Section to produce a significant committee report exposing the need for national and regional guidelines for engineering fees, salaries, and employment conditions.

The 1969 Regional AMF Conference pursued the "Unionism and Professional" theme, and generated strong support of the guideline proposal in California and elsewhere across the nation. This swell of membership demand was definitely a factor in the action undertaken in 1970 at the national level of ASCE leading toward greater emphasis on salary and employment problems.

But the California AMF campaign had even greater impact, as the result of a combined effort of the four Forums to elect a Director to represent District 11 in the Board of Direction who was considered knowledgeable about the unionism problem and aware of the need for prompt action. Such a candidate was nominated, and he lost to the official nominee by only 127 votes in almost 4,000 ballots cast. The incident drew attention to certain shortcomings in the nomination and election procedures, however, which also prompted the Board of Direction to explore reform measures.

Opinion is divided in the Society as to the wisdom of an organizational pattern that sets the younger members apart from their elders. Many hold the view that such polarization actually complicates and delays the progression of the young member into the mainstream of Society activity and leadership. The Forums are intended, however, to add a new dimension of interest and challenge for the younger member, without interfering in any way with his contact and interaction with his older associates. Many leaders in ASCE found their first opportunities to serve the Society through their participation in the younger member groups.

The Associate Member Forums have proved to be a vigorous and dynamic force for young member expression and action in their local sections, and they promise to exercise a measurable influence upon ASCE in the future.

Women in ASCE

The first woman to become a member of the American Society of Civil Engineers was a fitting exponent of the so-called "liberation" movement, albeit somewhat before her time. She was Nora Stanton Blatch De Forrest Barney, said to be the first woman in the United States to receive a civil engineering degree (Cornell, *cum laude*, 1905). The granddaughter of Elizabeth Cady Stanton, a pioneer in the women's suffrage crusade, Nora Blatch became a Junior of the Society in 1906. While working in the New York City Engineering Department she contributed a technical discussion to the 1906 volume of *Transactions*. In 1909 she married Dr. Lee De Forrest, inventor of the vacuum tube, but the marriage was of short duration.

When Miss Blatch sought to advance to the corporate grade of Associate Member in 1915, her application was declined. She responded, however, with a petition for a peremptory mandamus that would have required the Society to grant her admission to the Associate Member grade. The petition was denied, and Miss Blatch was dropped from membership for failure to qualify for advancement to corporate status in the required time. She later became an architect.

Eleven years later, in 1926, Mrs. Laura Austin Munson was elected to the grade of Junior. The occasion was marked by a special item in *Proceedings*:

> ". . . The question of opening Society membership to women has received considerable attention in the past and opinion has been divided. Some have argued the question from the standpoint of expediency and some from justice, but none from exclusiveness. Putting it in another way, frequently members took the view that women might be engineers—in fact, that some already were engineers—but that the number was small from the physical limitations usually required of one in the

practice of the profession. Thus, in the past, the discouragement of women as members of the Society was considered to avoid inconvenience and embarrassment to all concerned.

"This action by the Board is the result of the usual complete scrutiny given all applicants and is significant of the trend of the day. Thus are women accorded full equality with men in professional as well as public life."

Miss Elsie Eaves, admitted as a Junior in 1926, was the first woman to progress through the grades of Associate Member, Member, and Fellow. In 1962 she achieved Life Member status, at which time she was at the height of her career as manager of the business news department of the McGraw-Hill Publishing Co.

A 1949 *Civil Engineering* item noted that about fifty women had been admitted to membership up to that time, and that 43 were then enrolled. Most of these, however, had joined since 1940, possibly as a result of the emphasis on technology during World War II. The majority of women members were employed in federal and state agencies, about a dozen in private and industrial practice, and two in teaching.

Although the sex of members is not noted in ASCE records, at least 226 women could be identified in the roster as of March 1974. This was but 0.3% of the total membership.

There was a noticeable increase in the enrollment of women in civil engineering programs in the 1970's, and it was not uncommon to find young women serving as officers in the student chapters. Women were rarely involved prominently, however, in the activities of the Society at the local section and national levels.

Problems in Professional Practice

A professional society must be prepared to identify and to seek the solutions to any situations that might interfere with the effective performance of its practitioners. Some of these problems may be circumvented by such measures as the Code of Ethics, described in Chapter III. Others may be managed through administrative procedures, as in the Committee on

Professional Practice, wherein equal representation insures recognition of the interests of engineers in both public practice and private practice.

ASCE has been confronted with many unusual professional practice problems through the years—far too many to report here. A few typical examples will illustrate the readiness of the Society to cope with the wide range of incidents and trends that affect the service of its members in the public interest.

Early in 1923 Arthur P. Davis, Past-President ASCE, was requested by Secretary of the Interior Hubert Work to resign his post as Director of the United States Reclamation Service, a position Mr. Davis had filled with distinction for many years. Secretary Work justified his replacement of Mr. Davis with a former Governor of Utah, a non-engineer, on the grounds that the engineering phases of the Service were superseded in importance by the business and economic aspects. The summary manner in which Mr. Davis was dismissed, together with the implication that engineers were not capable of managing important business functions, elicited a strong and determined reaction from ASCE.

A special committee of the Board of Direction was formed to investigate the circumstances of the case, with the result that a strong letter of protest was sent to Secretary Work and given public circulation. Local sections were informed and urged to make known their displeasure to their legislators. The National Civil Service Reform League entered the fray on the side of ASCE, giving emphasis to the fact that Secretary Work had abolished the Civil Service position held by Director Davis, creating a new position with the same responsibility under the title of Commissioner outside the pale of Civil Service.

The storm grew to such proportions that Secretary Work found it necessary to set up a special Fact-Finding Commission "to investigate the whole system of Government methods in reclaiming arid and semi-arid lands by irrigation." By this gambit he was able to allay much of the criticism, but in the spring of 1924 he appointed Elwood Mead, M. ASCE, to succeed the ex-governor as Commissioner of Reclamation. The Society graciously commended Secretary Work for this action.

To be sure, the magnitude of the reaction to Mr. Davis' dismissal may in some measure have reflected his stature as a Past-President of the Society. This was definitely not a factor, however, when in 1935 the Board took note again of the unfair treatment accorded a number of members employed in public agencies who were deprived of their jobs for doubtful cause and without adequate hearing.

In January 1935 the Board of Direction approved the principle that the Society should defend its members in such cases. This was followed in April by adoption of detailed conditions and procedures to be followed in implementing such defense of its members "against unjust accusation or dismissal, or oppression, without justification or proper hearing." The procedure involved the local sections, a personal investigation by a Vice President of the Society, with delegation of powers of adjudication and action to the Executive Committee of the Society.

Occasion to apply the procedure arose immediately after its adoption, when a state director of the Federal Public Works Administration was asked to resign and a former associate in consulting practice was declared ineligible for participation in PWA projects. The Society acted on behalf of both these members to obtain a rehearing and then to prove that the actions against them had been founded on false information. This resulted in rescinding of the punitive rulings by the Federal Administrator of PWA.

The defense process was again implemented in 1936 in the interests of members employed by state agencies in Pennsylvania and Delaware. But then both the policy and the implementation process appear to have been lost, only to be unearthed in the course of this historical research. Neither of the two executive secretaries who served through the period 1945-72 were aware that the policy existed, which was unfortunate as there were several instances during those years in which it might have been exercised. It was restored to official notice in 1973.

The Society has also acknowledged responsibility of the profession for actions of its members that may have been contrary to the public interest. In 1923 a number of suits were filed by the government, with members of the Society among those indicted, for alleged fraud in connection with certain contracts for the construction of World War I National Army Cantonments. By formal resolution of the Board of Direction the Society urged the President, the Congress and the Attorney General to bring these charges immediately to trial "in order that the guilty be punished and that the innocent may be freed of the serious accusations which have been made against them . . ."

An unusual membership service was rendered by the Society in 1931, when the Great Depression in the United States brought about interest in possible employment opportunities for civil engineers in Russia. A special committee study produced a very useful report for the guidance of any members involved in contract negotiations. The report confirmed the responsibility of the Russian government in meeting its obligations, but advised that professional service contracts include appropriate provisions covering health services, duty-free importation of necessities, and similar items.

ASCE has consistently advocated the principle of Civil Service, and on numerous occasions has made representations on behalf of civil engineers employed under that system. In 1935, for example, there was cause for concern with erosion of Civil Service standards and procedures. Specific recommendations were conveyed to the proper authorities through the Society's Field Secretary in Washington. This action was pursued further a year later, in a formal resolution reaffirming endorsement of the merit system and also adopting the U.S. Civil Service classification of positions for use in future ASCE studies of civil engineering employment and compensation.

The projects of the consulting engineer keep him in the public eye, and at times he is subjected to unfair criticism. As early as 1914, when the integrity of three reputable consultants to the New York City Board of Water Supply was impugned by an overzealous Grand Jury, the Society reacted crisply. A sharp rebuttal and condemnation of the Grand Jury action was transmitted to the authorities involved and to the press.

Civil engineers in public service are sometimes victimized by irresponsible collectors of political campaign funds. In 1926 the Society condemned this practice, and termed such solicitation of funds or endorsement thereof on the part of a state department head to be "wrong and reprehensible."

Publicity of a highly unfavorable nature has resulted from time to time as the result of questionable practices in the promotion of professional services by consulting engineers. In 1956 there was considerable activity on the part of commission agents, who offered to secure engagements for consulting firms. A terse and timely pronouncement was made by the Society to the effect that such procedure was in conflict with the Code of Ethics provision declaring it unethical "to use influence or offer commissions or otherwise to solicit professional work improperly, directly or indirectly." This action sufficed to resolve the problem.

A related situation arose in 1970 as the result of a growing effort on the part of political entities to solicit contributions from professional engineers and architects. The problem reached public attention when a Nassau County, N.Y., newspaper reported that such contributions were possibly associated with the allocation of professional service contracts.

Under ASCE leadership, representatives of architectural and other engineering societies promptly developed a set of guidelines that clarified the issue beyond question:

> "An engineer who makes a direct or indirect contribution in any form under circumstances that are related to his selection for professional work shall be *(a)* subject to disciplinary action by the Society, and if appropriate, *(b)* reported to the public authorities."

A policy statement embodying this forthright admonition was adopted by the Board in 1971, providing specific documentation for guidance of both consultants and political campaign fund solicitors.

The civil engineering profession suffered great loss of public esteem in 1973 when a federal investigation of corruption in Maryland implicated three members of the Society in the irregularities resulting in the resignation of United States Vice President Spiro T. Agnew. The investigation focused, in part, upon the contribution of funds for political purposes in return for favors in connection with the award of contracts for engineering services.

The three members were expelled from the Society. In addition, a Task Committee on Professional Civic Involvement initiated a series of actions, including condemnation of immunity concessions to transgressors who implicate others, affirmation of the 1971 guidelines previously mentioned, endorsement of legislation requiring full disclosure of all campaign contributions, and support of state registration boards in suspension or revocation of licenses as punishment for professional misconduct.

The very nature of the development of public works projects insures a continuing parade of challenging professional practice problems for the civil engineer. Management of these problems will demand a high order of discipline, sensitivity, and integrity. ASCE has proved that it is willing and able to serve as a medium for the prevention and control of such obstacles to the effective performance of the practitioner.

Honors and Awards

Peer recognition of competence and noteworthy accomplishment is cherished by all professionals. The identification and rewarding of those whose works have been exemplary enhance the development of any profession by the encouragement of its members toward meritorious service. a significant awards program is, therefore, an essential feature of a professional society. ASCE has met this need more than adequately.

Honorary Members

Held to be ASCE's highest honor, the requirement for Honorary Membership has remained essentially unchanged. In 1874 it was "acknowledged eminence in some branch of engineering" with at least 30 years experience; in 1974 "acknowledged eminence in some branch of engineering or in the arts and sciences related thereto, including the fields of engineering education and construction."

Since the first five Honorary Members were elected on March 2, 1853, only 259 have been so designated through the next 121 years. The complete roster, given in Appendix IV, is replete with the names of famous civil engineers. Several personages of international stature are included, such as Herbert Hoover, Field Marshall Ferdinand Foch of France, Baron Christian Phillipp von Weber of Germany, Sir William Henry White of Great Britain, Karl Imhoff of Germany, Baron Koi Furuichi of Japan and Prof. Luigi Luiggi of Italy.

Medals, Prizes, Awards and Fellowships

In 1872 George H. Norman, M. ASCE, contributed about $1,200 to fund an annual award of a gold medal for "the best essay on engineering subjects." Rules were drafted and dies were prepared for the medal according to Mr. Norman's design. The first award was made in 1874 to J. James R. Croes for his paper, "Memoir of the Construction of a Masonry Dam."

The 1881 Annual Report of the Board of Direction noted that the Norman Medal was then the only Society prize, and invited gifts to fund additional awards. Thomas Fitch Rowland responded immediately by endowing, in 1882, the prize which bears his name for papers on engineering construction. This was followed, in 1894, by the Collingwood Prize, endowed by Francis Collingwood, M. ASCE (Secretary 1891-94) and to be awarded to Junior (later Associate) Members in technical-paper competition.

The J. James R. Croes Medal and the James Laurie Prize, established by the Society in 1912, and the Rudolph Hering Medal, endowed by the Sanitary Engineering Division in 1924, brought to a total of seven the number of awards being made by the Society a half century after the first presentation of the Norman Medal. In that year the first ASCE fellowship was funded by Past-President John R. Freeman, Hon. M. ASCE, to encourage young engineers in research.

By 1960 there were 21 prizes, medals and fellowships being awarded by ASCE. Efforts in the next several years to encourage additional awards were quite successful, and by 1974 the list had grown to 39. Most of the new awards initiated in the 1960's were endowed through the activity of the Technical Divisions. Reserves being held by the Society for prizes, awards and trusts totalled almost $600,000 in 1974.

In addition to the awards of the Society were the four important awards made jointly with other societies. These included the John Fritz Medal, instituted in 1902, the Washington Award (1916), the Hoover Medal (1929), and the Alfred R. Noble Prize (1929). Four of the Founder Societies are involved with all these awards, the Western Society of Engineers with the Alfred R. Noble Prize and the Washington Award, and the NSPE with the Washington Award.

A complete listing of all awards and their recipients from 1872-1974 is given in Appendix V.

* * *

During the 80 years or so that were needed for ASCE to prescribe the direction and scope of its professional policy it is not surprising that its early efforts toward creation of a professional environment for its members were both impromptu and disconnected. This was not true, however, after 1930—the era of the Great Depression. Suddenly there was almost too much planning of programs that the Society was not fully equipped to administer.

Nevertheless, there was enough positive accomplishment to produce a creditable record. And in so doing there was not loss of public respect or stature by reason of resort to pressure tactics inappropriate to a professional society.

Most significant are the forcefulness and specificity of policy and planning in this area in 1974. With the benefit of many years of experience, coupled with an unprecedented commitment of resources in funds and personnel, a productive future seems assured.

CHAPTER V

ADVANCEMENT OF THE CIVIL ENGINEERING ART

Article 2 of the 1852 Constitution of the American Society of Civil Engineers and Architects stated that the objectives of the Society shall be:

> "2. The professional improvement of its members, the encouragement of social intercourse among men of practical science, the advancement of Engineering in its several branches and of Architecture, and the establishment of a central point of reference and union for its members."

This statement could readily be interpreted in a broad sense. Article 3, however, indicates that this was apparently not intended by the authors:

> "3. Among the means to be employed for attaining these ends, shall be periodical meetings for the reading of professional papers, and the discussion of scientific subjects; the foundation of a library, the collection of Maps, Drawings and Models, and the publication of such parts of the proceedings as may be deemed expedient."

All of these activities are directed toward advancement of the technical art, and the majority of early members felt that this was the sole concern of the Society. Such emphasis is not surprising, in view of the relatively underdeveloped status of applied science in 1852.

As has been brought out in preceding chapters, passage of time has seen a shift of emphasis toward professional development in other than technical matters. Even so, ASCE has always devoted a high order of its interest and resources to furtherance of the civil engineering art.

Significant contributions of the members have been encouraged and recognized by many honors and awards. A continuous effort has been made to offer meetings that would attract the greatest possible attendance and participation of members. The publications of the Society have achieved worldwide renown. The specialized Technical Divisions have fostered hundreds of committees in their attack on specific technical problems. The Engineering Societies Library has provided "a central point of reference" for the entire engineering profession—not just civil engineering alone.

We Stand on Their Shoulders

By the time ASCE was organized in 1852 the Canal Era was declining, and a memorable group of civil engineering pioneers had already passed from the scene. Included in this group were such greats as the Loammi Baldwins, father and son, Canvass White, Benjamin Latrobe, Sr., William Roberts, Lewis Wernwag, James Finley, James Geddes, William Howe, and Benjamin Wright. Benjamin Wright had been a participant in the 1838 effort to form a national society, but he died ten years prior to the 1852 movement.

Three Honorary Members of ASCE—John B. Jervis, Moncure Robinson, and Stephen H. Long—were leaders in the transition from the canal to the railroads as were Past-Presidents James Laurie, Horatio Allen, W. J. McAlpine, W. Milnor Roberts, and Octave Chanute. Other members of the Society who contributed to the early development of the railroad included Joseph G. Totten, Theodore M. Judah, and Albert B. Rogers.

Among the famous pioneer bridge-builders were Honorary Member Squire Whipple and Past-Presidents Albert Fink, Julius W. Adams, Elmer L. Corthell, George S. Morison, Henry Flad, and Thomas C. Clarke. Many innovations in bridges originated with ASCE members, some of which were the first use of wrought-iron members by Jacob H. Linville and

L. G. Bouscaren; the steel-wire suspension bridge by John A. Roebling; the pneumatic pile by William Sooy Smith and the pneumatic pier by James B. Eads; the arch bridges of Lefferts L. Buck and the cantilever designs of Charles Shaler Smith and Charles C. Schneider. Eads earned fame for both his Mississippi River Bridge at St. Louis and his daring execution of the South Pass Jetties in the Mississippi, which made New Orleans a major port.

Sanitary engineering as a branch of civil engineering was also identified by the organizers and first leaders of ASCE. Past-Presidents James P. Kirkwood, Alfred W. Craven, Julius W. Adams, Rudolph Hering, and Ellis S. Chesbrough were the forerunners of a distinguished cadre of innovators in water supply, sewage works, and public health technology. Kirkwood introduced the art of water filtration to America, Adams and Chesbrough were the first designers of public sewer systems for Brooklyn and Chicago, respectively, and ASCE was organized in the office of Alfred W. Craven, chief engineer of New York's Croton Aqueduct. Chesbrough was also experienced in water supply, and may have been the first city engineer in his service to Chicago in that capacity. Charles Payne in Providence and John P. Davis in Boston also provided the function of the city engineer, though not under that title.

Just as the canal builders transferred their talents to the rising needs of the railroad industry, so did they move on farther to face the challenges posed by urban growth in the late 19th century. Laurie, Adams, Kirkwood, Craven, and Chanute typified this generalist characteristic of the early civil engineer. The demands of burgeoning America were so diverse that the luxury of specialization could not then be afforded.

From the days of Hadrian in ancient Rome, intense public works activity has been plagued by political chicanery, exploitation of public rights and property, and fraudulent schemes by some promoters of commercial and industrial enterprises. This was, indeed, the case during both the Canal Era and the Railroad Era in America.

Many of the early leading members of ASCE were brought

into the execution of projects being conceived by quick-profit entrepreneurs, and instant conflict was often the result. At the 1873 Annual Meeting of ASCE, Martin Coryell pointed to the mismanagement of railroads by political manipulators and Wall Street speculators who had no interest in the scientific management of transportation. John B. Jervis, Mendes Cohen, Alfred F. Sears, and William R. Hutton were also among the ASCE members who exposed the incompetence and corruption in railroad management. Cohen and Hutton resigned their top-level administrative positions rather than remain privy to practices that were unprofessional or worse. This spirit of integrity and rugged independence extended into the public sector as well, as exemplified in 1860 by the public resistance of Alfred W. Craven to interference by the mayor of New York City in the engineering aspects of the development of the city's water supply.

The general ineptness of railroad and industrial management prior to the Civil War created a vacuum that was filled by a cadre of civil engineers who adapted their disciplined education and training toward the development of the scientific management process. Engineers knew the importance of sound basic data, and were able to compile statistical information in such manner that it could be analyzed, interpreted, and then applied to management decisions. Ironically, in years to come engineers would often be required to defend their capacity to administer important functions beyond the realm of technology, despite the fact that professional management was actually "invented" by the civil engineers of this era.

Again, leaders in ASCE were prominent among the pioneering professional managing executives. In 1861, the ubiquitous John B. Jervis published his "Treatise on Construction and Management of Railways." During the Civil War, Montgomery Meigs and Samuel Felton were among those who employed the science of logistics to the allocation and movement of men, materials, and weaponry.

Foremost in this group, however, was Albert Fink, whose contributions to statistical and economic analysis of railroad construction and operation far transcended in importance the early bridge truss design for which Fink is best known to engineers. As vice president of the Louisville and Nashville Railroad he published a classic report in 1873, entitled "Cost of Transportation," in which he developed, scientifically, several of the theories and indices that are fundamental to transportation economics.

Octave Chanute demonstrated dramatically the efficacy of the engineering approach in his 1873 rescue of the Erie Railroad from financial disaster. As general manager, Chanute assumed the almost impossible task of reorganizing and rebuilding the road, which he did most successfully.

Another member of the Society eminent in this field was Arthur M. Wellington, whose 1877 book on *The Economic Theory of the Location of Railways* became a standard reference on the subject. Ashbel Welch was also an accomplished railroad executive, who further made significant contributions to railroad safety.

By 1880 the role of the civil engineer as a professional executive and administrator was well established. A number of ASCE members were being brought into service as railroad presidents, among them Ashbel Welch, Mendes Cohen, Samuel Felton, Sidney Dillon, Eckley B. Coxe, and Alexander J. Cassatt. Many more distinguished themselves in such top-level managerial positions as vice president, general manager, chief engineer, etc.

The engineer-manager also found opportunities in areas of industrial enterprise other than railroads. Alexander L. Holley, James B. Eads, and Martin Coryell became widely known as such consultants. In the public sector, James B. Francis, Alfred Noble, Frederick P. Stearns, and others filled responsible administrative posts in various governmental agencies.

There were some highly respected civil engineers who practiced during the period 1850-80 who did not choose to become affiliated with ASCE, but the number was small. Several of them were truly outstanding, including Herman Haupt, J. Edgar Thompson, Benjamin H. Latrobe, Jr., John

A. Devereux, George Stark, and Charles Ellet. These men exhibited the same dedication to duty, professional conviction, and courageous initiative that characterized the top civil engineers of the time, and they, too, made noteworthy contributions to the advancement of the art and the profession.

This abstract of the works of the founders and early movers of ASCE is by no means complete. Its purpose is only to establish that the American engineering art was cultivated and nurtured by a relatively small group of rugged individuals who constantly sought better ways to do what had to be done in providing the public works and services that were so sorely needed by a young nation.

Any instrument that would have brought these talented men together for purposes of exchange of information, the coalescence of ideas, the evaluation of innovative proposals—even for debate over differences in judgment—would have been beneficial in enhancing their individual capabilities. It would also have fostered joint enterprises in which special knowledge and skills were combined to great advantage. It was as such an instrument that ASCE was to become a national influence in the building of America.

The functions of ASCE that were dedicated to the furtherance of knowledge of the art and science of civil engineering were commingled in a complex system. Its principal elements were the many kinds of meetings, the wide range of publications, the multitude of technical committees and councils, the growing list of specialty fields, and the several major administrative committees (Technical Activities, National Meetings Policy and Practice, Publications, etc.). All these elements were interrelated; changes occurring in any one were almost certain to have some effect upon the others.

Meetings, Conventions, and Conferences

During the struggling pre-Civil War days of the Society the only real activity was ". . . the encouragement of social intercourse among men of practical science . . . ." that occurred by way of the nineteen meetings that were held in the years 1852-55. All of these gatherings took place in the office of Alfred W. Craven, chief engineer of the Croton Aqueduct Department, Rotunda Park in New York City. Attendance was woefully small, averaging only six for the eight meetings held in 1853 and even less for the six held in 1854.

Although prescribed for the first Wednesday of every month, the meetings were actually less frequent. President Laurie's description of an elevated railway plan for "The Relief of Broadway," given on January 5, 1853, was the first technical program. Similar informal presentations at subsequent meetings covered a wide range of subjects, such as the extension of Church and Mercer Streets in New York City, "The Use and Abuse of Iron as Applied to Building Purposes," a suspension bridge over the Kentucky River, rebuilding of a Morris Canal aqueduct, comparison of materials used for carrying water, economics of inclined planes and locomotives on railroads, "Ball's indestructible water pipe," a new process for blasting rocks, a comparison of English and American wire ropes, "Recent Inventions for Economizing Fuel in Generating Steam," and others.

The first formal paper, ordered placed in the files of the Society, was read on March 2, 1855, by William H. Talcott under the title "Results of Some Experiments on the Strength of Cast Iron." Regrettably, this was the last meeting to be held until the Society was "resuscitated" in 1867.

The revitalization meeting on October 2, 1867, was held in the office of C. W. Copeland at 171 Broadway, and was adjourned to a session at 76 John Street on October 9. It may well have been the most important meeting in the lifetime of ASCE.

On November 6, 1867, the Society met—for the first time—in its own headquarters, located in the Chamber of Commerce Building at 63 William Street. The next meeting, on December 4, 1867, was noteworthy for several reasons: 54 new members were elected, a semi-monthly meeting schedule was adopted, and James P. Kirkwood gave his "Address of the President," which was to become the first paper to be published by the Society (*Transactions*, Vol. I, p. 3).

While an increase now took place in meeting attendance, with occasional participation by non-resident members, the gatherings were still quite local in character. The following notice was to have the effect of giving a new measure of emphasis to "the encouragement of social intercourse among men of practical sciences:"

May 10, 1869

"Dear Sir:

"Our Society has of late been impressed with the importance of annually gathering its members in Convention for the interchange of ideas and the discussion of professional subjects, as well as those relating to the extension of its own field of usefulness.

"These objects cannot be attained by means of the twenty-three regular meetings, held during the year, on account of the distant residence of many of its members, but they require the appointment of some day on which all can meet and strengthen that bond of sympathy which should animate hearts engaged in the furtherance of a common cause.

"It is proposed, therefore, to hold the first of such a series of conventions on the 16th day of June, 1869, commencing at 10 a.m.

"Through the courtesy of the Chamber of Commerce, the meeting will be held in their Assembly Room. It will occupy most of the day, and will be followed by a suitable collation.

"The Society relies on the individual efforts of its members for sustaining the interest of the occasion, and desires that you will either come prepared to read a brief professional paper —in case of unavoidable absence will forward such to the Secretary—or else be prepared to make some statement of engineering experience calculated to prove of general interest . . ."

Fifty-five members registered for this first of many ASCE conventions. The day was devoted to five formal papers and two informal statements by Honorary Members John B. Jervis and Squire Whipple, and closed with a dinner. This convention pattern was followed for a good many years.

The "suitable collation" mentioned in the first convention notice was evidently of some importance to the format. In 1876 a Bylaws amendment provided that:

> "The officers of the Society may give a reception at the Society's rooms on the evening of the second and fourth Wednesday of each month, between October and April, for the purpose of informal professional conversation and social intercourse. Regulations concerning refreshments and the invitation of guests may be made by a committee appointed for the purpose, but shall be without expense to the Society."

When the second Annual Convention failed to achieve the success of its predecessor, a landmark decision was reached to hold the annual conventions in other parts of the country, and thus give positive emphasis to the national character of the Society. The wisdom of the move was immediately apparent. After a successful experiment in Chicago in 1872, the fifth Annual Convention in Louisville drew 79 Members and Fellows, the largest number of civil engineers ever to be assembled in North America up to that time.

A complete record of the dates and locations of all of the 204 full-scale national conventions and meetings of the Society is provided in Appendix XI. After the adoption of the Annual Convention of the membership, in 1869, as a supplement to the Annual Business Meeting in January and the regular monthly meetings at Society headquarters, other important policy changes concerning the conduct of meetings were implemented in 1922, 1929, 1950, 1960, and 1962.

Dual emphasis was given to the technical activities of the Society in 1922 when a schedule of three Conventions per year was adopted instead of the one held previously. In this same year the specialized areas of civil engineering (sanitary, structural, highway, construction, etc.) were first authorized to organize as separate Technical Divisions, and the two actions were no doubt related. In 1929 the regular monthly meetings that had been held since the creation of the Society were discontinued. The schedule at that time comprised the Annual

Meeting in January, with the Spring, Summer and Fall Conventions.

Frequency of national gatherings was cut back in 1950, when the Annual Meeting was changed to October, and only two conventions, in February and June, were provided for. It should be noted, however, that all previously authorized meetings were not actually held. In the period 1932-1961, no conventions were held during the World War II years of 1944-45, only one convention was held in four of the years, two conventions in thirteen of the years, and three conventions in eleven of the years.

Until 1930 the programs of the meetings were limited as to geographical coverage. In that year the concept of regional programs was introduced at the July convention in Cleveland. A regional committee handled the program and general management, with the business management left to the host Local Section. By this time, of course, the Technical Divisions were the primary source of program material. This pattern continued for about thirty years.

In the late 1950's, when the three national conventions were annually attracting a total registration of 4,000 to 5,000, a new trend began to develop. The fourteen Technical Divisions (Fig. 14) had recently acquired a major degree of autonomy in the periodic Journals made available to each Division for the publication of its technical papers and discussions. The national meetings produced about 450 to 500 papers per year, but the capacity of the Divisions was much greater. To provide further outlet for their publications, the Divisions began to hold secondary conferences, sometimes alone and sometimes jointly with other Divisions or other engineering organizations. These Division conferences were usually devoted to a specific problem area, such as electronic computation, weather modification, jet age airports, flood plain regulation, shear strength of soils, pollution abatement, economics, etc. Some Divisions, however, established a pattern of periodic conferences devoted to the complete spectrum of their interests. By 1960, six to eight Division conferences drawing an aggregate registration of 1,200 to 1,500 were being held annually.

A new format was initiated in 1962 in order to give renewed emphasis to the national meetings and conventions. Instead of attempting to cover the entire range of all thirteen Divisions in every national meeting—which could not be done in depth—each of the national meetings was concentrated in one of the four areas of civil engineering: Structural, Water Resources, Transportation, and Environmental Engineering. The plan envisioned the classification of the Technical Divisions into these four topical areas; the Structural Engineering Conference would serve the Structural, Engineering Mechanics, and Soil Mechanics and Foundations Divisions; the Water Resources Engineering Conference would accommodate the Hydraulics, Irrigation and Drainage, Power, Waterways and Harbors, and Sanitary Engineering Divisions; the Transportation Engineering Conference would include the Highway, Air Transport, Waterways and Harbors, and Pipeline Divisions; and the Environmental Engineering Conference would feature the programs of the City Planning and Sanitary Engineering Divisions. It was also intended that the broad functional Construction and Surveying and Mapping Divisions would offer papers under any of the topics.

Hopefully, the plan would have provided very strong, in-depth programs that would have given every Society member at least one meeting each year that would have justified his attendance, regardless of location. The four basic annual conferences were to be supplemented by an Annual Meeting program devoted to professional affairs and to major public projects and issues of civil engineering import. Also, the Divisions were encouraged to hold Specialty Conferences, but only in very specific problem areas.

Although formally adopted by the Board of Direction, the plan was never actually implemented in the manner intended. Some of the Divisions insisted on participation in every national meeting without regard for the relevance of their sessions to the conference topical theme. Other Divisions continued to hold their own general program conferences, which were essentially in competition with the basic national meetings. The Annual Meeting concept was also not implemented as envisaged.

By the early 1970's, the four annual topical meetings were aggregating only about 4,000 registrants. Seven to twelve Division conferences were drawing an attendance of 1,500 to 2,500 yearly, and the Divisions were also participating with other organizations in many joint specialty conferences. The Committee on National Meetings Policy and Practice was engaged, in 1973, in a review and appraisal of the situation, with a view toward increasing attendance while reducing the number of meetings.

Before 1957 the only exhibits displayed at ASCE meetings were non-commercial, consisting mainly of models and pictorial representations of local projects. In that year a trial commercial exhibit was a feature of the Annual Meeting in New York. The showing of new developments in civil engineering equipment and materials added to the value of the meeting, and at the same time income was derived from the sale of exhibit space.

The 1957 Civil Engineering Show was modestly successful—sufficiently so, at least, to merit its continuance in ensuing Annual Meetings. After a few years, however, the Federal Internal Revenue Service alleged that the exhibit income was unrelated to the tax-exempt purposes of the Society, and filed a claim for $47,978 in income taxes for the years 1969-70. The claim was successfully contested, and the complaint withdrawn, but the incident raised questions as to the relative advantages and disadvantages of the exhibit. It was concluded that the values hardly justified the effort, and the Board of Direction acted in 1965 to discontinue the holding of commercial exhibits in conjunction with basic or specialty conferences, while permitting any educational or other non-commercial exhibits arranged by the local committee.

There was a renewal of interest in commercial exhibits a few years later, however, when the Water and Wastewater Manufacturers Association offered to manage an equipment exposition at the 1973 Annual Meeting. The Board amended its policy to sanction such functions at national meetings, when so requested by the local committee. The 1973 Civil Engineering Show in New York City was only moderately successful. Participation by exhibitors and attendance at the show were not up to expectations, although there was a slight surplus of income over expenses. It was decided on the basis of this experience to explore other approaches to the management of exhibits, both as an adjunct to national meetings and specialty conferences, and as a potential source of Society income.

Many national meetings of the Society were noteworthy for one reason or another, but a few of them stand out in the record. The 1876 American Centennial Exposition in Philadelphia provided an unique opportunity for directing public attention to the promising young engineering profession, and the ASCE leadership took full advantage of the situation. The 1876 Annual Meeting was held in Philadelphia in conjunction with the Exposition, under the direction of a special ASCE Centennial Commission and a Resident Secretary. A general exhibit in the main exposition building featured 1,900 displays, including pictures, books, manuscripts, maps, drawings, and even full-scale models. The exhibit was the means of "bringing together, expeditiously and effectively, engineers from all parts of the world, placing them in commumication with manufacturers and supplying them with letters of introduction and data concerning such engineering works as they were desirous of visiting." Many foreign engineering displays were also shown in the Exposition, which was a fitting stage for the first international recognition of American engineering.

A highlight of the program of this 1876 Annual Meeting was a paper by ASCE Vice President Theodore G. Ellis. His review of engineering progress and prediction of the future of aluminum as an engineering material attracted wide attention:

"In the short space of a century, American engineering has

risen to the highest standard, and ranks with that of any nation in the world. We have the longest railways, the best equipment and the most comfortable conveyances. We have by far the finest passenger steam boats in the world. We have the largest bridges, the greatest spans, and the deepest foundations, and the only successful railway suspension bridge in the world.

"The question remaining to be solved is what are we coming to? When will our improvements end? Can it be possible that a century from this time the American engineer can look back to such a record of progress as now appears to us? I reply, it is not only possible but probable.

"With a metal only 2½ times as heavy as water, 6 times as strong as steel, as easily worked as iron, and almost indestructible in air or water, conceive what would be the possibilities of engineering skill, to say nothing of the revolution in all the tools and utensils of ordinary life. Bridges of a mile span would be entirely possible, railway trains for passengers could be reduced to at least one-fourth their weight, and the navigation of the air (all except the alighting, in which I should never have great confidence), would become practicable. With a motive power which would not be obliged to carry so great weight of fuel as our present steam engines require, speed of navigation might easily be doubled. But these are merely visions in which a practical engineer of the present day has no right to indulge; he cannot yet afford to wait for the new discoveries that are to throw our works in the shade; like the builders of the pyramids, we must go on plodding in our old-fashioned way, leaving to those of our profession who shall look back upon us from the standpoint of the next century, to say if any of us shall be ranked among the 'old masters.' "

So outstandingly successful was the 1876 Annual Meeting that the Society was conferred an award by the United States Centennial Commission "for the very large and important exhibition, and for the great service rendered by the Society to the art and science of engineering."

ASCE also was awarded a Diploma of Honor for its exhibit at the Paris Exposition in 1878, and was given responsibility for the highly successful Civil Engineering Division program in the International Engineering Congress featured at the Columbian Exposition in Chicago in 1893.

ASCE did not exhibit at the Paris Exposition in 1900, although the possibility was considered. (In the discussion the comment was made that, "There is a great craze on this subject of automobiles.") Instead, the beginning of the new century was commemorated by holding the 1900 Annual Meeting in London in July. The facilities and hospitality of the Institution of Civil Engineers were made available for the occasion, and 68 members attended.

At the request of the Louisiana Purchase Exposition, ASCE financed and conducted an International Engineering Congress in conjunction with the 1904 Annual Meeting of the Society. This significant event was a feature of the St. Louis World's Fair, attracting a registered attendance of 876. The program was truly international in character, including 51 papers from the United States with others from France (18), England (10), Holland (7), Japan (5), and single papers from Austria, Belgium, Canada, Denmark, Russia, and Switzerland. Volume LIV of *Transactions,* comprising six parts, was devoted entirely to the papers and discussions presented at the Congress.

The 1924 Annual Convention in Pasadena, California, is remembered for its tragic aftermath. En route to the East after the meeting, a railroad train carrying a number of Society officers and members, including Secretary John H. Dunlap, was involved in a collision at Buda, Illinois, near Chicago. Secretary Dunlap was so severely injured that he died a month later, on July 29, 1924. He was survived by his wife and three sons, and the Society assumed a deep and lasting interest in their welfare.

Appropriately, the 100th anniversary of the founding of America's first national engineering society was celebrated with great pomp and fanfare. However, the Centennial of Engineering, held in Chicago in September 1952, was much more than a birthday celebration for ASCE. It was actually the occasion for a worldwide commemoration of the first century of engineering as a profession in America.

As developed under the direction of Major Lenox Lohr, the Centennial had for its aims: (1) an accounting of the contributions of the engineer in peace and war, (2) the "personalizing" of the engineer, (3) encouragement of young people to pursue engineering careers, and (4) to depict the role of technology in the enhancement of the standard of living and national prosperity. These objectives were fulfilled by a program of several hundred professional and technical papers, a dynamic exposition of exhibits, encouragement of other engineering organizations to hold meetings in Chicago during the ten-day celebration, and a strong program to bring these activities to public attention. Total registration reached 27,964, including 2,600 ASCE members.

FIG. 12. — ASCE CENTENNIAL COMMEMORATIVE STAMP AND FIRST-DAY COVER

President Carlton S. Proctor represented the Society in distinguished fashion throughout the complex and extensive program, and on its behalf he received 66 gifts and expressions of greeting and congratulation from engineering societies throughout the world. An outstanding feature was the Centennial Day luncheon address by former President Herbert Hoover, Hon. M. ASCE.

As a tribute to the Society on its 100th anniversary a special commemorative stamp was issued by the Post Office Department during the Chicago Convocation. The stamp (Fig. 12) depicts progress by comparing a typical 1852 wagon bridge with the George Washington Bridge in New York City. The service of George Washington as a pioneer in civil engineering is also symbolized. An official engraved cover was also provided for first-day mailing of the commemorative stamp.

Publications of the Society

When James P. Kirkwood assumed office as the second President of ASCE immediately following its rebirth in 1867, he lost no time in calling for the initiation of services that would insure the permanence of the Society. In his inaugural address he said:

> "The non-resident members may many of them connect themselves with our Society from a kind of esprit de corps . . . but this link of union between the Society and those distant members who can but rarely enjoy the privileges of its meetings or the convenience of its rooms or library when we shall possess one—this link of good-will, I say, must be maintained and nourished by some palpable food. The short papers to which I have alluded on subjects of professional interest or a selection from them, must be printed, to admit of the doings of the Society here, reaching and interesting the non-resident members."

Action to provide the "palpable food" was immediately forthcoming, for at that same meeting, on December 4, 1867, a Publications Committee was created to develop a feasible operation. (Supervision of publications was later combined with the duties of the Library Committee. However, there has been a standing Committee on Publications since 1891, except for the periods 1923-27, when there was a Committee on Technical Activities and Publications, and 1927-30 when there was a Committee on Meetings and Publications.)

A few months later a special fund was inaugurated for the sole purpose of printing and disseminating papers. After considerable effort fifty donors contributed $100 each in demonstration of "fellowship among the patrons and capitalists of public improvement and those interested in the onward progress of the sciences." In February 1870 the membership grade of Fellow was established to recognize all contributors of $100 or more. The Fellowship Fund more than met the need for publication costs. By 1882 the fund reached $10,000, and it was determined that all receipts above that level would be transferred to the general revenues of the Society. By 1891 the resources of the Society were adequate to accommodate publication expenses in the regular operating budget, and the Fellowship Fund was abolished.

From 1867 to 1873 those technical papers accepted for publication were issued separately and irregularly as *Transactions* of the Society. A total of 57 papers was produced and distributed in this fashion. The following resolution, adopted early in 1873, laid the foundation for the system of handling technical papers that was still being used a century later:

> "*Resolved*, that hereafter every paper presented to the Society shall be immediately examined by the Library Committee, who shall decide whether it shall come before the Society. If yes, it shall be printed in cheap form and distributed to the members, with notice that discussion, written or oral, will be received, within definite limits as to time; at the expiration of which, said discussion, with the original paper, shall be referred to a special committee, with instructions to examine and recommend a final disposition of the same, with reference to the permanent proceedings of the Society."

A second action, also in 1873, provided for the first regular publication of the Society. It was to comprise not less than 48 pages, and to be issued on the second Wednesday of each month. Included in the new monthly *Transactions* were all papers accepted by the Society together with all contributed discussions on them, as well as a section designated *Proceedings*, comprising notices and reports of meetings of the Society and the Board of Direction, news items, announcements, new book lists, membership actions, memoirs of deceased members, and library acquisitions.

The 1872 number of *Transactions*, Volume I, was issued in November 1873. In it and Volume II were carried the 57 "Transactions Papers" selected for publication during the period 1867-72. One of these was the 1867 Presidential Address of James P. Kirkwood, in which he charted so well the early course of the Society.

The volume of material available for publication was more than sufficient to support the authorized monthly issues of 48 pages, and *Transactions* was produced with this frequency from 1873 to 1895.

From 1879-1892 a few extra copies of each paper were printed and circulated in advance of presentation to selected reviewers as a means of stimulating discussion. In 1892 a *Bulletin* was published for circulation to all members. It carried all official notices and announcements, with abstracts of papers to be read at forthcoming meetings. Any interested member could request an advance copy of the complete paper. The *Bulletin* was discontinued in 1895.

January 1896 brought the first number of the new *Proceedings*, to be issued on the fourth Wednesday of each month except July and August. The *Proceedings* carried all items pertaining to the business of the Society, meeting notices, minutes, etc., as well as all papers accepted by the Publications Committee for presentation at subsequent meetings. The final drafts of these papers, with all discussion and correspondence, were then published in an annual volume of *Transactions* for the permanent record.

For the benefit of foreign members and others unable to attend meetings of the Society regularly a newsletter supplement was published as "Part 2" of *Proceedings* from 1924 to 1930. Content of this "informal sheet" was limited to notices, news items and short feature articles of general interest.

Except for the frequency of issue of *Transactions*, this publication program continued on a complimentary basis to all membership grades for the next 50 years. The growth in volume of published material made semiannual publication of *Transactions* necessary in 1887, and a quarterly schedule was required in 1909. Resort to the use of the best quality of low-bulking India paper in 1912 made possible return to the single annual volume without reduction in coverage.

A major development was the introduction—in October 1930—of the monthly magazine *Civil Engineering*, which provided valuable internal communication to the fast-growing Society (then 14,000 members). According to a notice in the Annual Year Book:

> "It carries the more animated and graphic articles submitted for publication, and many Society announcements of official and semiofficial character which formerly were contained in *Proceedings*. The *Proceedings* were thus simplified, making more readily accessible those contributions to engineering literature of the more studious nature. The change contributes largely to a wider spread of knowledge of both Society and technical matters, providing, with *Proceedings*, two mediums different in style of expression and character of content."

Civil Engineering met its editorial objectives most successfully. It was furnished without charge to all members, its costs being offset in sufficient part by advertising revenue to sustain it through the Great Depression of the 1930's. As "The Magazine of Engineered Construction," advertising income by 1955 not only covered all production and staff expense, but also began to contribute to general revenue. In 1972 a claim by the Internal Revenue Service that surplus advertising and subscription receipts were taxable "as income unrelated to the exempt purposes" of the Society was successfully resisted. In recognition of a trend in public concern, in 1970 the subtitle of *Civil Engineering* was changed to "The Magazine of Environmental Design and Engineered Construction."

After more than 30 years' usage of the name "*Civil Engineer-*

ing," a notice received in 1964 advised that a commercial periodical in England held prior claim to that title under copyright. The matter was promptly settled by a change to *Civil Engineering — ASCE*, with other nominal concessions. Oddly, not a single one of the then 70,000 readers gave any indication that the revision was noticed!

Rising printing and mailing costs in the 1940's created economic problems in the production of *Transactions* and *Proceedings*, both of which were almost completely subsidized from general operating revenues. The only income derived from these publications came from non-member subscriptions. Thus, a modest charge was imposed for *Transactions* in 1947 to avoid waste circulation and to reduce the subsidy.

By 1950 the amount of technical material being provided by the Technical Divisions for publication in *Proceedings* was accelerating. To reduce waste circulation, the production of *Proceedings* on a monthly schedule was discontinued, and beginning in February 1950 each technical paper or report was published as a *Proceedings-Separate*. Abstracts were listed in each monthly issue of *Civil Engineering*, with a convenient order form. Up to 25 *Separates* per year were furnished without charge to members, additional copies being priced at 25 cents each. In 1954 this was modified by permitting members to enroll in one Division and to receive automatically, and without charge, all the papers sponsored by that Division.

The system was not popular with the membership, and restiveness became evident in several Technical Divisions. The proliferation of technical specialty organizations throughout the engineering profession is often a result of a one-time need for a suitable publication to serve a specialist field. Murmurs as to the propitiousness of new organizations began to develop in several of the Divisions. A major adjustment was in order.

The situation was effectively resolved with the production of *Proceedings* in the form of a separate periodical for each

Technical Division effective in January 1956. These *Proceedings-Journals* assembled the papers of each Division under its own review, and every member was privileged to receive any two *Journals* without charge. Additional *Journals* and separate papers were available for purchase. The unrest in the Technical Divisions was satisfied, as the Divisions now had essentially all of the services and autonomy of an independent organization.

The *Proceedings-Journals* were still fulfilling their purpose in 1974, although economic considerations had required that they be put entirely on a subscription basis in 1966. Also the Air Transport, Highway, and Pipeline Divisions decided in 1969 that it would be advantageous to combine their separate aperiodical Division *Journals* into a single joint quarterly *Transportation Engineering Journal.* The publication frequency of the *Journals* varies from annual to monthly. Several *Journals*—including those of the Engineering Mechanics, Environmental Engineering, Hydraulics, Geotechnical Engineering, and Structural Divisions—have acquired worldwide eminence in their fields.

When the *Proceedings-Journals* were adopted in 1956, a *Journal of the Board of Direction* was provided to accommodate professional material of broad membership interest. Renamed the *Journal of Professional Practice* in 1958, it found limited demand even though a complimentary mailing of an issue was made to all members in an effort to promote circulation. In 1972 it was again renamed as *Engineering Issues —Journal of Professional Activities*, and assigned a staff editor. The content and readership of the publication were making good progress in 1974.

From the 1950's to the present time, considerable attention has been paid to the physical appearance of the Society's publications. When, in 1953, ASCE adopted photo-offset printing methods for its technical publications and became the first technical society to use typewriter composition, the appearance of the text was considerably less attractive than that produced by the hot-type methods previously used. Constant upgrading of the text composition quality served as a guide to other organizations that were interested in publishing economies along with acceptable quality standards. By the late 1960's computer-assisted photocomposition had become a reality, and ASCE used it to produce its 1968 *Directory* and 1971 *Official Register*. In 1972 the *Journals* were produced by computer typesetting with an attendant improvement in quality coupled with a stabilization of the growth rate of costs. All these advances in composition were achieved with no outside funding support; if they involved extra initial expenditures, these were shown to be economically supportable in a very short period of time.

Until 1956, all technical papers published in *Proceedings* were reproduced as finally revised, with all discussion, in the *Transactions*. The number of *Proceedings* (papers) at that time (Fig. 13) exceeded 250 per year. Double publication of this volume of material was not justified by demand and cost, and a decision was reached to limit *Transactions* papers to relatively few selected *Proceedings* papers. The plan was short-lived, however, because of the difficulty in determining just which papers merited the rating of "permanent reference value." It survived for only three years—1956-59.

FIG. 13 — ASCE TECHNICAL PUBLICATIONS, 1852-1973

Year	*PROCEEDINGS Papers*		*TRANSACTIONS*	*Cost of PROC. and TRANS.*	
	Submitted	*Published*	*Papers Published*	*Amount*	*Percentage of Total Budget*
1875	—	—	15	NA	NA
1900	NA	18	23	$ 17,862	32
1925	NA	38	23	59,479	19
1950	NA	53	34	89,709	12
1973	1,669	842	960*	766,500	17

*Digests of papers

Return to the publication of all *Proceedings* papers in *Transactions* during the period 1960-63 was equally unsatisfactory, for by this time the number of papers had risen to 330 per year. It was necessary to publish the 1961, 1962 and 1963 volumes of *Transactions* in five parts, and the operation was incurring an annual loss of $75,000.

After careful management review in the staff, it was decided to produce *Transactions* as a compilation of digests of all *Proceedings* papers, together with the complete "Annual Address of the President" and abstracts of the memoirs of deceased members. The paper digests, preferably prepared by the authors, were intended to be functional capsule presentations of the original paper, its discussion, and the closure. Hopefully, the abstract would of itself meet the reference requirements of the searcher in the vast majority of cases. Digests of all *Civil Engineering* articles were also included in the single *Transactions* volume. Thus, one compact bookshelf reference could provide a key to an entire year of the Society's output of publications. The vast majority of *Transactions* purchasers favored this move.

The technological "information explosion" had reached such proportions by 1960 that new measures were required to assist the user of the literature. In 1963 a system of abstracts and key-words was developed in accord with the EJC *Thesaurus of Engineering Terms*. This was followed, in 1966, by introduction of a new bi-monthly *ASCE Publications Abstracts*— a compilation of information-retrieval cards for all *Civil Engineering* and *Journal* papers, together with subject and author indexes. The cards could be conveniently clipped, permitting the reader to select those he was interested in saving in a personal reference file. The service as well received from the beginning and was finding a rising demand in the years following its introduction.

In 1970 ASCE resorted to the electronic computer in its effort to best serve the information needs of the profession. This made possible the production from a single input of information of all the abstracts and indexes that unify the many periodical publications of the Society into a single system. As a result of its favorable experience with computerized abstracting, indexing, and typesetting, the Society proposed in 1973 that a monthly *Civil Engineering Abstracts* be published jointly by all of the professional and technical organizations serving the various specialized areas of civil engineering practice. The idea was well received and appeared likely to develop successfully.

The growth of the ASCE technical publications program is shown in Fig. 13. Of particular interest is the substantial segment of the operating budget allocated to this important function of the Society.

Manuals of Practice

The *Manual of Engineering Practice* series was initiated in 1927 when the Board of Direction sought a publication format for a "Code of Practice," setting forth practical ethical guidelines for the professional engineer. This was Number 1 of the long series of monographs intended for ready reference. The distinctive mark of the Manual is its specific authorization by the Board of Direction.

The Technical Procedure Committee defined the Manual in 1930 as "an orderly presentation of facts on a particular subject, supplemented by an analysis of the limitations and applications of these facts." Also, it is "the work of a committee or group selected to assemble and express information on a specific topic." The Manual "is not in any sense a 'standard,' however; nor is it so elementary or so conclusive as to provide a 'rule of thumb' for non-engineers."

In 1962 the series was renamed *Manuals and Reports on Engineering Practice*, and it was required that manuscripts be published in a *Journal* and subjected to the critical review of the entire membership before being published in Manual form. A total of 53 numbered Manuals had been produced as of 1973, of which seven were on professional subjects and the remainder on technical matters.

On several occasions ASCE has collaborated with other technical and professional organizations in the production of special reports on subjects of mutual interest. Typical of these are the following:

Glossary: Water and Wastewater Control Engineering (AWWA, WPCF, and APHA)

Recommended Guide to Bidding Procedure on Engineering Construction (AGC)

Water Treatment Plant Design (AWWA and CSSE)

Design and Construction of Sanitary and Storm Sewers (WPCF)

Miscellaneous Publications

In recent times the basic technical publications program of ASCE has been augmented by a wide assortment of special-purpose books of various kinds. Some of these, such as the newsletters of the Technical Divisions and the Civil Engineering Research Letter, provide administrative communication to technically oriented activities. Others afford similar stimulus to internal functions and affairs, such as the newsletters concerned with sections, student chapters, and public relations.

As early as 1871, a "List of Members" was furnished as a membership service. In 1882 the Constitution and Bylaws were included in this annual booklet. Lists of committees, award recipients, local sections, and student chapters were added as time passed, with more and more general information for members. In 1918 the publication was designated the *Yearbook*. Beginning in 1950, the membership lists were produced in a separate *Directory*, the remainder of the material being retained in a booklet titled the *Official Register*. For reasons of economy the *Directory* was made a biennial publication in 1956, and a charge has been made for it since 1962

A constitutional provision since 1852 has required the Board of Direction to issue a report on the state of the Society at the Annual Meeting. There was no requirement that such a report be published, but from 1937 to 1949 it was reproduced in full in the *Yearbook* and a "Secretary's Abstract" of it was carried in the *Official Register* from 1950 to 1955. On the premise that an informed membership would be more interested, enthusiastic, and active — and thus engender the most productive

Society — an attractive, readable "corporation-type" Annual Report was mailed to each member beginning in 1956. Not only did the new format enhance readership within the membership, but it also brought widespread attention to the work of the Society on the part of nonmembers and other organizations. The report proved to be an extremely effective medium for membership promotion, and for public relations purposes. Its cost, only about 25 cents per member, was well justified.

The publications policy of the Society has accommodated the printing and distribution of many documents of interest and value to the civil engineer. The full range of "Miscellaneous Publications" extends from the Conference Proceedings volumes containing the papers from Specialty Conferences, to the ASCE Historical Publication Series, to the translation of the Russian monthly journal *Gidrotekhnicheskoe Stroitelstvo* (Hydrotechnical Construction.)

Organization for Technical Activities

For almost 70 years the advancement of the art and science of civil engineering in ASCE was generated primarily through the national meetings and publications functions. Special committees were formed from time to time to pursue studies in specific technical areas, but no attempt was made to administer technical activities under a formalized organizational pattern until well into the 20th century. The early technical committees were mostly concerned with methods for testing engineering materials, and with the application of such test data to rational design.

The Committee on Tests of American Iron and Steel was created in 1872 when General William Sooy Smith, M. ASCE, proposed:

> "That a committee of five be appointed to urge upon the United States Government the importance of a thorough and complete series of tests of American iron and steel, and the great value of formulae to be deduced from such experiments."

Although this committee was not ultimately successful in its efforts to enlist continuing support from Congress, it established a fitting example in its dedication of purpose for the many technical committees that were to follow.

A committee recommendation was adopted in 1874 to join with Stevens Institute of Technology in founding a testing laboratory "for making complete and impartial tests of the characteristics, value and strength of materials used in the arts."

In 1875 it appeared that the aims of the committee were to be completely realized when the following Presidential Order was issued from the White House:

Executive Mansion,
March 25, 1875

"In pursuance of the 4th section of the act entitled 'An act making appropriations for sundry civil expenses of the Government for the fiscal year ending June 30, 1876, and for other purposes,' approved March 3, 1875, a Board is hereby appointed, to consist of—

Lieutenant-Colonel T. T. S. Laidley, Ordnance Department, U.S. Army, President of the Board;
Commander L.A. Beardslee, U.S. Navy;
Lieutenant-Colonel Q. A. Gillmore, Engineer Department, U.S. Army;
David Smith, Chief Engineer, U.S. Navy;
W. Sooy Smith, Civil Engineer;
Alexander L. Holley, Civil Engineer;
R. H. Thurston, Civil Engineer;

who will convene at the Watertown Arsenal, Massachusetts, on April 15, 1875, or as soon thereafter as practicable, for the purpose of determining, by actual tests, the strength and value of all kinds of iron, steel and other metals which may be submitted to them, or by them procured, and to prepare tables which will exhibit the strength and value of said materials for constructive and mechanical purposes, and to provide for the building of a suitable machine for establishing such tests, the machine to be set up and maintained at the Watertown Arsenal.

"The funds appropriated for the purposes of these tests will be disbursed under the Ordnance Department of the Army, and the Board will receive instructions from, and make its report to, the Chief of Ordnance.

"Mr. R. H. Thurston, Civil Engineer, is designated as Secretary of the Board at an annual compensation of twelve hundred dollars.

"Actual traveling expenses, as provided by law, will be allowed the members of the Board.

"U. S. Grant."

At the next session of Congress, however, further appropriations to support the testing operation and the supervisory Iron and Steel Board were denied. The determination of the committee is manifest from the following extract from its report in April 1877:

". . . The Committee considers it necessary to more than merely allude here, to the want of knowledge of the characteristics of the new varieties of iron and steel offered for our use, of which we all are so painfully conscious, to bring the mind of each member of the Society to a realization of the value of the knowledge which seems just within our grasp. Shall we fail to attain it?

"We beg each member to put this question to himself. There is not a member of Congress of the United States who cannot be reached through some member of our Society, who is personally acquainted with him.

"Let us make a vigorous effort at once to get Congress to repeal the legislation discontinuing the Board when the money already appropriated has been expended and to appropriate the money which the Board may need to enable it to complete the very valuable work it has undertaken."

The testing machine erected in the Watertown Arsenal was turned over to the U.S. Ordnance Department in 1879. Although the 1881 appropriations bill contained a directive that the Chief of Ordnance "give attention to such programme of tests as may be submitted by the American Society of Civil Engineers," no systematic series of tests on engineering materials was ever accomplished. The analysis of iron and steel continued to receive specific committee attention in ASCE, however, until 1903.

Among other early technical committees that performed sterling service were:

Committee on Rapid Transit and Terminal Facilities (1873)
Committee on Railway Signals (1874 and 1875)
Committee on Cost and Work of Pumping Engines (1874)
Committee on Preservation of Timber (1880)
Committee on Proper Manipulation of Tests on Cement (1898)
Committee on Rail Sections (1875)
Committee on Concrete and Reinforced Concrete (1905)
Committee on Steel Columns and Struts (1909)
Committee on Materials for Road Construction (1910)
Committee on Bearing Value of Soils for Foundations (1913)
Committee on Stresses in Railroad Track (1914)
Committee on Bridge Design and Construction (1921)
Committee on Valuation of Public Utilities (1910)
Committee on Highway Engineering (1920)

The ubiquity of Octave Chanute (President, 1891) is noteworthy in reviewing the earliest technical committee activities. This capable and energetic engineer served on at least four committees operating in 1875, and he was chairman of two of them!

Origin of the Technical Divisions

The 1919 report of the Committee on Development, which foresaw such significant expansion of professional activities in ASCE, also envisioned growth in the technical area. Additional meetings and emphasis on the publication of papers and discussions were urged at both national and local levels. Further, it proposed that standing advisory committees to the Board of Direction be appointed "to promote the study of important engineering subjects and charged with outlining and coordinating the work of like committees of the Local Sections." This was the report that was received by the Board without action, although it was undoubtedly a factor in broadened aims and responsibilities that were accepted by the Society in the 1920's and thereafter.

Under date of February 4, 1921, Director John C. Hoyt addressed the following proposal to members of the Board of Direction:

"With a view to increasing the value of the technical activities of the American Society of Civil Engineers, it is suggested that the field of civil engineering be divided into several branches (for example, sanitary, hydraulic, structural, topographic, highway, railroad, irrigation, drainage, river and harbor improvement), and that a committee be appointed to follow the work in each branch, such committee to be composed of men who are active in the branch. Following are possible functions of each such committee:

"1.—To keep in touch with the work that is being done throughout the country in its branch.

"2.—To advise the membership in a general way, through the *Proceedings*, of activities in its branch.

"3.—To solicit papers for presentation to the Society, and to promote discussions of them to the end that the Society publications may contain a record of important current activities in engineering, as well as a history of completed engineering work.

"4.—To keep sufficiently in touch with the activities of other organizations that our work may be co-ordinated with others and duplication avoided.

"5.—To initiate the organization of sub-committees for work within limited fields in the branch if there is demand for them.

"6.—To assist the Committee on Publication, especially by reviewing manuscripts submitted for presentation, or to suggest names for this function.

"7.—To study systematically the needs for sub-committees on research problems in the branch and to make appropriate recommendations to the Board of Direction.

"The above plan of procedure should tend to create an interest by the individual members of the Society in its work, and the committees could be composed of the younger men located in various parts of the country. It could also be extended to promote activities in the Local Sections, and might eventually lead to the formation of technical sections."

Two months later the Board appointed a Committee to Promote the Technical Interests and Activities of the Society, under the chairmanship of Past-President Arthur N. Talbot. A crisp and action-oriented report from this body was adopted by the Board in January 1922, providing for creation of a standing Committee on Technical Activities and Publications, with advisory subcommittees covering the principal branches of the civil engineering field. Other recommendations called for greater emphasis on technical activities in the local sections, and for greater involvement of younger members.

No time was lost in effecting the necessary Bylaws amendments. When the sections providing for the Committee on Technical Activities and Publications, and the new Technical Divisions were approved on first reading, the officers were authorized to proceed with implementation without waiting for formal adoption at the next meeting. This enabled the Board to act forthwith (June 1922) to authorize formation of the Power Division, Sanitary Engineering Division, Irrigation Engineering Division, and Highway Division, in that order. The latter two authorizations were subject to the subsequent filing of the required petitions.

At this point Acting Secretary Elbert M. Chandler cited the existence of sixteen organizations having overlapping and duplicating interests in various aspects of civil engineering, and urged that action be taken to invite these bodies to become Technical Divisions of ASCE. A committee was authorized to promote such affiliations, but none was subsequently consummated.

As originally established, the Technical Divisions were administered by their own executive committees; they could enroll nonmembers of ASCE as affiliates, and they could assess dues to finance their operations. They were directly responsible to the Board of Direction. In 1926, a separate Committee on Technical Procedure was created to coordinate all technical affairs, its membership including the chairmen of the Publications Committee, Research Committee, and of all Technical Division executive committees, together with the President and Secretary of the Society and two members of the Board. Assessment of separate Technical Division dues was discontinued in 1943, when provision was made for the Divisions in the regular operating budget of the Society.

FIG. 14 — ASCE TECHNICAL DIVISIONS — 1922-74

Technical Division	*Year Established*	*1973 Enrollment*	*1973 Committees Number*	*1973 Committees Personnel*
Power	1922	2,992	13	73
Environmental Engineering	1922	10,031	20	168
Irrigation and Drainage	1922	4,660	22	129
Highway	1922	12,197	12	97
Urban Planning and Development	1923	9,215	22	129
Structural	1924	23,546	80	656
Waterways, Harbors and Coastal Engineering	1924	4,159	30	156
Construction	1925	18,774	23	188
Surveying and Mapping	1926	3,731	17	97
Engineering Economics	1931	(1)	(1)	(1)
Geotechnical Engineering	1936	15,247	24	179
Hydraulics	1938	11,029	39	220
Air Transport	1945	1,889	13	58
Engineering Mechanics	1950	6,098	18	150
Pipeline	1956	1,639	15	165
Urban Transportation	1971	1,027	13	56

(1) Engineering Economics Division dissolved in 1952.

All the Technical Divisions formed to 1974 are listed in Fig. 14, with pertinent information. Their influence has been extended into the local sections by way of technical groups in the sections and each Technical Division has its own newsletter to further membership interest and communication.

The comprehensive reorganization of the Society under the 1930 Functional Expansion Program classified all activities into three departments: Administrative, Professional, and Technical. The Technical Procedures Committee, Technical Divisions, and the Research Committee constituted the new Technical Activities Department. The following extracts from the Functional Expansion Program report are noteworthy:

> "The technical functions of the Society are well performed through the medium of Society committees and Technical Divisions. No change is proposed in them . . .
>
> "The program provides at once for the extension to members of an increased participation in Society affairs and for the development of that participation definitely along nontechnical or professional lines . . ."

While there had previously been occasional mention of "technical" activities as apart from "professional" activities, this was the first official recognition of such separation of interests. In subsequent years this dichotomy acquired something of a competitive aspect that went beyond administrative structure to become manifest in budgetary considerations, planning of meeting sessions, and even to the assignment of Board members to standing committees.

The Technical Department operated very effectively until 1950, when a reorganization eliminated the departmental arrangement altogether. However, Technical Division operations continued to enjoy relatively independent supervision under a Committee on Division Activities constituted of members of the Board of Direction, and a Committee on Technical Procedure including in its membership the chairmen of the Technical Division executive committees. A few years later the Divisions were authorized, if they chose, to name a Board of Direction "contact member" to meet on invitation with the Division executive committee, but without vote. In 1963 a Director from the Board was specifically designated to serve as a full member of each Division executive committee. Such Board contact members had been serving in professional committees for many years.

In 1963 all technical affairs of the Society were placed under the direction of a Technical Activities Committee (TAC) as a counterpart of the Professional Activities Committee (PAC). The Technical Divisions were represented by their Board contact members in TAC, just as the various professional committees were represented in PAC. The Division executive committee chairmen were no longer directly involved in the general administration of technical activities.

For almost thirty years after they were authorized, the Technical Divisions operated within themselves, with only nominal collaboration in areas of overlapping interest. An important step was taken in 1959, when the Hydraulics, Sanitary Engineering, Irrigation and Drainage, Power, and Waterways and Harbors Divisions set up a Coordinating Committee on Water Resources, as a device for advancing their mutual concerns. A similar Coordinating Committee on Transportation soon followed, bringing together the Air Transport, City Planning, Highway, Pipelines, and Waterways and Harbors Divisions; it was dissolved in 1962. Coordinating committees on Civil Defense (1960), Water Rights Laws (1964), and Structural Engineering (1967) also served effectively.

The benefits of cooperation among the Technical Divisions were so apparent that a more formal liaison was introduced in 1965 with the formation of the Technical Council on Urban Transportation. Technical councils followed on Ocean Engineering (1967), Aerospace (1970), Water Resources Planning and Management (1972), and Computer Practices (1973).

The pendulum made another arc after the 1968 Professional Activities Study Committee brought about the extensive reorganization of professional activities in 1970 (see Chapter III). Even as this new administrative mechanism was being implemented in 1971 the Board of Direction approved a restructuring of technical affairs upon recommendation of a Technical Activities Study Task Committee. The thrust of the change was to group the Technical Divisions and Technical Councils into five "Management Groups," comprising Board contact members and nonmembers of the Board. In essence, the move added another level of administrative review in that a proposal originating in a Technical Division or Council must be cleared by the Management Group and then by TAC before it could be considered by the Board. At the same time it relieved the Board of concern with many routine operations of the Technical Divisions and Councils, delegating this authority to TAC and the Management Groups. The arrangement is illustrated in Fig. 15.

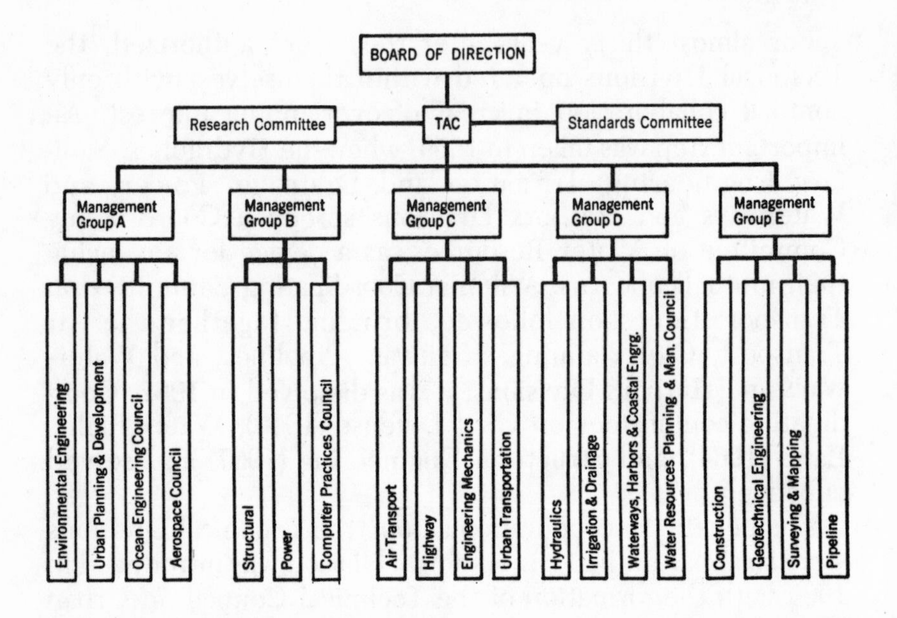

FIG. 15. — TECHNICAL ACTIVITIES ORGANIZATION CHART — 1974

"The technical and professional activities of ASCE are carried on in two separate spheres of the Society—united only at the Board of Direction level." This is the basic premise of a Task Committee on Cooperation Between PAC and TAC that was appointed in 1972 "to study cooperation and coordination between the Technical and Professional Activities of the Society." The 1973 report of the task force was more philosophical than pragmatic in presenting its recommendations for improving communication and interaction between the two areas of concern in the Board of Direction and staff, in meetings, in committee work, and in continuing education. A cogent point was the observation, "It is not necessary that TAC and PAC have the same organizational set-up . . . it is essential that TAC and PAC activities be thoroughly coordinated and tied more closely with local section programs."

Time has proved Director Hoyt's suggestion to the Board of Direction in 1921 to have been a major contribution to civil engineering, with impact on the entire engineering profession. His concept of the Technical Divisions as a means of advancing the art was fulfilled to the letter, with complete success. In a broader sense, the movement had two opposing effects on the trend toward proliferation into specialized technical organizations that was extant at the turn of the century. On the one hand, the Technical Divisions slowed the trend by supplying the need for identification and services in some specialties; on the other, several technical associations were actually created by ASCE members as a by-product of their activities in the Divisions.

Development of Civil Engineering Standards

The coordinated programs in ASCE directed toward advancement of engineering standards and research evolved from the same special committees that preceded the Technical Divisions. With a few exceptions, the standards activities have been completely cooperative with those organizations having the establishment of engineering standards as their main purpose. It has always been the policy of the Society that standards preferably should be founded upon the broadest possible base of acceptance, rather than the issue of a single organization.

The independent activities of ASCE in standards development extended over the early years of the Society, when there were few other engineering societies available for joint work. In the period 1875-95 important committee reports bearing upon standardization were produced on such subjects as tests of American iron and steel, railway signals, standard rail sections, uniform systems of tests on cement, compressive strength of cements and cement mortars, uniform methods of testing materials, and preservation of timber. To be sure, much of this work was in appraisal of the state of the art, but such information is basic to any consideration of standards.

Two early ventures of ASCE into the realm of standards merit documentation at some length: Its activities in connection with the metric system of weights and measures, and its role in the establishment of the present system of standard time. The reporting of these exercises will be followed by an account of the more formal standards programs that came later.

Consideration of the Metric System

The concern of ASCE with regard to adoption of the metric system of weights and measures goes back to 1876 when Clemens Herschel, (President, 1916) offered the following resolution:

> "*Resolved:* That the American Society of Civil Engineers will further by all legitimate means, the adoption of the metric standards in the office of weights and measures at Washington, as the sole authorized standard of weights and measures in the United States; that the Chair appoint a committee of five to report to the Society a form of memorial to Congress in furtherance of the object expressed."

After extensive discussion the resolution was submitted to letter ballot of the membership, and was approved by a vote of 138 to 73. The Committee on Metric System of Weights and Measures was promptly appointed. It was also decided that authors of papers presented before the Society be requested to include the metric equivalents of all weights and measures included in such papers.

The movement came upon strong opposition however, when the Congressional "memorial" drafted by the committee failed of adoption by a letter ballot in 1877, and the committee was not continued. In 1878 a resolution was adopted by membership referendum to postpone indefinitely further consideration of the metric system of weights and measures.

An inquiry in 1881 disclosed that only two authors had complied with the 1877 resolution calling for the inclusion of metric equivalents in ASCE papers. This led to immediate rescinding of that policy.

A special Committee on Status of the Metric System in the United States was appointed in 1907, and its informative report was published in *Proceedings*, Vol. XXXIV, p. 415. The movement was still strongly opposed, however, and the committee was discharged in 1910, without approval of its recommendation that copies of its report be sent to all members of Congress and the New York State Legislature.

In 1917 the Publications Committee was requested to consider again the question of an ASCE requirement that metric equivalents be used in its technical publications. The committee's recommendation, however, was negative.

Eighty-one years after ASCE resolved that "further consideration of the metric system of weights and measures be postponed," the Board of Direction acted in October 1959 to reaffirm the 1876 policy that "the Society further, by all legitimate means, the adoption of the Metric Standards . . . as the sole authorized standard of weights and measures in the United States." In 1969, recognizing the growing public interest in the metric system, the Committee on Standards created a Task Committee on Metrication to study the effects of conversion upon civil engineering and the construction industry in the United States.

The deliberations of the task force resulted in adoption by the Board of Direction in 1971 of the following metrication program:

1. To support conversion to the International System (SI).
2. To recommend that no additional metric units be incorporated into the SI System.
3. To begin immediately the use of dual units in all ASCE publications.
4. To adopt the ASTM Metric Practice Guide.
5. To recommend the use of SI units in all cartographic and geodetic projects.
6. To recommend revision of pertinent publications using SI units.

7. To recommend that instruction in SI units be initiated immediately in the teaching of civil engineering subjects.

A brochure, entitled "Metrication and the Civil Engineer," was published by the Society in 1972 for use as part of an information packet, in partial implementation of this program.

ASCE Fosters Uniform Standard Time

By far the most important and exciting venture by ASCE into the realm of standards was the successful espousal by the Society of a uniform comprehensive system of standard time, to replace the multiplicity of unrelated local time systems throughout the world. It all began when Sandford Fleming, M. ASCE, a Canadian, presented a paper "On Uniform Standard Time for Railways, Telegraphs and Civil Purposes Generally" at the 13th Annual Convention in June 1881 (*Transactions*, Vol. III, p. 387).

Mr. Fleming described the time notation situation then prevalent, which was the cause of great confusion especially in North America:

> "According to the system of notation which we have inherited from past centuries, every spot of earth between the Atlantic and Pacific is entitled to have its own local time. Should each locality stand on its dignity, it may insist upon its railway and its other affairs being governed by the time derived from its own meridian. The smaller and less important localities, however, as a rule, have found it convenient to adopt the time of the nearest city. The railways have laid down special standards which vary, as has been held expedient by each separate management. In the whole country (U.S. and Canada) there is so far, an irregular acknowledgment of more than one hundred of these artificial and arbitrary standards of time. The consequences of this system are unsatisfactory. They are felt by every traveller, and in an age and in a country when all, more or less, travel, the aggregate inconvenience and confusion is very great, and it will be enormously multiplied as time rolls on. If the system already results in difficulties to trouble our daily

life, and to lead to embarrassments which often occupy our courts of law, which, indeed, too often are the cause of loss of life, what will be the consequences in a few years, when population will be immensely increased and travel and traffic indefinitely multiplied, if no effort be made to effect a change?"

The problem had already been recognized by the American Meteorological Society, the Imperial Academy of Science of Russia, the Royal Society of London, and the Canadian Institute. The American and Canadian organizations had joined in recommending a system under which 24 standard meridians would be designated, with the prime or zero-time meridian to pass near the Bering Strait, 180 degrees from the Greenwich Observatory. Also, consecutive numbering of the 24 hours was proposed, instead of the 12 hours A.M. and P.M.

Mr. Fleming reported that these proposals were being brought to public attention, and he sought an expression of support from ASCE as he felt that such endorsement "must carry with it great weight, and will exact respect in every quarter." His eloquent plea found receptive ears. The study group that he recommended was immediately appointed, with himself as chairman.

At the next meeting of the Society, in June 1882, the need for a standard time system was confirmed, and the principle of the 24-hour universal day to be reckoned from a prime meridian of longitude was endorsed. Reporting this progress at the January 1882 Annual Meeting, the committee requested authority "to invite the cooperation of other scientific associations, and that of other bodies in the furtherance of this important object, and that all such societies and government departments interested be invited in the name of the Society to attend a general convention to meet at New York or Washington . . . for the purpose of determining the Time System advisable to adopt." Not only were these recommendations approved, but also the President of the Society was authorized to invite representatives from Canada and Mexico, as well as appropriate state and federal agencies, to the proposed conference.

Chairman Fleming and his associates lost no time in developing a brochure and supporting documentation outlining the standard time plan, and the Society gave this material wide circulation among railroad and telegraph company officers, leading scientists, and public officials. The response was almost unanimously favorable. This led to the decision that the necessity for time reform in North America was so urgent that action should be taken at once to establish a zero meridian for the reckoning of time, as basis for "a system capable of extension to the whole globe." The committee forthwith was authorized to petition the Congress of the United States to take the necessary steps to have a prime meridian established.

Such interest and momentum had been developed by distribution of the Society's time reform literature that there was no patience for the ponderous movement of Congress. In his report at the January 1884 Annual Meeting Chairman Fleming stated:

> "On the 11th October last, the railway authorities met in convention at Chicago and determined, without further delay, to take energetic action. They decided to adopt the hour standards, and they fixed upon the 18th of November (1883) as the day when they would generally begin to operate their lines by the hour meridians. The public with great unanimity acquiesced to the changes. It is now generally and universally admitted to be a great public boon."

Thus did the present system of Standard Time come into being in North America, with its Atlantic, Eastern, Central, Mountain and Pacific time zones reckoned from the Greenwich anti-meridian. This action did not, however, encompass the so-called "24-hour notation."

Meanwhile, Congress had also responded to the ASCE petition by adopting on August 3, 1882, a Joint Resolution authorizing the President "to call an International Conference to fix on and recommend for universal adoption a common prime meridian to be used in the reckoning of longitude and the regulation of time throughout the world." The International Meridian Conference was held in Washington in October 1884, with representatives of 26 nations participating. The deliberations of the conference resulted in essentially complete acceptance of the concept of uniform standard time advocated by ASCE, as manifest in "the system of regulating time which has been adopted with signal success in North America . . ." The following resolutions were adopted:

> I.—"That it is the opinion of this Conference that it is desirable to adopt a single prime meridian for all nations, in place of the multiplicity of initial meridians which now exist.
>
> II.—"That the Conference proposes to the Governments here represented the adoption of the meridian passing through the centre of the transit instrument at the Observatory of Greenwich, as the initial meridian for longitude.
>
> III.—"That from this meridian longitude shall be counted in two directions up to 180 degrees, east longitude being plus and west longitude minus.
>
> IV.—"That the Conference proposes the adoption of a universal day for all purposes for which it may be found convenient and which shall not interfere with the use of local or other standard time when desirable.
>
> V.—"That this universal day is to be a mean solar day; is to begin for all the world at the moment of mean midnight of the initial meridian, coinciding with the beginning of the civil day and date of that meridian; and is to be counted from zero up to twenty-four hours.
>
> VI.—"That the Conference expresses the hope that, as soon as may be practicable, the astronomical and nautical days will be arranged everywhere to begin at mean midnight.
>
> VII.—"That the Conference expresses the hope that the technical studies designed to regulate and extend the application of the decimal system to the division of angular space and of time shall be resumed, so as to permit the extension of this application to all cases in which it presents real advantage."

Two months after the Washington conference, on January 1, 1885, the hour zone system was adopted at the Greenwich Observatory. This began a slow but steady movement toward worldwide acceptance. In 1891 Chairman Fleming reported that:

"The hour zone system has been adopted for ordinary use in portions of the three continents of Asia, Europe and America. In 1887 an imperial ordinance was promulgated directing that on and after the first day of January of the year following, time throughout the Japanese Empire would be reckoned by the third hour meridian. The reckoning in England and Scotland is by the twelfth hour meridian; in Sweden the eleventh hour meridian is the standard, and quite recently it has been resolved in Austria-Hungary to be governed by the same meridian. Efforts are now being made to follow the same course in Germany and in other European countries . . ."

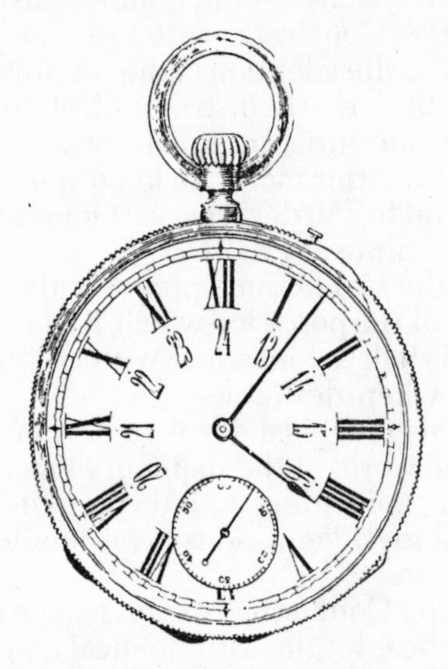

FIG. 16. — THE 24-HOUR CLOCK (From ASCE *Proceedings*, June 1884, p.77)

The crusade for adoption of another recommendation of the international conference—the "24-hour clock"—was not, however, to meet with such success. This reform would have been effected by universal use of the watch dial illustrated in Fig. 16. The concept was applied to the publications and meetings of ASCE in 1883. Also known as the "Universal Day," the notation was subsequently adopted by the Greenwich Observatory, the railways of Canada, the Eastern Telegraph Company (serving parts of Asia, Africa, Australia, Europe, and New Zealand), and by several national governments. Because one of the international conference recommendations referred to the possibilities of a decimal time system, however, the issue became controversial. While apparently favored by a wide majority of railroad managements in the United States, the 24-hour clock was not adopted by them as was done with the hour zone system in 1883.

When President Grover Cleveland in his 1886 message to Congress urged adoption and implementation of the recommendations of the International Meridian Conference, there were high hopes that Congress would give legal sanction to these proposals. Chairman Fleming suggested that it would be appropriate and feasible for the "Universal Day" of 24 consecutively numbered hours to be formalized on January 1, 1900. But this was not to be.

Several surveys and mailings of brochures aroused sufficient interest on the part of railroads, telegraph companies, and the public to cause the Board of Direction in 1890 to produce a "memorial" urging that Congress adopt the recommendations of the 1884 international conference. The resolution was approved by a mail ballot of the membership, and was conveyed to President Benjamin Harrison and to both Houses of Congress. Bills "respecting the reckoning of time throughout the United States," setting forth the principles of time reform advocated by the Society, were introduced before both Houses in January 1891. Neither bill, however, achieved passage.

The procrastination of Congress did not, however, deter progress elsewhere in the world, notably in Europe. The hour zone system was adopted in Belgium, the Netherlands, and Germany, and the 24-hour notation by India. France would not accept the Greenwich Prime Meridian, but adopted the reckoning of Paris as the time for the entire nation—a difference of only nine minutes from Uniform Standard Time!

The issue came back to Congress in 1896, in connection with the International Conference recommendation that astronomical time be abolished and replaced by civil time in nautical almanacs for the purpose of navigation. Six nations publishing nautical almanacs were prepared to accept this provision, but only upon condition that the United States would do so simultaneously. A most convincing resolution by the Society again petitioning Congressional approval of the International Conference findings was no more successful in overcoming the intransigence of Congress than its predecessors.

After these disappointments, the efforts of Mr. Fleming and his committee were centered upon the 24-hour notation system, but with a gradual decline in emphasis. In 1899, a small but vocal band of members protested the continued use of the 24-hour system in the notices of the Society because it was "not likely to accomplish the end for which it was adopted . . ." Outcome of the discussion was an action to request the Committee on Uniform Standard Time to make its final report at the next Convention.

Chairman Fleming made his last report by letter dated January 13, 1900. He reviewed the accomplishments of his committee since its appointment in 1881, noting that the hour zone system of Standard Time was by now effective on five continents. Although disappointed in the limited acceptance of the 24-hour clock, he still expressed hope in these words:

> "The American Society of Civil Engineers took a leading part in initiating Standard Time, and it has continuously stimulated the development of a great reform in time-reckoning, not on this continent alone, but throughout the world. It must be recognized to be desirable that the Society should participate in the complete fruition of the movement. At one time some were sanguine enough to think it possible that the final step would be effected on the opening of the coming new century, but whether then or later, I am satisfied that like the Gregorian reform, the modern time-reform must in the end become an accomplished fact."

With this report, 19 years of dedicated, significant activity was terminated by a terse expression of thanks to the committee for its "long-continued service." Sandford Fleming served as chairman of the committee throughout its existence, from 1881-1900; others who served on the committee for 19 years were Thomas Eggleston, Charles Paine, and John M. Toucey. Theodore G. Ellis, Theodore N. Ely, J. E. Hilgard, and Frederick Brooks served for shorter periods. A much more fitting acknowledgement of their signal contribution is the action on the committee's progress report in June 1883:

> "*Resolved*, That the American Society of Civil Engineers hereby acknowledges the extent and value of the work accomplished to date by the Committee on Uniform Standard Time, and tender to that Committee hearty thanks and earnest congratulations for the diligence and the intelligent and fruitful labors of which the results have been so well exhibited . . ."

Evolution of "American Standards"

By 1916 the proliferation of engineering standards activities had reached such a state of complication that ASCE, AIME, ASME, AIEE and the American Society for Testing Materials (ASTM)—which had been founded in 1898—collaborated in the creation of the Joint Committee on Organization of an American Engineers Standards Committee. The state of affairs at the time is summarized in the preamble of that committee's report to the participating societies in June 1917 as follows:

> "At the present time many bodies are engaged in the formulation of standards. There is no uniformity in the rules for such procedure in the different organizations; in some cases the committees engaged in the work are not fully representative, and in a considerable proportion of cases they do not consult all the allied interests. The present custom results in a considerable duplication of work, and there are in some fields several "standards" proposed for the same things that differ from each other only slightly and that often in unimportant details. It is very much more difficult to obtain agreement between the proposers of overlapping standards after they have been published than it would be to get the proposers to agree before they had committed themselves publicly."

The joint study resulted in a recommendation for the creation of a permanent American Engineering Standards Committee, to be constituted of representatives from the five sponsoring societies and others admitted later. Specialist "Sectional Committees" would develop specific standards, with input from interested "Co-operating Societies," including government agencies and others. Standards approved by the main committee were to be designated first as "Recommended Practice," and, when suitably established, as an "American Standard."

The American Engineering Standards Committee (AESC) was established in October 1918. Its membership was soon expanded to include government and industrial bodies concerned with standards, and it performed an important service in coordinating the work of the standards-producing committees of the many "sponsor" bodies. For some years the AESC operated from offices in the Engineering Societies Building in New York.

In 1928 the AESC was reorganized "to permit broader participation in its activities by trade associations and other bodies having an important concern with National standardization work," and became the American Standards Association (ASA). Procedures in ASA were broadened to accommodate alternative methods of approach. The Sectional Committees were modified to comprise representatives of all interested entities.

By 1930 ASA had 320 projects on its agenda, and 155 American Standards were in use. The association was affiliated with the International Standards Organization (ISO). ASCE was participating through its own representatives in 22 Sectional Committees concerned with a broad range of subjects.

ASCE Standards Activities in Recent Years

Through the next 35 years ASCE concern with standardization was satisfied through its close cooperation with ASA and ASTM. From time to time there was occasion to collaborate with other bodies—such as the United States Bureau of Standards and various industrial associations—on special problems.

In the early 1950's it was noted that some of the ASA Sectional Committees were not as active as they might be, and a means for selecting and reviewing ASCE appointments to these committees was sought. To this end a Committee on Standards was appointed in 1956, with the intent that it work closely with the Technical Divisions and serve the Board of Direction in an advisory capacity. Several of the Divisions set up their own committees on standards to better support the Society committee in its efforts to involve effective personnel and thus expedite the operations of ASA in its handling of civil engineering standards.

In 1961 the Society accepted an invitation to name representatives to the U.S. National Committee of the ISO.

The ASA underwent a second major reorganization in 1966, emerging as the USA Standards Institute (USASI). ASCE continued to be represented at the administrative level through the Member Body Council, and carried on its participation in some 35 Sectional Committees and Correlating Groups. In 1970, USASI was renamed as the American National Standards Institute (ANSI).

The standards program assumed new dimensions in the mid-1960's with a rising interest in building codes. This was first actuated through representation in committees of the U.S. Chamber of Commerce, and later through participation with other interested professional organizations in the Model Code Standardization Council. In 1968 the Committee on Standards joined in an effort to explore with the National Fire Protection Association and American Institute of Architects the best approach to development of a model national building code that would find universal acceptance. Although this complex issue has not been resolved, these studies have instigated several actions at national level toward consistency in building codes.

In 1974 about 150 members represented the Society in some 75 standards and codes activities, including seven Technical Advisory Boards and 28 committees of ANSI, 17 committees of ASTM, three U.S. National Committees of the ISO, various committees of the U. S. Bureau of Standards and other government and private organizations, and in the Model Codes Standardization Council. ASCE was cooperating with other major engineering societies in support of federal legislation on metric conversion. The recently enacted regulations of the federal Occupational Safety and Health Act were found to conflict with long-established local building codes, and these divergences were being reviewed by the Task Committee on Building Codes. New task committees on nuclear standards and on building code requirements for excavations and foundations were also becoming operative under the aegis of the Committee on Standards.

Advancement of the Art Through Research

Research services in ASCE have generally been carried on under the direction of the Technical Divisions, with emphasis upon the stimulation of research effort in the most urgent problem areas. The Society has sought to provide two-way communication between the field of civil engineering practice and the research institution, both in identifying research needs and in making the results of research available for practical application. It has thus provided the catalytic function that is necessary for effective interaction between science and art.

In the preceding accounts of the historical backgrounds of the ASCE Technical Divisions and the standards activities of the Society, reference has been made to a number of very early ad hoc technical committees. Some of these committees were engaged with the testing of engineering materials; others were addressed to the review of the state of the art in various aspects of civil engineering practice. While these projects do not represent fundamental research in a strict sense, they are strong examples of the kind of applied research that was in such great need in the latter part of the 19th century.

Arthur N. Talbot (President, 1918), in a review of early research activities (*Transactions*, Vol. 86, p. 1280), cited as particularly significant the committee work on uniform tests of cement (1912), concrete and reinforced concrete (1917), steel columns and struts (1919), bearing values of soils, and stresses in railroad track. Past-President Talbot provided much of the leadership toward the formalization of research programs in the Society. He considered that, among other things, a national engineering society should be ". . . a stimulator of research, and of the progress of engineering science . . ."

Some consolidation of research-related activities in ASCE and the other Founder Societies occurred in the 1914-20 period when Ambrose Swasey, a Past-President of ASME gave endowments totalling $500,000 to the United Engineering Society (later United Engineering Trustees) for "the furtherance of research in science and engineering, or for the advancement in any other manner of the profession of engineering and the good of mankind." This was the beginning of Engineering Foundation, which has achieved a distinguished record in the advancement of engineering research in all fields. Administered by representatives of the Founder Societies, the Foundation has financed—wholly or in part—many millions of dollars of important technological pioneering. ASCE has been a strong supporter and a frequent applicant for research funding. ASCE projects in the early years of the Foundation included concrete and reinforced concrete arches, steel columns, soils and foundations, and arch dams.

When the National Research Council was formed by the National Academy of Sciences in 1916, as a war-time advisory body of scientists and engineers, Engineering Foundation offered the funding and staff needed until the Council was permanently authorized in 1918. All of the Founder Societies are represented in the NRC Engineering Division.

It was not until 1920 that research activities in ASCE were identified as a specific area of concern, and given administrative direction. In that year the Engineering Foundation invited ASCE to name representatives of its Research Committee "to co-operate with other Founder Societies and with Engineering Foundation as to the best possible service in their joint instrumentality for research . . ." As no such committee existed at the time, the Board of Direction designated the chairmen of the Committee on Bearing Power of Soils, the Committee on Stresses in Railroad Track, and the Committee on Specifications for Bridge Design and Construction to serve as a Conference Committee on Research.

Just two years later the committee was expanded to nine members, and given advisory status to the Board of Direction in connection with the appointment of new special committees engaged with research problems. In 1923 the Research Committee was given direct supervision over nine special committees—those on Bearing Values of Soils, Stresses in Railroad Track, Concrete and Reinforced Concrete Arches, Flood Protection Data, Hydraulics Phenomena, Impact in Highway Bridges, Irrigation Hydraulics, Steel Column Research, and Stresses in Structural Steel.

This arrangement prevailed for the next 15 years, in which period the Technical Divisions came into being and gained in strength. In 1938 it was decided to abolish the Committee on Research, and to assign the special committees for which it was responsible to the Technical Divisions. When it became apparent a few years later that coordination of research operations in the Divisions was essential, the Committee on Research was reconstituted for that express purpose. The committee has been continuously in existence since 1946.

A significant spinoff of ASCE research operations occurred in 1947 when a committee in the Structural Division proposed formation of the Reinforced Concrete Research Council (RCRC), an independent entity supported by Engineering Foundation with ASCE sponsorship. With the further cooperation of the American Concrete Institute this council introduced the ultimate-strength design concept that is widely applied in structural engineering.

In the late 1950's the Research Committee foresaw great opportunities for ASCE in the advancement of civil engineering research. In 1958 it held the first of a series of research conferences designed to illuminate and to delineate these opportunities. In 1959 the committee initiated a move providing for ASCE administration of research councils such as RCRC, which are formed for the conduct of research in a particular field by the voluntary association of all those interested in the problem, with operating funds being contributed by interested groups. A new Research Council on Pipeline Crossings of Railroads and Highways was formed soon thereafter, and the RCRC transferred from Engineering Foundation to the administrative umbrella of ASCE. These and other research councils to be created in the next 15 years were to become a major segment of ASCE research activity.

The year 1962 brought important developments. With the aid of a $10,000 grant from Engineering Foundation a professional Research Manager was added to the staff. At the same time the Research Committee broadened its program to encompass the following objectives:

1. To review and clarify the research goals of the Society.
2. To formulate a plan to up-grade engineering research and the image of civil engineering in terms of research.
3. To obtain increased support for civil engineering research at educational institutions.
4. To establish suitable methods of coordination.

Under these guidelines the new ASCE Research Department opened an era of productiveness never before approached.

These accomplishments began with the generation of research activities in the Technical Divisions at an unprecedented level. Increased interest was stimulated particularly by research conferences—some general and some on specific problems. A 1963 publication, entitled "Advancing Civil Engineering Techniques Through Research," stimulated new projects. In 1965 a Task Committee on Study of National

Requirements for Research in Civil Engineering began its work by requesting input from every Technical Division. Each passing year brought at least one new research council.

Dismantling of the New York World's Fair in 1966 afforded a unique opportunity for destructive testing of structures. With funding of about $250,000 from several government agencies and private foundations, ASCE joined with the Building Research Advisory Board of the National Research Council in carrying out such a project. This highly successful project led to further exploration of opportunities for full-scale testing.

It has always been the policy of the Society to give all possible aid to the financing of civil engineering research in qualified institutions, and to accept grants for direct administration of research projects only when this procedure is prefered by the grantee or justified for other reasons. A typical example of this was the Combined Sewer Separation Project undertaken in 1966 and reported in 1970, which was funded by the Federal Water Pollution Control Administration. In 1969 the Research Committee was given broader authorization to engage in projects in which the Society would accept research funds for direct administration. An ASCE Research Fund was established in 1962 as a reserve source of funds for special research purposes.

The public concern with the social impact of technology was reflected in ASCE research operations in the early 1970's. Indicative of this influence was the 1971 Conference on Goals of Civil Engineering Research — Its Responsiveness to the Needs, Desires and Aspirations of Mankind. The Society also undertook a project covering "Case Studies of the Impact of Civil Engineering Projects on People and Nature." Other research conferences sponsored during this period by the Technical Divisions and Engineering Foundation also pursued this theme.

The Research Committee instigated several measures to stimulate general membership interest in research. Five annual research prizes with a modest stipend were initiated in

1946, and were endowed in 1965 in the name of Past-President Walter L. Huber. In 1958 an annual Research Luncheon was authorized at a national meeting of the Society, to draw attention to the accomplishments of the winners of these prizes. A $5,000 Research Fellowship was funded annually by the Society from 1961-66. Other research awards include the O. H. Ammann Research Fellowship in Structural Engineering and the Raymond C. Reese Research Prize in Structural Engineering.

The scope and pace of research activity in 1973 presage a high priority to these functions in the future. The 1971 report on "Research Needs in ASCE Relevant to the Goals of Society" estimates such needs to require funding on the order of $3 billion annually for the next 10 years, Research councils administered by ASCE were working on Pipelines, Reinforced Concrete, Air Resources Engineering, Urban Hydrology, Environmental Engineering, Coastal Engineering, Urban Transportation, Performance of Structures, Urban Water Resources, Construction, Expansive Soils, Underground Construction, Structural Plastics, and Computer Practices.

ASCE also has cooperated freely with many other organizations in their research councils and related activities. The Column Research Council, the Riveted and Bolted Joints Research Council, and the Steel Structures Painting Research Council are just a few examples.

In the decade ending in 1973, collaboration with Engineering Foundation was never more energetic and productive. More than 50 of the research conferences held by the Foundation were on topics of primary interest to civil engineers, and projects of the ASCE Technical Divisions and research councils were being widely supported.

More than 400 members of the Society were engrossed in 1974 with service in some 40 research committees and 13 research councils. Their projects—principally aimed toward initiation and stimulation of study in new areas of research—involved more than $500,000 of outside funding. Obviously, soundly conceived and executed research programs are a vitally essential element in the pursuit of a learned art in the modern world.

* * *

While not all of the early leaders of ASCE were agreed on the nature of its role in effecting the "professional improvement of its members," there seems always to have been unanimity of opinion that advancement of technology was fundamental. Hence, the development of meetings, publications, technical committees, standards, and research services came much more rapidly and easily than did growth in the professional domain.

All the benefits from technological progress that may have originated or been furthered in the Society cannot possibly be enumerated or quantified. Let it simply be recorded that the American standard of living—unmatched in the world—has been made possible at least in part by the advances in structures, transportation, environmental management, and in the conservation and utilization of natural resources that have been aided in realization through the functions of the American Society of Civil Engineers.

CHAPTER VI

IN THE SPIRIT OF PUBLIC SERVICE

Inherent among the founders and early builders of the American Society of Civil Engineers was a strong sense of responsibility for the safety, welfare, rights, and financial interests of the public. This quality comes through powerfully in the actions, writings, and biographies of such leaders as Laurie, Craven, Kirkwood, McAlpine, Chanute, Adams, Jervis, Allen, Fink, Welch, Chesbrough, Greene, Roberts, and many others. Public service was not set out as a specific obligation of the civil engineer, but it was implicit in the integrity and concern for the public interest that characterized the performance of these men in the execution of their important projects.

Neither did the constitution of ASCE state or even imply that the Society was to be used as an instrument for protection or advancement of the public welfare. Within the first ten years of its effective existence, however, there were several ventures into the domain of public affairs. True, these incidents encompassed technical overtones relevant to the "advancement of the profession," as then interpreted, but they also represented public service of a high order. The propriety of ASCE concern with this area was often questioned by conservative members, and definitive policy was slow in formulation.

After its first two decades of existence, ASCE involvement in public affairs ranged beyond the investigation of national disasters and the encouragement of technical services by government agencies. Advice on major public works programs and conservation issues was given to Congress—sometimes by invitation and sometimes not. Likewise, recommendations on administrative procedures were offered to government bureaus with which civil engineers had to work. Later, during both World Wars, assistance was extended to federal authorities on technical manpower problems, wartime construction, and postwar planning. Through the years the Society has offered advice and guidance to elected officials, legislative bodies, and agencies on public policy related to the development of natural resources and environmental quality.

Administration of Public Affairs Activities

Many actions were taken by ASCE in the public interest prior to 1919, but it was not until that year that such service was expressly identified as an institutional responsibility. When Chairman Onward Bates presented the report of his 1919 Committee on Development, the Board of Direction endorsed the premise ". . . that the time has now come when this society should adopt the principle of becoming an active national force in economic and industrial and civic affairs." Moreover, although the Board did not adopt the recommendations of the Committee on Development, it was sufficiently impressed by the section on "Relations to Public Affairs" to authorize, in 1921, an ASCE Committee on Public Relations.

The name of this committee was misleading, because it was actually an instrument of the Board of Direction intended to advise the Board on "matters of public policy and professional relations." In its operations the committee concerned itself almost exclusively with national legislation. It was commendably effective in this capacity until its functions were divided in the Functional Expansion Program of 1930 between a Committee on Legislation and a Public Educational Committee, both as parts of the new Professional Department. The Committee on Legislation was charged "with being alert in the matter of legislation, other than registration, and to advise the Board promptly on needful decisions or actions." After only three years, however, this approach was found to be "impracticable" and the committee was discharged, with its duties to be assumed by the Board of Direction as a whole. Actually, so demanding a function as identifying, evaluating, and obtaining informed consensus on legislative issues is properly a staff responsibility. It was also too much for the Board, and in the absence of sufficient staff at the time, the cycle was repeated when, in 1946, a Committee on National Affairs was appointed to "advise and guide the Society in activities regarding all national legislation concerned directly with the welfare of the profession." Again, this was too much to expect from volunteers as a continuing service, and the committee was discharged in 1954.

In 1940, before the entry of the United States into World War II, the Committee on Society Objectives recommended that a member of the staff "spend as much time as necessary in Washington, particularly during the next few months, to look out, in all practicable ways, for the interest of civil engineers." High interest at the time in the developing National Defense Construction Program resulted in April 1941 in the designation of a full-time staff representative in Washington. It was intended that this arrangement would be maintained only for the duration of the national emergency.

With the dissolution of the American Engineering Council in 1941, and closing of its Washington office, ASCE invited the other Founder Societies to enter upon a joint venture because of the national emergency and the need ". . . for a definite measure of solidarity and concurrent action through representation at Washington of the four Founder Societies." AIME, ASME, and AIEE all refrained from accepting the invitation.

The Washington Field Office of the Society was continued for 14 years. Four staff members successively filled the post of Washington representative. All of them were civil engineers experienced in the legislative process, and at least two were registered lobbyists. Highest priority was given to legislative developments related to effective use of civil engineers in military service during World War II, to public works planning and construction programs, and to national policy on military training and on peacetime reconversion.

There was also interaction on regulations and procedures

of federal agencies engaged with engineering and construction, contracts, civil service, research activities, and education. Liaison was maintained with AIA, AGC, and other engineering bodies based in Washington. The office also extended its cooperation to EJC in the handling of national affairs.

By 1955 transportation and communication facilities in the Eastern megalopolis had progressed to the point that the economics of the Washington office were no longer favorable as compared to the handling of national affairs from the New York City headquarters of the Society. The office was closed in January 1956 ". . . in the interest of more efficient operation," its functions to be assumed by the headquarters staff and Society officers. From 1955 to 1972 all public affairs activity was handled through the staff, with assistance from several sources. A law firm in Washington was engaged to screen all national legislative proposals, and to refer to Society headquarters any bills that might be of interest. About 150 to 200 bills were referred annually, of which 25 to 40 were the subject of action of some kind. Legislative concerns were also brought to attention by members, local sections, technical or professional committees, or other societies. A professional journalist was retained to furnish a monthly Washington news column for *Civil Engineering*.

When a situation or problem was judged appropriate for ASCE attention the first need was to determine the Society's official position in the matter. This might be found in an established policy; if not, it was necessary to secure direction from the best available authority (Board of Direction, Technical Division, or special committee having specific competence in the subject area of the legislation). The Water Policy Committee, the Land-Use and Environmental Systems Policy Committee, and Transportation Policy Committee were particularly effective in this function.

It is implicit in the statement of a Society policy that it represent the consensus of viewpoint of a majority of the membership. This was theoretically the case until 1922, when attitudes of the Society generally were affirmed by membership referendum. The process was questioned from time to time, as in 1890 when it became necessary to establish clearly the policy of the Society on adoption of the 24-hour clock as the standard for time notation. Extensive debate ended with a call for a letter ballot, which overwhelmingly supported the proposal.

As the Society grew and public issues became more numerous and complex, a faster, more flexible, and less costly procedure for policy determination became desirable. This came about with constitutional changes in 1922 that gave the Board of Direction broader powers. Expedience and timeliness of action were gained, but occasionally at the expense of clear-cut majority representation on highly controversial questions.

Any question as to the authority of constituent units of the Society to speak or act in the name of ASCE was clarified by a policy resolution in 1939. This provided that no pronouncement concerning national policies was to be made by a committee or Technical Division without specific prior clearance by the Board of Direction.

Once the attitude of the Society on a legislative proposal was determined, it was possible for the staff to communicate with the sponsors of the legislation to offer comment or suggestions, or with the legislative committees engaged with the issue. Frequently, arrangements were made for competent witnesses to appear before legislative committees. Upon the enactment of legislation, advice was offered the implementing officials and agencies concerning appointments, regulations or procedures that might have engineering implications.

A provision in the Internal Revenue Service regulations applicable to the exemption of educational and scientific organizations from federal income taxes has imposed some measure of constraint upon ASCE and other engineering societies. This is the requirement that no "substantial part" of the activities of such organizations "may be devoted to the carrying on of propaganda or otherwise attempting to influence legislation." It has been the attitude of the Society in recent years that it will act in legislative matters on behalf of the public whenever this is justifiable as a professional responsibility, thus in some measure earning the privileges that are enjoyed as a tax-exempt body.

In deference to the IRS rule, ASCE has limited its legislative activity to measures with a clear-cut public interest dimension, acting in the capacity of a competent adviser rather than as a lobbying agency. Some of the bills acted upon were more or less self-serving in aspect, of course, but action on them was held to be justified on the premise that every nation needs a strong engineering profession in order to develop properly its natural and human resources. This logic might be paraphrased as "what is good for engineering is good for the country!"

Members were not exhorted to "write their Congressmen," although they were free to endorse the position of the Society as individuals. Political-action techniques were avoided, as was any kind of activity in political campaigns.

The special role of the National Capital Section must also be acknowledged with reference to legislative and bureaucratic activity in Washington. The Section has always maintained close liaison with the headquarters office, and cooperated frequently in obtaining information and in making personal contacts.

Public affairs activities at the state and local levels have been almost entirely carried on through the sections of the Society and their branches. In states encompassing several sections—as in California, Ohio, and New York—there has been joint action in state affairs by councils made up of the sections.

Society policy has also favored cooperation at all levels with other professional and technical engineering organizations in public affairs of mutual concern. Collaboration was particularly effective with the National Society of Professional Engineers (NSPE) and with the American Consulting Engineers Council (ACEC) after 1955.

The sweeping reorganization of ASCE professional activities in 1972 resurrected two operations from past approaches to the administration of public service functions:

(1) the use of a general committee and (2) the restoration of the Washington Field Office. Actually, not one but two committees were created, both within the new Member Activities Division.

The Committee on Legislative Involvement was charged:

> ". . . to stimulate and encourage participation by individual members in legislative activities on local, state and national levels. It shall inform itself on legislative problems and assist in developing Society policies in regard to them. It shall continuously monitor proposed legislation in collaboration with others to determine its possible effect on Society activities and recommend corrective actions where necessary."

The Committee on Public Affairs was given an equally wide-ranging assignment:

> "The Committee shall encourage professional involvement by members of the Society as concerned citizens in local, national and international affairs. This will include assistance to government agencies and community action groups. It shall find ways for the civil engineering profession to better serve the public in the improvement of man's welfare and his environment."

These purposes went far beyond those of the two predecessor committees of the 1930's and 1940's. The new goals were ambitious, to be sure, but were commendable in their reflection of the professional ethic as it was applicable in a period in which the impact of technology upon the quality of life was being subjected to searching questions. The availability at this time of a capable professional staff gave promise of greater success for these committees than was the case with their earlier counterparts.

Reestablishment of the Washington office was effected in 1972, this time staffed by an attorney knowledgeable in the ways of the national capital. Close rapport was promptly established with AIA, NSPE, ACEC, AGC and other Washington-based organizations. The operation showed much promise, and its early success with legislation dealing with professional-contract negotiation and with pensions was highly gratifying.

Public affairs activity in the sections was given new emphasis in 1972-73. A Legislative Involvement Handbook, released in 1972, furnished guidance in "organization and procedures to assist members involved in legislative activities at Local Section level." Early in 1973 a conference of section representatives was held in Chicago for the specific purpose of encouraging the sections to assume leadership in bringing about the enactment of state laws requiring that engineering services for public works projects be engaged by professional negotiation procedures instead of competitive bidding. The meeting "generated considerable enthusiasm for state level legislative-action programs," and appeared likely to be followed by similar gatherings as there might be occasion in the future.

Motivation Toward Public Service

It is relevant here to delve into the motivation that spurred ASCE toward ever-increasing emphasis on public service as a responsibility of the Society. Did this trend stem from self-interest or altruism? More than likely it was a combination of the two.

A common thread throughout all studies of ASCE aims and activities was the striving for greater professional status, public respect, and a higher order of economic welfare for the civil engineer. As time passed, however, a growing sense of public responsibility became evident in the ASCE leadership.

The 1875 report of the Committee on Policy of the Society observed that:

> ". . . The influence of the Society upon the public is unfortunately not very *rapidly* developed by the mere professional character of the members, although this is the ultimate basis of its influence and usefulness. There must be some connecting links between the Society and the public."

At that time such "links" were considered possible through publications and special committee studies.

In 1919 the Committee on Development stated forthrightly:

> ". . . The engineering profession owes a duty to the public which it is believed can best be discharged by every engineer in the civic work of his community."

The very purpose of the Functional Expansion Plan in 1930 was "to promote the desirable functions of the Society to best serve the public and the membership." Likewise, the 1938 Committee on Professional Objectives referred to "the responsibility of the engineer to society," and to ". . . increasing the usefulness of the profession to society."

A milestone in the movement toward altruistic motivation, however, was the emphasis given to concern for the public by the 1962 Committee on Society Objectives and Tax Status. This body considered modification of the objective of ASCE in Article I of the Constitution to stress "service in the public interest." The amendment was not recommended, however, on the reasoning that actions meant more than words, as follows:

> ". . . Such a reference would reflect the importance to the national economy and standard of living that is inherent in the proper determination of policies with respect to such matters as natural resources, transportation, urban and suburban planning and housing, including business and industrial housing. The Committee concluded, however, that the principle of 'by his works you shall know him' was of more consequence than a change of the wording in the stated objective."

As a result of the input of this committee the public interest was specifically recognized in 1962 in the preamble to the Code of Ethics, and this factor became thereafter a basic criterion in the evaluation of all Society programs.

Most significant is the action taken in 1973 to undertake as the first goal of the Society (page 75) ". . . To provide a corps of civil engineers whose foremost dedication is that of unselfish service to the public." It was clear by this time that the self-serving enhancement of professional standing and economic welfare—although given unprecedented emphasis in the agenda of the Society—was henceforth to share priority in ASCE with service on behalf of the public.

There was never, however, any groundswell of membership concern for the public welfare. When Presidential Nominee Mason G. Lockwood, in 1955, sought guidance from the local section presidents and secretaries as to deficiencies in society programs, there was strong demand for attention to such internal needs as the status of younger members, economic betterment, public relations, and membership apathy, but there was no mention whatever of outwardly directed services in the public domain. As the leadership realized that professional status and public respect had to be earned by deeds — not by rhetoric — policy, aims, and programs of the Society were modified accordingly.

This conclusion applies only to the Society in a composite sense. There have always been countless members of ASCE who were outstanding individually as public-spirited citizens, more than generous in their contribution of time and talent to the public weal. Former President Herbert Hoover, Hon. M. ASCE, heads the long list that cannot be given here.

A Review of ASCE Accomplishment in Public Affairs

Reference has been made previously to several ventures of the Society of important public concern, *e.g.*, the 1872-77 effort to develop a federal testing facility for iron and steel (page 215) and the leadership of ASCE in establishing the present system of standard time (page 228). There have been literally hundreds of public service contributions through the years, some by formal action of the Board of Direction and others upon the initiative of the staff in its day-to-day operations. Some noteworthy examples are documented here, grouped in the categories of public safety, public policy, wartime services, service to government agencies, and public communication.

Public Safety

In its early years ASCE did not hesitate to become involved publicly and energetically in the analysis of any national disasters of civil engineering import. Further indication of this attitude was the appointment of a committee, in 1892, to urge upon the members their duty to present papers recording "for the benefit of all not only their successful works but the experience not infrequently acquired by failure."

As early as 1869 there was concern about the safety of bridges, which were failing at a rate of 25 or more annually. At the 1873 Annual Convention it was resolved:

> "In view of the late calamitous disaster of the falling of the bridge at Dixon, Ill., and other casualties of a similar character that have occurred and are constantly occurring, a committee . . . be appointed to report at the next Annual Convention the most practical means of averting such accidents."

The committee, chaired by James B. Eads, set its own objectives:

> "It would seem, therefore, to be our duty as a Society to establish in a few general terms—such as can be readily embodied in a law—a standard of maximum stresses and a table of least loads for which bridges should be designed, and to add thereto a practical suggestion as to the necessary legislation required to give the public that protection which an adherence to this standard would afford."

The final report of the committee, published in 1875, consisted of four parts, each composed by one or more members of the committee. There was some disagreement, both on technical details and on the propriety of the Society's preparing a model bridge law. Although never actually formalized in legislation, the findings of the committee provided badly needed guidance for bridge designers.

During discussion of the Ashtabula, Ohio, bridge tragedy in 1877, Director C. Shaler Smith moved that the Society appoint a committee to draft a law covering the points outlined in the above report and to include in addition "the necessary provisions to secure the inspection by experts of all questionable bridges now in existence." The motion provided further that the law so drafted be recommended for adoption by the state legislatures, and that the members lend their support to this aim. At the same time Clemens Herschel (President, 1916) urged appointment of a committee "to draft a law requiring tests of finished bridges, before, and at stated times after, their opening for public travel."

Both of these proposals were put to letter ballot of the membership, as required at the time, and both were soundly defeated. The deciding argument appeared to be that "The Society, *as a body,* is not competent to set forth opinions on any special points of practice." This was not meant to imply that the Society did not have in its membership engineers competent to evaluate "special points of practice;" it did question the propriety of setting forth an opinion in the name of the entire Society which would be voted upon by some members not expert on the matter at issue. The dilemma arose from the attempt to produce an authorative standard before there was a suitable procedure for doing so.

The Board did not always wait, however, for membership approval. The Committee on Tests of American Iron and Steel had been quite active since its beginning in 1872 (page 215). Through its efforts, a bill calling for a national commission of experts to execute tests on structural materials and to "deduct useful tools therefrom" was introduced into Congress in 1882. When the legislation failed to pass, the Board of Direction moved unilaterally in 1883 to select a qualified panel "to prepare and promote such a programme of tests of structural material as to secure the best results possible from the Watertown Arsenal Experiments."

The number of bridge failures declined after 1900, but a most spectacular one was the collapse of the Tacoma Narrows Bridge ("Galloping Gertie") in 1940. Soon thereafter President John P. Hogan was authorized to appoint an investigating committee, but the ensuing discussions produced the view that the study would be better undertaken as a joint project including other interested organizations. An appeal by ASCE to the Federal Public Roads Administration was successful in enlisting the sponsorship of that agency, and the Advisory Committee on Investigation of Long-Span Suspension Bridges was the result. This body developed valuable guidelines for the testing and research then needed for suspension bridge design. ASCE served as a catalyst in this instance.

The importance of regulation of engineering practice in the public interest was starkly exposed by the failure in May 1874 of the dam on Mill River, in Connecticut. This tragedy, costing 143 lives, was investigated by an ASCE panel headed by James B. Francis (President, 1881).

Three extracts from the committee's report *(Transactions*, 1874, p. 118) tell the story:

> "The plan adopted, or rather the specifications (for it does not appear that any plan was drawn), were prepared by Mr. Lucius Fenn, Civil Engineer, now of New Britain, Conn., who, according to the evidence reported to have been given by him at the coroner's inquest, claims to have written them under the direction of the Directors,—'he acted only as the attorney of the company in drawing up the specifications' . . .
>
> "In the construction of the work by the contractors, it appears that there was no sufficient inspection, so peculiarly important in a work of this description, and during part of the time none at all, except by the Directors of the Company or their building committee, during their occasional visits. The remains of the dam indicate defects of workmanship of the grossest character . . .
>
> ". . . it is obvious that this cannot be called an engineering work. No engineer, or person calling himself such, can be held responsible for either its design or execution . . ."

The disastrous failure of the Lynde Brook Dam at Worcester, Mass., in 1876 was followed promptly by appointment of a special committee, chaired by Past-President Theodore G. Ellis, to examine and report upon the incident. The three-man committee visited the scene and reviewed the available data, noting that it "was obliged to procure the information desired from other sources" when it was denied access to the plans from which the dam was constructed. The report of the committee (*Transactions,* 1876, p. 244) contained the following cogent conclusion:

> "In the opinion of your committee there is no doubt as to what was the immediate cause of the failure of this dam. There was evidently a stratum of porous material lying under the upper gate-house and the upper end of the pipe vault, partaking of the nature of a quicksand, which should have been removed, and greater precautions taken to prevent the access of water from the reservoir, than appears to have been the case in the foundations above described . . ."

Discussion of the report was concluded by Past-President William J. McAlpine, who had been engaged by the City of Worcester as a member of a panel to investigate the failure. He stated tersely that he could not "assent to the conclusions of the report," and that he hoped to be able to report later on his own conclusions. He did not, however, take issue with the committee report in publishing his recommendations for repairing the dam (*Proceedings,* Vol. II, p. 97).

The devastating Johnstown Flood in 1889 gave rise to appointment of a special working party "to visit the scene of the disaster and report to the Society the cause of the calamity." Six months later the Committee to Investigate the Failure of the South Fork Dam was ready to report but stated that:

> ". . . the Committee have considered that the presenting of the report at this time would not be a proper thing. Lawsuits involving claims for very heavy damages have been instituted by many of the sufferers from the disaster against the South Fork Fishing and Hunting Club, and these suits are now pending in the courts of Pennsylvania, and the Committee, therefore, thought that to give publicity at this time to their conclusions might prejudice the case and have a tendency to affect the issue of the trials. The Committee have, therefore, agreed to recommend to the Society that the report . . . fully endorsed and sealed, should be placed in the custody of the Chairman of the Committee, Mr. James B. Francis, and kept

> there to be called for at any time after the issue of the trials in court has been decided. . ."

Calling for presentation of the report at the 1891 Annual Convention, the Secretary noted that:

> "The time for personal liability is now past and the principal suit has been decided that it was absolutely impossible to have foreseen this disaster; in other words that it was the act of God . . ."

The exhaustive report of the committee was published in *Transactions,* Vol. 24, p. 431. Following are extracts from its conclusion:

> ". . . There can be no question that such a rainfall had not taken place since the construction of the dam . . . The spillway, however, had not a sufficient discharging capacity; contrary to the original specifications of Mr. W. E. Morris, requiring a width of overflow of 150 feet and a depth of 10 feet below crest, which would have been a sufficient size for the flood in the present case—it had only an effective width of 70 feet, and a depth of about 8 feet; the accumulated water rose to such a height as to overflow the crest of the dam and caused it to collapse by washing it down from the top."
>
> ". . . There are today in existence many such dams which are not better, or even as well provided with wasting channels as was the Conemaugh Dam, and which would be destroyed if placed under similar conditions. The fate of the latter shows that, however remote the chances for an excessive flood may be, the only consistent policy when human lives . . . are at stake, is to provide wasting channels of sufficient proportion and to build the embankment of ample height."

Widespread incidence of "floods of exceptional magnitude" in the spring of 1913 resulted in authorization of a Committee on Floods and Flood Prevention. The committee very properly pointed out that first attention must be given the gathering of fundamental data, and it encouraged systematic collection of such information by federal and state agencies.

In 1922 the American Red Cross sought and received the cooperation of ASCE in its flood prevention efforts. The national impact of flooding in the Mississippi River valley brought specific action by the Board of Direction in 1928, when it urged President Herbert Hoover, Hon. M. ASCE, to create a professional Board of Review on the Mississippi River flood control program, and to seek Congressional authorization for funds needed for surveys and compilation of data on the controversial aspects of the plan. A strong committee was delegated to convey the recommendations to the President and the Congress.

Need for continuing pressure and guidance to sources of flood information elicited a new Committee on Flood Protection Data in 1923, and a Committee on Flood Control in 1936. Both of these committees were discharged after reporting in 1940, when their functions were relegated to the Hydraulics Division for "continuing study of flood protection data and determination of principles of practical application of such data to the design and operation of flood control works."

Urbanization of America was occurring so rapidly in the 1880's that provision of safe, potable municipal water supplies became a real problem. ASCE found an opportunity for public service in this situation.

The following resolution was adopted at the 1889 Annual Convention:

"*Whereas,* It is a well known fact that many cities and towns on the Atlantic Coast have suffered very greatly from impurities in their water supplies, due to various causes, and that no adequate remedy meeting all conditions, has been found therefore; and

Whereas, these impurities are often due to natural causes which have not been adequately investigated on account of the difficulty of centralizing the individual efforts of all parties engaged in such investigations,

Resolved, That a Committee . . . be appointed to ascertain the best means of concentrating all available information in such a manner as to secure useful results and to report . . . what further action should be taken . . ."

In January 1890 the Committee on Impurities in Domestic Water Supplies recommended that a complete bibliography of physical, chemical, biological, and public health information be developed according to a devised outline, and that such references be assembled in the Society's library; further that a system be set up to gather data continuously from boards of health, waterworks managers, federal bureaus, and researchers throughout the country. Its report ended with the statement:

> "If the engineering profession desires to retain the right to act as the final arbiter in the selection of sources of water supply, it must resolve to take the lead in the intelligent application of the results of scientific inquiry into the conditions affecting the purity of stored water and the subtle differences that separate waters of good quality from others, equally good in appearance and chemical composition, that do not stand this test."

Chairman A. Fteley supplemented this report with the information that the AWWA had recently established a committee of similar purpose, which was already gathering information. He had communicated with that body, and he was authorized to exercise his discretion with regard to joint or independent action on the part of ASCE.

Communication was initiated with the AWWA group, but an effective joint activity did not develop. The independent studies of the ASCE working party eventually indicated that it was not feasible to attempt to assemble and collate the large amount of analytical data required without substantial outside funding. The correspondence and circulars distributed by the committee to Boards of Health and various water-supply agencies had been beneficial, however, in emphasizing the need for higher standards of domestic water quality. In 1893 the committee concluded its efforts to bring about consolidation of all information on the quality of domestic water supplies. However, the exercise was no doubt contributory—in conjunction with contemporary activities of the American Academy for the Advancement of Science, the American Public Health Association, and AWWA—in bringing to fruition the first USPHS Drinking Water Standards in 1914.

Catastrophic earthquakes, causing great loss of life, have plagued the world from its beginning. They are ever a challenge to the civil engineer, who strives to build his structures to withstand the damage and hazard to human life that come with major seismic disturbances.

The San Francisco earthquake in April 1906 brought action to appropriate $1,000 for the prompt relief of the engineers of the city. More important, however, was the offer of the San Francisco Association of Members to serve the stricken city in any way possible. The mayor responded promptly by appointing seven ASCE members to his Committee of Forty on Reconstruction. The Association also created six technical subcommittees to study the effects of the earthquake on all kinds of engineering structures. The comprehensive report resulting from this work was published in March 1907 (*Proceedings*, Vol. XXIII, p. 299). The information and conclusions it contained provided invaluable guidance to designers and public authorities on earthquake and fire-resistant construction.

Again in 1923, following the disastrous Japanese earthquake, the Society acted quickly to communicate with its members in that country. The Board of Direction also authorized a committee of experts to conduct an "investigation of the effects of earthquakes on structures in Japan and elsewhere." The panel included engineers from both sides of the Pacific, and a voluminous report was submitted to the Society in 1929. Financial stress incident to the national economic depression prevented publication of the report, but it was made available in the Engineering Societies Library as a prime reference on design of structures subject to seismic forces. By this time the Structural Division was well established, and the analysis of seismic and other dynamic forces has received continuing study through the years. Typical examples are the investigations of the effects of the Florida hurricane of 1926 and of the Santa Rosa earthquake in 1971.

The Long Beach, California, earthquake of 1936 prompted the Los Angeles Section to join with the Structural Engineers Association of California and the Los Angeles Engineering Council in a useful study of damaged structures. The joint task force report included both general and specific recommendations on earthquake-resistant design.

Since 1900 ASCE has recognized the necessity for restraint in undertaking investigations of disastrous incidents, because of possible legal implications incident to liability suits and damage claims. Members of the Society are frequently involved as individuals in investigative commissions, but in recent years the major contributions of the Society toward public safety have come about through the work of the technical divisions. These studies cover the full range of such topics as flood control, structural failures, fire protection, public health, highway safety, earthquakes, hurricanes and tornados, landslides, and similar hazards to the public safety.

Public Policy Guidance

Because civil engineering practice is primarily applicable to public works, which are, in turn, heavily dependent upon public policy, ASCE has always been engrossed with such decision-making. The range of its interests encompasses transportion, conservation, water resources, environmental quality and the full spectrum of public works administration, planning, financing, and management.

ASCE was still in its infancy when it moved, in 1870, to appoint an ad hoc committee to be "charged with collecting such documents and information in relation to inter-oceanic communication between the waters of the Atlantic and Pacific as they can obtain." Financial commitment to the study was limited to $250. President Craven himself chaired the committee.

Just a month later, apparently upon the recommendation of the task committee, the Society adopted the following resolution:

> "That in the opinion of the Society the relations of the Isthmus of Tehuantepec to the United States are so peculiar and so different from those of any other route for canal or transit, that it is highly important that the patronage and influence of the government. . . . should not be committed to any ship canal or transit enterprise until the practicality or non-practicality of a canal across the isthmus has been determined by a survey made for that purpose."

The resolution was transmitted to both Houses of Congress.

Over the next decade many papers and discussions were published by the Society on the feasibility of an Isthmian canal at various locations. These important contributions were climaxed when the Society sponsored a public meeting in the theater of the New York City Union League Club on February 26, 1880, at which Ferdinand de Lesseps led a general discussion on "Inter-Oceanic Canal Projects." According to the account published in *Frank Leslie's Illustrated Newspaper,* he "delivered an address which for style, charm, interest and brilliancy should serve as a model for scientific lectures in general, and those in this country in particular."

It is not unlikely that an 1873 survey of national problems by the Committee on Library was intended to generate public policy guidance as well as technological study. Of 37 subjects "relating to the practice of engineering, its connection with kindred arts and public affairs, . . ." the majority dealt with transportation. Included in the list were three items still of major concern a hundred years later:

> "On rapid transit for passengers in large cities.
> "On the prevention of pollution of rivers by, and the utilization of, sewage.
> "On the production of mineral oils."

In view of the universal and complex problem of urban transportation across the United States in 1974, the following resolution—adopted a century earlier—attracts special notice:

> "*Resolved,* That a committee . . . be appointed . . . to investigate the necessary conditions of success, and to recommend plans for:

> *First*—The best means of rapid transit for passengers, and
> *Second*—The best and cheapest methods of delivering, storing and distributing goods and freight, in and about the city of New York, with instructions to examine plans, and to receive suggestions such as parties interested in the matter may choose to offer, and to report on or before the first day of December, 1874."

The directive having been enacted on September 3, 1874, only three months were allowed for its fulfillment! Although not directly a policy determination, the 80-page report published a year later by this committee *(Transactions,* 1875, Part II, p. 1) provided urban transportation policy guidance for larger cities for some years thereafter. Chaired by Octave Chanute, the committee introduced a number of new ideas, such as the suggestion by Richard P. Morgan, Jr., that an elevated rapid transit trackway be constructed over the center of the street.

Strangely, there appears to have been no official ASCE involvement with railroad transportation policy, despite the explosive development of railroads from 1850 to 1900. However, national policy concerning railroads was forthrightly and critically discussed by prominent members of the Society, in its meetings and publications. The personal contributions in this area by Albert Fink, Arthur M. Wellington, Octave Chanute, Ashbel Welch, John B. Jervis, Martin Coryell, and Albert Sears were especially noteworthy.

After 1950 there was strong involvement with national highway and urban transportation policy. A special task committee provided input to the National Highway Bill in 1956 and 1959; a report on "Principles of Sound Transportation Policy" was adopted in 1963; recommendations were submitted by correspondence and hearings on the Federal Aid Highway Program (1970), highway safety (1971), rapid transit service to airports (1971), the Highway Trust Fund concept (1972), and a proposed reorganization of federal transportation agencies (1972).

Considerable membership reaction was aroused in 1973 by a pronouncement opposing the use of funds from the Federal Highway Trust Fund for support of urban mass transit studies and facilities. Many members protested the validity of this position, which originated in the Transportation Policy Committee, as not being truly representative of the views of a majority of the membership. There was some demand for membership ballot on policy issues, as practiced in the early years of the Society.

Concern in ASCE for conservation of national resources is long-standing. The following resolution was among the incentives that led President Theodore Roosevelt to call a White House Conference of Governors in May 1908:

> "*Whereas* the timber resources of this country are being rapidly diminished owing to unscientific methods of forestry, to the prevalence of forest fires and a wasteful use of lumber . . . which may result moreover in the diminution of the natural storage capacity of our streams, and increasing the irregularity in the flow, and consequent impairment of the value of our water powers:
>
> "*Resolved,* That . . . every endeavor should be made to further the introduction of principles of scientific forestry and the creation, preservation of National and State forest preserves . . . and the Board of Direction approves and urges the passage by Congress of a bill providing for national forest preserves in the Appalachian and White Mountains."

The presidents of all four Founder Societies were invited to participate in the conference.

The declaration issued by the state governors following the conference confirmed the need for conservation, urged formulation of policies, and enactment of legislation to insure best use of forest, water and mineral resources, and recommended that a national Commission on Conservation of Natural Resources be established. President Roosevelt was completely receptive to these recommendations, and invited ASCE President Charles MacDonald to serve in the eminent national commission that he appointed.

A series of public meetings on conservation of natural resources was sponsored jointly by the Founder Societies during 1908-09 as an aftermath of the White House Conference.

The Weeks Law, which provided for federal appropriations to purchase land for public use in the conservation of forest and water resources, was strongly supported by ASCE in 1923-24. Again, in 1927, legislation fostering purchase of land for forest development was endorsed.

Although ASCE had a Committee on Impurities of Domestic Water Supply from 1890-93, it was more interested in the gathering of data relevant to the improvement of water potability than in the broader aspects of water resources policy. The latter intent, however, was implicit in the following resolution adopted in May 1913:

> "*Resolved,* That the Board of Direction . . . appoint a Special Committee to investigate the advisability of drafting a National Water Law applicable to all navigable interstate and other waters within the jurisdiction of the United States, and embracing all uses of water, and that such committee be directed to prepare a preliminary draft of such a law . . . if, in their judgment, it appears advisable."

Apparently it was found infeasible at the time to draft a national policy. This committee was replaced in 1917 by the Committee on Regulation of Water Rights, which functioned until 1920 and was concerned more with state control than with water as a national resource.

Stream pollution as a national problem came to official attention in 1922 when a communication from Charles Haydock, A.M., ASCE, urged that the subject of pollution of streams by industries be carefully considered and the need for corrective legislation be studied. The Board of Direction responded by taking the following action:

> "*Resolved:* That the United States Government be requested to undertake, through the Department of Commerce, a complete investigation of the cause, extent and effect of pollution of waters by industries, that methods of mitigating such evils be investigated, and that existing legislation be reviewed in order to determine what if any legislation is required to cope with the situation."

This seed, unfortunately, fell upon barren soil. The only federal official to respond was the Secretary of War, and he merely referred to the modest appropriation provided the Public Health Service for studies of stream pollution.

One of the most significant contributions ever made by the engineering profession to national water management was the report, "Principles of a Sound National Water Policy," produced in 1951 by the EJC National Water Policy Panel, and the 1957 "Restatement" of that report. This merits mention here because the EJC Panel was largely composed of eminent members of ASCE.

By 1960 ASCE had reassumed direct involvement in water-resource matters. A top-level Committee on National Water Policy was set up in 1961 to devote its full attention to that purpose, and it found much to do. In 1963, for example, of 39 Congressional bills of major concern to the Society, 19 were referred to this committee. Because of its sharply focused objectives, the committee was able to maintain effective communication with all federal water agencies, such as the National Water Commission and the Water Resources Council.

Water pollution legislation was the major area requiring attention through the 1960's and early 1970's. Many other problems were also considered, however, including water-resource planning, watershed protection, flood prevention, flood insurance, wild and scenic rivers, estuarine preservation, reclamation, and coastal management. Particularly important was the committee's analysis, in 1972, of the National Water Policy Commission's report, "Principles and Standards for Planning Water and Related Land Resources."

When the International Joint Commission of the United States and Canada recommended, in 1922, that a special technical board be created to investigate and report upon certain proposed improvement of the St. Lawrence River as a waterway, ASCE acted vigorously to endorse such a procedure. A proposal that legislation be enacted to authorize a commission "to give further study to the essential facts with reference to a St. Lawrence Waterway, to guide the action of Congress with respect to its construction" was communicated to the appropriate committees of Congress. The other Founder Societies and the Federated American Engineering Societies were urged to assist in the venture. By October 1924, a new International Board of Engineers was engaged in appraisal of earlier conclusions.

Some public-policy concerns of the Society were not limited to engineering import. In this category was the resolution adopted at the 1924 Annual Meeting supporting federal legislation that would reduce personal income taxes. The action was taken unanimously and without discussion!

One of the most significant and successful ASCE public-service efforts arose from the depths of the Great Depression. A group of 16 members of the Society, headed by John P. Hogan (President, 1940) drafted a memorandum entitled, "A Normal Program for Public Works Construction to Stimulate Trade Recovery and Revive Employment," which proposed creation of a federal credit corporation. The principle was adopted on May 9, 1932, by the resolution:

> "*Resolved,* that the American Society of Civil Engineers, through its Executive Committee,
> (1) Approves in principle a normal program of public works construction as the most effective immediate means of increasing purchasing power, stimulating recovery and reviving employment; and
> (2) Urges on the Congress of the United States the enactment of the necessary legislation to extend Federal credit facilities to solvent states, counties and municipalities to enable them to carry out their normal programs of necessary and productive public works."

A special committee with Colonel Hogan as chairman composed a communication to President Herbert Hoover, and on May 19 the committee presented it to the President in person. He welcomed the proposal, inviting the cooperation of the Society in drafting the necessary legislation, and also asked the Society to provide data on the nature and cost of public works construction that was being deferred for economic reasons. A survey made with the aid of the local sections identified a backlog of more than $3,000,000,000 of such delayed construction. This information was quickly transmitted to the President and to the Congress.

On July 16, 1932, the Emergency Relief and Construction Act of 1932, known as the Wagner Bill, was passed by both Houses and was signed by President Hoover on July 21. The financing principle espoused by the Society was embodied in the Act by reconstituting the Reconstruction Finance Corporation.

Further to the credit of the dynamic Committee on Public Works was its intensive support of the National Industrial Recovery Act of 1933, Title II of which authorized establishment of the Public Works Administration. These laws were of immeasurable value to the national welfare, and more than 40,000 engineers were estimated to have found employment through their implementation.

In addition to Colonel Hogan, the Committee on Public Works included Harrison P. Eddy (President, 1934), Malcolm Pirnie, (President, 1944), Alonzo J. Hammond, (President, 1933), and Joseph Jacobs (Vice-President, 1940-41).

Only a few members could recall the leadership given by ASCE in the 1880's to the establishment of a Uniform Standard Time System when the National Committee on Calendar Simplification sought an expression of interest and support from the Society in 1928. Realizing full well that such calendar reform would be a long and arduous undertaking, the Board of Direction approved the proposal to urge the Secretary of State to involve the United States in the international deliberations on the question then under way.

Five years later, in June 1933, the Board adopted the following resolution:

> "*Resolved* that the American Society of Civil Engineers, in view of its resolution adopted in 1929 in favor of improving the calendar, notes with satisfaction that in 1931 an international conference of the League of Nations (Fourth General Conference on Communications and Transit), in which the United

> States Government participated, officially considered this question, and that the conference, although it decided that the time was not then favorable to proceed with a change, recognized that the simplification of the calendar was certainly desirable, and gave to the governments a survey of the question for their future decision.
>
> "*Resolved* that the Society is of the opinion that further steps to secure this much needed improvement in our measure of time, especially from the economic standpoint, should no longer be delayed, and expressed the hope that the League of Nations will not let the opportunity pass to invite the governments to a further consideration of this question at the next meeting of that same conference in 1935 with a view to conclusive action.
>
> "*Resolved* that the Society call the attention of our government to this resolution and express to the Secretary of State the hope that our government will indicate to the League of Nations a desire to have this question again taken up at this conference and to again participate in the discussion.
>
> "*Resolved* that the Society is of the opinion that a reform based on a division of the year into 13 equal months would best adjust the calendar to modern conditions."

This policy of the Society still prevailed in 1974.

The four Founder Societies combined in 1946 in an important expression of national policy when their presidents communicated a joint statement to the Senate Committee on Atomic Energy urging that responsibility for nuclear energy development be placed in a civilian commission. Such a body was created in the Atomic Energy Commission (AEC), and in 1952 ASCE formed a Committee on Atomic Energy to study and recommend policies and actions relevant to the interests of the Society in the development of nuclear energy. For several years liaison was maintained with AEC in reviewing professional problems of manpower and contract negotiation as well as technical problems of power plant design and environmental impact.

By the early 1970's, ASCE was closely attuned to federal policy determination over a wide spectrum. Reference has been made in this abstracted review to national policy positions and actions in recent years.

Two timely and significant ventures were undertaken in 1973 in the Land Use Policy Committee and the National Energy Policy Committee. The land-use-policy action was in anticipation of the early enactment of federal land-use-policy legislation, and was aimed toward stimulation of grass roots implementation at the state and local levels. The national energy policy panel sought to establish a data base and to devise a coordinated energy-conservation program to manage short-term dislocations in the energy supply. Other topical areas of legislative and bureaucratic interaction on the part of ASCE in 1974 included weather modification, negotiation of professional service contracts, environmental quality education, housing, transition to the International System of units of measurement, and ocean resource development.

Relationships with Government Agencies

Many civil engineers are employed by agencies of government at all levels. Also, a large number of government bureaus, particularly federal, provide services that are closely related to the professional practice of civil engineering. Thus, ASCE is keenly interested in the laws, regulations, and procedures that dictate the operations of these entities. Only a few representative actions of the several hundred that have been taken in this domain, officially and unofficially through the years will be summarized here.

In 1878 a "memorial" to Congress gave ASCE endorsement to the extension of the national system of triangulation by the U.S. Coast and Geodetic Survey into all states where surveys have been authorized by the state legislature. Triangulation was especially important at the time for accurate and convenient location of the roads, railroads, waterways, and other public works construction. Many later actions dealt with various kinds of surveying services—triangulation, topographic, seismic, oceanographic, etc; they reached a peak during the depression years of the thirties as a source of productive employment for engineers.

Being so familiar with the Coast and Geodetic Survey, the Society was on solid ground when, in 1934, it opposed transfer of the Survey from the Department of Commerce to the Navy Department. The transfer did not occur.

Obviously, the operations of the Bureau of Reclamation are hard-core interests of ASCE, and many engineers employed in the Bureau are members of the Society. The vigorous reaction that accompanied the politically-inspired dismissal of Arthur P. Davis (President, 1920) as Director of the Bureau in 1923 is related in Chapter III. The services of 12 staff engineers and 15 consultants were also discontinued in this purge by Interior Secretary Hubert Work, who was forced to set up a special Fact-Finding Commission in an effort to justify his actions. Noting that ". . . no evidence has been provided sufficient to warrant so phenomenal an overturning in the Reclamation Service," the Society made clear its unhappiness with the whole situation.

Taking note of certain expressed aims of President Warren G. Harding and, later, President Calvin Coolidge, to streamline the federal bureaucracy, the Society embarked upon an ambitious undertaking when the Board of Direction took the following action on April 16, 1923:

> "*Resolved:* That the Board of Direction of the American Society of Civil Engineers endorses and commends the recommendations of the President of the United States and his Cabinet, that the military and non-military engineering activities of the Government be separated, and that the design, construction and maintenance of non-military public works be assembled as far as practicable in one department, under one head, and that only those activities closely related be included in that department. We also commend the effort to apply similar principles to all the departments, and to allocate the numerous independent offices to appropriate departments so far as possible. We believe such action will tend to eliminate duplication, to co-ordinate public activities, and in many ways to promote economy and efficiency in the public service.
>
> "*Resolved:* That the President of this Society be empowered to appoint a committee of five members of this Society of which he shall be the Chairman, to present the above resolution to the President of the United States, and to appropriate officials of the Congress, and of the Executive Departments, and to

> take such other action as it deems wise in the furtherance of the principles above endorsed."

The sum of $1,000 was appropriated to cover any expense incident to execution of the resolution.

Complying with its directive, Society President Charles F. Loweth and his Committee on Federal Reorganization presented in person to President Coolidge, on September 30, 1923, the documentation setting forth the position of ASCE on the several bills dealing with reorganization. To bring about profession-wide support, the committee also enlisted the sponsorship by Federated American Engineering Societies of a Conference on Public Works in Washington, in January 1924. About 60 engineering and architectural organizations participated and reacted favorably.

This venture was pursued with full realization that it would be difficult to accomplish. Naturally, there was outspoken and vigorous opposition by the affected federal bureaus, highlighted by a personal appearance before the Board of Direction by General Lansing H. Beach, Chief of Engineers. A bitter exchange took place between the Society and General Beach, following the hearings before the Congressional Joint Committee on Federal Reorganization. He and other ASCE members in the Corps considered the position of the Society to be contrary to their interests.

The reorganization legislation failed of enactment. Nevertheless, the Society was consistent when it responded to the Wyant Public Works Bill, in 1928, by adoption of the principle:

> ". . . that reorganization and concentration of the engineering functions of the Federal Government is desirable and advisable and the Board of Direction favors such legislation as will produce these results . . ."

and, again in 1937, when it supported the recommendation of President Franklin D. Roosevelt that a federal Department of

Public Works be established. But, as before, Congress did not agree.

ASCE was forced to take sides in 1928 when there was a movement to put a proposed National Hydraulic Laboratory under the direction of the Corps of Engineers after the Senate passed a bill authorizing the laboratory in the National Bureau of Standards (NBS). The Society supported operation of the laboratory in the NBS, but the legislation did not pass. An appropriation by Congress in 1931 for such a laboratory in the NBS triggered an immediate offer of cooperation by the Society to advise and assist in its design to insure that it would be "fully adapted to the study of the problems which the tremendous natural resources and great industrial progress of this country must of necessity bring to it for solution." A blue-ribbon committee was set up to implement the offer.

Through the initiative and persistence of its then youthful Sanitary Engineering Division, ASCE pressed for years (1920-24) for the enactment of legislation authorizing the commissioning of sanitary engineers as officers in the USPHS. This was essential to full collaboration of these engineers with other commissioned professionals, particularly medical personnel.

A 1924 "Bill to Promote the Efficiency of the Public Health Service" failed to gain the support of the Budget Director. In ensuing years other similar measures also failed of enactment, for various reasons. It was not until 1943—19 years after the original proposal—that a suitable USPHS reorganization measure became law. Aggressive action by the Society through its Washington staff was responsible for two important amendments favoring the status of sanitary engineers.

A regrettable sequel to this exercise came in 1971 when the Public Health Service lost most of its responsibility in public health engineering matters to the new Environmental Protection Agency (EPA). Commissioned status for sanitary engineers was essentially discontinued at that time, despite protest by the Society.

The Society's endorsement was helpful to the U.S. Geological Survey, in 1925, in bringing about enactment of legislation authorizing a continuing inventory of national water resources. A strong resolution urged appropriation of the necessary funds without a time limitation on the work, but recommended that the program not interfere in any way with the regulation by the states of any streams wholly within their boundaries.

It was not at all unusual for the Society to appeal to the President of the United States for action on matters of importance. In 1935 the Public Works Administration (PWA) sharply amended its procedure for project review with the result that several thousand projects—prepared in compliance with prescribed requirements—were summarily rejected. The following telegram of protest was among those that brought about restoration of the original PWA operation:

"The Honorable Franklin D. Roosevelt
"Hyde Park, New York
"The wholesale rejection of PWA projects prepared in accordance with the recommended and approved procedures breaks faith with applicants who under previously announced terms have advanced time and money in the preparation of plans for sound and useful projects (stop). Stoppage of proposed worthwhile PWA projects denies opportunity to engineers, architects, skilled workers, and business organizations in the construction industry to continue along lines of their normal training and operations thus throwing more into the ranks of the unemployed (stop). American Society of Civil Engineers strongly endorses the basic principles of PWA procedure (stop). You are urged to correct the injustice of the present situation.

"George T. Seabury,
"Secretary, American Society of Civil Engineers"

A special ASCE Committee on Atomic Energy was formed in 1953 to cooperate with the AEC in the advancement of nuclear technology. Liaison was promptly established, and the AEC welcomed the Society's interest and collaboration.

Numerous requests for improved or expanded service in technical agencies of the government originate in the Technical Divisions. Typical was the work done by the Irrigation Division in the mid-1920's, directed toward enhancement of cooperation between state agencies and the several federal bureaus in the Department of the Interior and the Department of Agriculture dealing with all aspects of irrigation. A report published in 1927 offered several useful recommendations. Another example was the strong plea in 1936 for an expanded program in the Weather Bureau to provide basic data required for development of water and agricultural resources, aviation, and public works construction. The resolution drafted by the Committee on Meteorological Data and adopted by the Board was specific in its recommendations, and was undoubtedly helpful in bringing about the increase in appropriation that was needed.

Communication with federal agencies is a continuing day-to-day part of the ASCE headquarters operation. The relationship extends to all bureaus that have any responsibility with construction or with engineers and their activities. Subject areas range from cooperation in technical affairs to personnel problems, to engineering contract negotiations, to public works administrative procedures, and beyond. The number and complexity of the issues arising in this sphere may be expected to increase, in accord with the trend toward growth and expansion of the federal bureaucracy.

Service in Time of War

ASCE lost no time in mobilizing for emergency duty in both World Wars.

In February 1917, before the entry of the United States into World War I, ASCE President George H. Pegram joined the leaders of the other Founder Societies in pledging to President Woodrow Wilson their united support:

> ". . . in the stand for freedom and safety of the seas, and we are confident that we represent the membership of the four societies in offering to assist toward the organization of engineers for service to our Country in case of War."

Only three months later America was embroiled in the war, and the Board of Direction expressed itself on the danger of "relying on volunteer armies and navies for the national defense" by resolving:

> "That Congress be petitioned to pass at once a bill providing for universal military training and service, which we hold to be the only proper, democratic and efficient way for creating the public defense."

At the same time, it was urged that all possible consideration be given to students in recognized engineering schools to enable them to continue their education.

The War Revenue Act of 1917 created a problem for ASCE when the professions were interpreted to be included in the "trade" or "business" categories that were subject to an 8% excess profit tax under the law. Two local associations openly opposed the effort of the Society to obtain exemption of professionals from the tax, holding the action to be unpatriotic. From the record, however, the Society appeared to be well justified in its objection to a tax clearly improper and unfair in its application to professional fees.

About 15% of the membership of the Society was in some branch of military or related wartime service during the period 1917-19.

At the 1919 Annual Meeting, R. S. Buck, M. ASCE, proposed that the Society urge upon all public works authorities

> ". . . that public works should be carried forward to the fullest possible extent consistent with sound judgment, not only for fundamental economic reasons, but for humanitarian reasons, to furnish employment for all who can properly claim employment, especially returning soldiers."

A post World War I proposal (1923), upon which the Society acted negatively, is of interest because of the widespread anti-war sentiment generated by the Viet Nam war in the late 1960's. The following resolution offered at the 1923 Annual Meeting was promptly tabled:

"*Whereas,* Many engineers, including members of this Society, actively supported the recent World War, which was entered by the United States with the avowed purpose of preserving civilization in 'a war to end war' and 'to make the world safe for democracy'; and

"*Whereas,* It is now evident that the World War did not end war, and that present conditions in Europe may soon lead to another war; and

"*Whereas,* the next war between civilized nations will in all probability result in the complete destruction of so-called Christian civilization; and

"*Whereas,* The Engineering Profession is devoted to the service of humanity in directing the great sources of power in Nature for the benefit of the human race, not for its destruction; therefore

"*Be It Resolved,* That the members of the American Society of Civil Engineers in Annual Meeting assembled this 17th day of January, 1923, hereby declare their opposition to human warfare and their refusal to support war for any purpose or at any time; and

"*Be It Further Resolved,* That the Board of Direction be requested to take such action as will aid in the establishment of an international economic commission for the purpose of regulating international commerce and directing the development of the resources of the world for the benefit of all mankind."

Quite different was the response when Secretary of War John W. Weeks sought the assistance of the Society in connection with War Department construction operations. Five representatives were promptly appointed to attend a conference, in January 1925, aimed toward the cooperation of the entire construction industry with the War Department.

Several members of the Society were indicted in 1922 for alleged improprieties in connection with the construction of World War I military cantonments. In fairness to the defendants, a strong ASCE resolution demanded prompt action by the Attorney General in order that his allegations be sustained or disproved. The accused members were completely vindicated when the courts found no criminal conspiracy had been involved in the negotiation of cost-plus construction contracts instead of the customary competitive bids, in the interest of expedience in the war emergency.

Responding to a petition signed by 44 members, in July 1940 the Board created a Committee on Civilian Protection to be concerned with "all matters relating to civilian protection; the safeguarding of life and of civilian activities in general and the protection of public utilities." The services of the committee were made available to federal authorities, and 62 local civilian defense committees were formed as a result of its efforts.

In July 1941 a resolution endorsing President Roosevelt's program and efforts in the national defense was adopted by both the Board and the membership in business session. These actions were reiterated in January 1942 just after Pearl Harbor. By this time the new Washington Field Office was well established, presumably for the duration of the war emergency but actually to continue until 1955. It rendered sterling service in the critical areas of military and civilian manpower, wartime construction controls, and planning for postwar construction.

In 1943 there was concern in the Board of Direction as to the degree to which the fullest capabilities of engineers were being utilized in the armed services. Discussion of the means by which personnel having special qualifications by reason of civilian training and experience might be assigned to services that would use these abilities to best advantage in the war effort resulted in the resolution:

> ". . . that the Board urges (1) that a procurement board for engineers similar to that now in operation for procurement of medical officers be established; and (2) that it be conducted and operated in like manner to that for medical officers."

The suggestion was forwarded to the War Manpower Commission.

Through the initiative of the Construction Division, a Committee on Postwar Planning (later renamed the Committee on Postwar Construction) was authorized in 1943 to study means of expediting conversion to peacetime industrial and public works activities. Within a few months a significant report, which outlined the needs and resources available in the construction industry for the transition to peacetime, received approval by the Board of Direction (*Civil Engineering*, September 1943, p. 439).

The committee study provided the basis for a draft public works bill that was submitted to Congress by President Ezra B. Whitman in January 1944. The bill proposed that all federal postwar construction be placed under the direction of a single agency, that planning funds be provided and that federal construction grants be made available. Congress still declined to accept the concept of a single public works agency, but the other recommendations of the Society carried into subsequent legislation. The Committee on Postwar Construction was discharged in 1945.

As the end of the war appeared to be approaching, President Malcolm Pirnie joined with the presidents of AIME, ASME, AIEE, and AIChE in offering the assistance of the engineering profession in solving the problems related to the demilitarization of Germany and Japan. Programs for postwar industrial control in the two countries were submitted to the Secretary of State in October 1944. Secretary Stettinius promptly welcomed the proffered aid, and invited the Engineers Joint Conference Committee to pursue its studies in detail, and to develop its program of work in consultation with his office and the Foreign Economic Administration. This commitment was carried to successful conclusion by the Engineers Joint Conference Committee, which evolved into Engineers Joint Council in 1945.

A membership questionnaire survey, in 1943, disclosed that more than 5,000 members—about 27% of the total—were then in the armed services. About 93% of all military personnel were commissioned officers, with approximately 35% in the Navy and 60% in the Army. In addition, the survey indicated that 40% to 50% of the membership was engaged in wartime construction, industry, and government service.

Any differences that may have existed in peacetime relationships between the Corps of Engineers and civilian groups in the Society were forgotten now. Every assistance was given the Corps in its recruitment effort, and many future leaders of the Society served in its ranks. Similarly, in the Navy, the Air Corps, the Sanitary Corps, and other branches of the military members of the Society were generously represented. The Construction Battalion (Sea Bees) of the Navy built a glorious wartime record, with ASCE members filling many leadership posts.

Deeply conscious of the need to preserve the national defense in the age of nuclear weapons, the Board of Direction resolved in October 1945:

> "That the establishment by the Congress of the United States of a system of universal military training be urged, to effect adequate National Defense and to promote the security and well being of the nation in time of peace and, if need be, to provide a maximum degree of protection and security in time of war.
>
> "That such system be so developed as to integrate civilian and military training for maximum effectiveness . . . and
>
> "That such systems be so devised as to give minimum interference with peacetime civilian functions . . ."

This expression was acknowledged gratefully by the Secretary of War and many members of Congress.

Since 1962 the Society has been continuously represented by the Executive Director in an advisory committee to the Secretary of Defense on "The Design and Construction of Public Fallout Shelters." In 1973 the Civil Defense Preparedness Agency of the Department of Defense entered into an agreement with ASCE to develop a program to train engineers and architects in the causes of building collapse and other structural failures.

Public Appointments

Representing such a great resource in public works expertise, ASCE has on many occasions been asked to nominate candidates to serve on technical or other public commissions, or to render special professional services. In 1875, at the request of the President of the United States, the Society named representatives to the U.S. Commission on the Improvement of the Mouth of the Mississippi, and three members of the U.S. Commission to Test Iron, Steel and Other Metals were nominated by invitation of the Secretary of War.

Questions of policy arose when the City of Providence requested ASCE to examine and evaluate a proposed sewerage plan for that city. It was decided that:

> ". . . the Society as such cannot undertake to nominate committees to serve private interests, but the Board of Direction [may] be requested to transmit to the mayor the names of a number of experts from whom he may select gentlemen to serve on the committee referred."

Policy determination was complicated by failure to distinguish between appointments to public commissions and advisory bodies as apart from appointments to public office or to perform consulting services for a fee. In any event, it was resolved at the 1875 Annual Convention that:

> ". . . it is inexpedient for the officers of the Society to take action upon applications for members to perform professional services."

Apparently this ruling was not considered mandatory, however, as lists of names were furnished in 1879 when the cities of Holyoke, Massachusetts, and Milwaukee, Wisconsin, sought consultants, and again in 1882-83 when panels of experts were set up to study water supply and street improvements for the City of Philadelphia.

Delineation of policy was again attempted in 1905, in the following words:

> ". . . that the nomination by the American Society of Civil Engineers, or by its officers, of persons to serve on technical commissions, or to render special professional services, does not come within the purposes for which the Society was or-

> ganized, and exists, and that the making of such nominations is not advisable."

But again the language was not construed to be prohibitive, and nominations were usually provided when requested.

Because of the several or even many members of the Society who might be qualified for any given appointment, it was customary to supply lists of names. Official endorsement or evaluation of a single candidate was seldom given, unless requested by the appointing authority for a specific individual.

No reason is recorded for the action taken in 1917 with regard to policy on appointments to public office when it was resolved:

> "That no officer of the Society shall officially recommend any one for any office or position."

This move was reconsidered by the Board of Direction in 1922, and after deliberation, it was "tabled." Having been adopted in 1917 and never rescinded, however, it would still seem to prevail in 1974 as Society policy.

The aborted effort in 1923-24 to do away with engineering leadership of the Bureau of Reclamation must have been at least partly responsible for the following declaration in 1926 concerning the qualifications of appointees to public positions requiring engineering training:

> "*Whereas:* Persons untrained in technical matters are frequently appointed to fill public positions and offices whose incumbents should have knowledge and experience along engineering lines;
>
> "*Whereas:* The practice is already well established in regard to other professions to recognize special qualifications, as, when a member of the Legal Profession is called upon to fill an office or position requiring a knowledge of law, or when a member of the Medical Profession is called to an office requiring medical training;
>
> "*Therefore Be It Resolved* by the Board of Direction of the American Society of Civil Engineers, that public positions and

offices whose incumbents are charged with duties requiring engineering or other technical training should be filled with properly qualified persons, and that whenever positions or offices are to be filled which are concerned with the administration of engineering matters, every reasonable effort should be made to fill the same with competent members of the Engineering Profession."

An action by the Board of Direction in 1953 affirmed this premise, and also formalized for the first time the manner in which the Society would nominate candidates for important public positions. It was voted:

"That the Society continue to support the principle that engineers be appointed to fill engineering positions in government . . . That current Society policy be continued in recommending slates of names for such possible appointments and that such questions be referred to the Executive Committee for action. When such slates are submitted, it should be stated that they are not to be considered all-inclusive."

Regrettably, the tendency toward appointment of unqualified political activists to posts requiring engineering judgment and expertise appears to be growing with the passage of time. The Bureau of Reclamation has been particularly subject to such treatment, beginning with the ouster of ASCE Past-President Arthur P. Davis in 1923 (page 181). A similar situation in the Bureau drew fire from the Society again in 1934, and some later BuRec appointments were highly controversial.

In 1927, a determined effort was made by Past President John F. Stevens, Hon. M. ASCE, to bring about appointment of competent civil engineers to the Interstate Commerce Commission. When legislation was proposed in 1941 to limit practice before the ICC to members of the bar, the Society communicated with the Judiciary Committees of both Houses of Congress to support the contention of the ICC that "the work of the Commission will not be improved by excluding non-lawyers from practice before it."

Since 1955 ASCE has encouraged the appointment of qualified civil engineers to posts in many federal agencies, such as the St. Lawrence Seaway Advisory Board, Department of Housing and Urban Development, Department of Transportation, Department of the Interior, and the several interstate boundary commissions. The extensive reorganizations triggered by the environmental-quality movement in the late 1960's demanded a great deal of attention, because of the claims of conservation extremists that engineers were not sufficiently aware of the ecological impact of their works. In consequence, lawyers and other non-engineers were often appointed to policy-making posts in agencies oriented to engineering output. The capricious and unrealistic standards and objectives issued under such leadership were frequently inhibitory rather than beneficial to the cause of environmental protection and enhancement.

This pattern for assisting public agencies and private interests in the identification of qualified consultants has prevailed in much the same form through the years. Since 1955 requests for the names of consultants were almost invariably handled by simple referral to the Professional Directory featured in *Civil Engineering*. Shorter lists were supplied only when a very high order of specialization was involved.

A comprehensive position paper on "Policy Regarding Government Agencies Employing Professional Engineers" was adopted in 1948 and up-dated in 1959. It reinforced the importance of qualified engineer-administrators, and set forth guidelines for the professional relationship between engineers employed in government and those who serve public agencies as consultants. This fundamental statement covered a highly sensitive area, and was still effective in 1974.

Further reference to the policies of the Society with regard to the procedure for selection of professional consultants will be found in Chapter III.

Public Relations and Information Services

Upon assuming the Chair as the third President of ASCE in

1868, William J. McAlpine listed as the first of the three greatest needs of the Society:

> "To make its advantages more generally known to the profession and to the public, and thereby obtain for it a higher standing and more influence with both."

However, except for the commendable participation of the Society in several international expositions (page 202), no real effort was made toward public education and recognition until the 20th century was well underway.

A communication to the Board of Direction in May 1921 from W. M. Hoyt, M. ASCE, stated that he did not see ". . . why a campaign of advertising cannot be undertaken to let the public know something of what the Society stands for and make the term, Engineer, mean something." Specific suggestions were invited from Mr. Hoyt, but there is no record of his reply.

Even while this exchange was taking place, however, another significant secondary impact of the 1919 Committee on Development was taking form in the new Constitution and Bylaws that were adopted in 1921. Among the areas of expanded activity proposed by the Committee on Development were the two items "Publicity" and "Public Affairs." The overall recommendations of the committee were never adopted, but the provision for a Public Relations Committee in the 1921 revised Bylaws must have originated with the Development Committee report and its accompanying debate.

The terms of reference for the new Public Relations Committee are interesting, because of their emphasis on public-affairs functions and their lack of application to publicity and public education:

> "The Public Relations Committee . . . shall consider and report to the Board of Direction upon such matters of public policy and professional relations as the Board may refer to it and shall call the attention of the Board, from time to time, to such matters affecting the welfare of the Society, or its members, or the Engineering Profession, as in its opinion should receive consideration or action . . ."

Thus, under the banner of public relations, the committee served commendably as a legislative review and advisory body. Through the nine years of its operation under the stated charge the committee was of great assistance to the Board of Direction in consideration of public-policy legislation ranging from conservation of natural resources to government reorganization. The committee performed so well that in 1923 the Board authorized formation of subcommittees in the Local Sections, and in 1924 the Board instructed the committee to initiate collaboration with the other Founder Societies to engender joint public-affairs action.

Notwithstanding the energetic work of the Public Relations Committee in legislative matters, Secretary John H. Dunlap found it necessary in 1922 to address the Board about his ideas for securing publicity about the Society.

The Functional Development Plan in 1930 brought about replacement of this Public Relations Committee by two new bodies—a Committee on Legislation and a Committee on Public Education. The charge given to the latter was quite in keeping with its name:

> "The Committee on Public Education . . . shall put into effect, methods and procedures calculated to acquaint the general public with notable achievements in engineering; the advantages of applying engineering principles to industrial and civic problems; and the necessity of placing the administration of engineering matters in the hands of men of suitable qualifications and experience. Such educational work shall be carried on at all times in a manner consistent with the dignity of the Society."

Under this committee the public-relations function in ASCE assumed unprecedented direction and scope. A program including radio broadcasts, syndicated articles for the popular press, slide presentations, film bibliographies, and public education committees of the local sections was initiated. *Civil Engineering* magazine was developed as an external relations medium as well as the primary means of communication within the membership. An inter-society task force was set up with the other Founder Societies to develop a plan for commercial publication of selected material too expensive to be published by the individual societies. Unfortunately implementation of this program was only nominally successful.

In 1938 the Committee on Public Education was renamed the Committee on Public Information, with the same charge. The committee was discontinued in 1941, the victim of an extensive reorganization.

For the next 33 years all public-relations services were directed and managed within the staff, without any administrative committee. Another extensive reorganization in 1971 recreated the Public Relations Committee under a new Administrative Division. The committee was charged to:

> ". . . develop a public relations program for the Society with the objective of improving the public understanding of the civil engineer and increasing his leadership in national, regional and local affairs. The public relations program shall be carried out at the national level through an in-house professional staff supplemented by consultants as appropriate, and at the regional and local level through effective liaison and assistance from the sections and branches . . ."

The need for an adequate staff facility to carry on a productive public-relations program was first realized in 1935, when authorization was given for the engagement of "experts" to prepare and distribute publicity material to the press. The first in-house Department of Public Relations was set up in 1936 under the direction of a professional journalist. This was primarily a mechanism for newspaper publicity and occasional articles for the popular press.

The staff operation continued in this general pattern until about 1953. Productivity must have been somewhat variable, however, because action was taken by the Board of Direction in 1938 to reinforce the public-relations staff, and again in 1945 to recreate the "Department."

The Centennial of Engineering in 1952 marked the effective

revival of the broad program of public education that was envisioned in 1935 but was limited in execution because of lack of resources. Emphasis shifted from newspaper publicity alone to such media as radio, television, special brochures and films, recognition of civil engineering projects and accomplishments, and to enhancement of references to civil engineers and their activities in books, encyclopedias, and the popular press.

In 1962 the paraphrase "By their works you shall know them" expressed the objective of the ASCE public relations effort. For the first time a professional specialist in oral, visual, and graphic expression was engaged to direct operations in the staff, with an assistant to handle routine publicity. The program was very much of the "soft-sell" type, but it was, nevertheless, dynamic and fruitful. Not only was the public reminded of the contributions of the civil engineer, but also were the members of the Society given cause for prideful projection of the profession in their own contacts and activities.

By Their Works You Shall Know Them

A recounting of specific public-relations undertakings in ASCE must begin with the "Seven Civil Engineering Wonders" project of 1955 — a spinoff of the Centennial celebration in 1952. At that time the Local Sections were encouraged to select and publicize outstanding civil engineering works in their areas. So much local interest was aroused that in 1954 the Board of Direction set up machinery for designation of the Seven Modern Civil Engineering Wonders of the United States, which were announced in 1955 as (1) Chicago's Sewage Works, (2) the Colorado River Aqueduct, (3) the Empire State Building, (4) Grand Coulee Dam, (5) Hoover Dam, (6) the Panama Canal, and (7) the San Francisco-Oakland Bay Bridge. The response aroused in the news media and among public-relations firms, advertising agencies, and the general public was far beyond all expectations. *Readers Digest* carried an 8-page feature given worldwide circulation. Generous space was devoted to the announcement in *Time* magazine and in leading newspapers throughout the nation. Thousands of reprints were distributed, and extensive radio and television coverage resulted.

As an afterthought, it was decided to prepare bronze plaques for mounting at each of the project sites to inform the thousands of visitors each year that they were viewing the work of the civil engineer. Each plaque was unveiled in a public ceremony, generating even more publicity. The cost of the Seven Wonders project was nominal; its benefits were inestimable.

The success of this venture suggested that the idea merited development in a continuing format. Thus came into being the annual "Outstanding Civil Engineering Achievement Award," established in 1959 for recognition each year of the project demonstrating the greatest engineering skills and the greatest contribution to engineering progress and mankind. By recognizing a project the award emphasizes the accomplishment of the civil engineer while acknowledging the input of all who take part in its execution—not just a few leading figures.

The first OCEA Award was conferred upon the St. Lawrence Seaway in 1960, with three bronze plaques unveiled at key structures for the information of future visitors. Subsequent selections are listed in Appendix XII. All nominated projects in the national OCEA yearly contest received public recognition and a number of sections initiated local OCEA programs for their own geographical areas.

The magic of the Seven Wonders theme was still extant in 1962, when the *Rotarian* magazine carried the story to its 22 million readers. A few months later *Readers Digest* published an article, "Five Future Wonders of the World," based on documentation supplied by the ASCE staff. In 1966 the staff gave encouragement and assistance to the author of a book, *Wonders of the World.*

In 1965 the Committee on History and Heritage of American Civil Engineering was established as part of the public-relations program. One of its first efforts was the National Historic Civil Engineering Landmark program, recognizing early projects of significance to the advancement of the profession and the development of America. The first such landmark was the Wendell Bollman Truss Bridge at Savage, Maryland, which was marked with a bronze plaque unveiled with appropriate ceremony in 1966. By 1974 a total of 36 National Historic Civil Engineering Landmarks had been so identified (Appendix XIII).

It is noteworthy that two ASCE members were named in 1935 to act with similar representatives of other bodies to consider the formation of a Joint Committee on the History of Engineering. The venture did not materialize. Also, the earliest historic landmark plaque had been mounted by the Connecticut Section in 1937. This was placed on the Quinnipiac River Bridge near New Haven, Connecticut, in memory of Clemens Herschel (President, 1916). Another special memorial plaque was unveiled at Wethersfield, Connecticut, in 1970, commemorating Benjamin Wright, 1770-1842, as "The Father of American Engineering."

In 1970 ASCE joined with the National Park Service and Library of Congress to create the Historic American Engineering Record, to insure documentation of important early projects. Working liaison was established with the National Geographic Society, the Society for the History of Technology, the American Historical Association, and the Smithsonian Institution. The Civil Engineering Museum and the Archival File maintained by the latter agency are particularly valued by the Society.

The History and Heritage Committee introduced in 1970 the first of a series of "mini-histories," entitled "The Civil Engineer: His Origins." Other issues on canals, bridges, railroads, etc., were planned. Also, in 1972 the committee produced its first *Biographical Dictionary of American Civil Engineers.*

Noting that the American Bicentennial Celebration in 1976 would offer an unique opportunity for favorable reference to civil engineering accomplishments, the committee assumed leadership in the bicentennial committee of EJC. It also undertook to launch an ambitious theater-scale audio-visual presentation that would portray the role of the engineering profession in the past and future development of America.

Advantage was taken of the growth of television as a communications medium during the 1960's. National coverage was given a one-hour show in 1963 on civil engineering aspects of the New York World's Fair. Collaboration and guidance were provided for the major network documentaries—"Essay on Bridges," in 1965, and "They Said It Couldn't Be Done," in 1970. A video-taped presentation on "Metropolitan Planning and Design" was aired by 78 stations.

Televised use of motion picture films prepared by ASCE was productive. The 1968 half-hour film, "The Invisible E—the Civil Engineer," was shown by 250 stations to more than 8 million people in its first year. By 1973 it had been telecast 744 times to a total estimated audience of over 14 million persons and shown some 418 times at group meetings to an audience of over 16,000 persons. A youth-oriented, career-guidance film, titled "Beginnings," was released in 1974 for showing on television and to student groups. Most promising in this area, however, is the series of one-minute non-commercial spot films for televising on public-service time. The first of these, titled "Water," was released in 1971, and followed by similar films on structural engineering, "people-movers," transportation and environmental engineering, and professionalism in the community. By 1973 five of these films had been telecast 6,700 times on 791 stations before 262 million viewers.

The public relations staff was also responsible for the production of career-guidance materials after 1960 (page 97). This covered an array of special brochures, a number of sound slide presentations, and the film "A Certain Tuesday" for showing in schools. As of 1973, the latter 14-minute film had been shown 367 times to an audience of 14,000 persons.

Encouragement and direction were provided continuously to the officers and sections in public-relations procedure. A "Public Relations Guide," first issued in 1958, served as a detailed manual for section personnel. At least since 1938 some form of public-relations newsletter has been circulated to the sections, under such titles as "Headliners," "PR Bulletin," "ASCE Information Bulletin," "The Section Leader," and "PR Newsbriefs." Some of the sections were outstanding in their conduct of timely news conferences and their public identification of newsworthy members and their accomplishments.

Advantage was taken of every occasion to assist authors of engineering books and career-guidance manuals, as well as editors of encyclopedias and dictionaries. An effort in 1957 to develop a cartoon strip featuring a civil engineer as the leading character was unsuccessful. In 1966, however, an educational ASCE cartoon, captioned "In This World," was released, ultimately being syndicated on a continuing basis to some 4,500 weekly newspapers. Two more such cartoon series, entitled "Building a Better World" and "News of Ecology," were also released.

Some unusual opportunities arise for public-relations benefits. One such situation was the request of the Boy Scouts of America, in 1958, for assistance in updating its Surveying Merit Badge. This was gladly provided, with a gratuitous recommendation that there should also be an Engineering Merit Badge. After ten years of persuasion, this new scouting award was initiated in 1969. Another one-time venture was the exhibit on "20th Century Engineering" at the New York City Museum of Modern Art in 1964. The Society supported the New York display of artistic photographs of engineering works, and also collaborated in arrangements for exhibitions elsewhere.

An overriding concept of the ASCE public-relations program is the realization that the members of the Society are themselves the most effective resource for building public respect and good will for the profession. Two media intended to engender professional pride and attitude were the corporate-type annual report issued since 1956 (page 215) and the editorials contributed by the Executive Secretary-

Director to *Civil Engineering* since 1962. Both of these devices were meant to inform the membership, thus to arouse interest, enthusiasm, and participation by the members in the advancement of their profession and the Society.

Upon its recreation in 1971, the Public Relations Committee obtained authorization for an evaluation study of the overall public information system by a professional public-relations firm. The consultants found the operation to be "thoroughly professional," and ". . . that it is making use of all methods of communication—written, audio, visual and audio-visual—and that it is aware of the latest techniques and uses them imaginatively to cover a multitude of audiences." The same consultants made a detailed proposal for an expanded program in 1972-73. Most of the recommendations called for allocation of additional funds to strengthen ongoing activities, and some new coverage was also urged.

Delayed though it was in getting under way, the ASCE public relations effort made commendable progress after 1935. There was every indication in 1974 that this responsibility of the Society would be given even greater emphasis in the future.

CHAPTER VII

ASCE AND THE ENGINEERING PROFESSION

Any meaningful analysis of engineering as a single profession must be prefaced by noting that engineering is a highly heterogenous and nebulous entity at best, not nearly as clearly delineated and compact as the classic professions of medicine and law. The difference is a result of both the systems of education and the modes of practice.

In medicine and law a basic professional education of at least seven years is common to all, with specialization coming thereafter. This is followed by formal internship in medicine and pre-professional training in law, again common to all beginners. In engineering, on the other hand, the "common core" of education is usually no more than two years, after which the student moves into his chosen branch of engineering specialty and thence into narrower specialization within that branch if he pursues graduate study. There is no formal internship in engineering except in the cooperative or "sandwich" plans offered by some schools.

By far the majority of doctors and lawyers are in private practice, with relatively few employed in government and industry. In engineering about 30% of civil engineers are principals or employees in private practice (consulting) firms and 40% are employed by government. The other branches of engineering are practiced predominantly in industrial organizations, with well under 10% of all practitioners engaged in private practice.

Thus, the medical and legal professions are relatively close-knit, by reason of the mutuality of education, training, and form of practice of their members. They are also well-defined by licensing procedures that are exacting and universally applicable. Engineers have very little in common in their education experience, and there is little real community of interest in a practice spectrum covering such diverse specialty fields as agriculture, electronics, construction, mining, chemicals, mechanical, electrical, public works, aerospace, etc. Moreover, the engineering professional is poorly defined by a licensing arrangement that is largely voluntary and not restrictive in its requirements.

As the first personification of the American engineering profession, ASCE understandably became subject to a variety of stresses and pressures when other professional groups began to emerge. Being the first and oldest national engineering society, it was only natural that ASCE in its early years would be cautious about yielding or sharing its identity, stature, autonomy, and resources with younger, less-respected bodies. The public-political orientation of civil engineering practice fostered different interests and attitudes than was the case for the engineering branches geared to the industrial-commercial world. This polarization was often the source of misunderstanding and disagreement in intersociety deliberations.

At the same time, ASCE did not attempt to become an island unto itself. In most cases the Society did participate readily in joint and group liaisons, and there were numerous occasions when it provided leadership in initiating, encouraging, and preserving intersociety collaboration.

As originally conceived, the American Society of Civil Engineers and Architects was intended to foster professional improvement and social intercourse "among men of practical science" and "the advancement of engineering in its several branches and of architecture."

The circular notice addressed to prospective Charter Members following organization of the Society in November 1852 envisioned a unified profession, with one organization serving "Civil, geological, mining, and mechanical engineers, architects and other persons who, by profession, are interested in the advancement of science." A note of realism is found, however, in the observation that:

> "It is anticipated that the union of the three branches of civil and mechanical engineering and architecture will be attended by the happiest results, not with a view to the fusion of the three professions into one; but as in our country, from necessity, a member of one profession is liable at times to be called upon to practice to a greater or less extent in the others, and as the line between them cannot be drawn with precision, it behooves each, if possible, to be grounded in the practice of the others; and the bond of union established by membership in the same Society, seeking the same end, and by the same means, will, it is hoped, do much to quiet the unworthy jealousies which have tended to diminish the usefulness of distinct societies formed heretofore by the several professions for their individual benefit."

This optimism proved, unfortunately, to be but wishful thinking. Only a handful of architects ever became members of the Society, largely because the American Institute of Architects (AIA) was organized in 1857 while ASCE was in the midst of its organizational hibernation from 1855-67. Reference to architects was deleted from the name of ASCE in 1869.

The American Institute of Mining Engineers (AIME) was organized in 1871 to identify the second branch of American engineering, followed by the American Society of Mechanical Engineers (ASME) in 1880, and the American Institute of Electrical Engineers (AIEE) in 1884. Some of the earliest engineering-related technical associations were also formed during this period, such as the American Iron and Steel Association (circa 1875) and the American Water Works Association (1881).

It is not altogether accurate to refer to these movements as a "splintering" of ASCE, as there were never significant percentages of mining, mechanical, and electrical engineers in the membership of the Society. Breakdown data are not available, but review of the index to *Transactions* for the period

1867-1920 is revealing. No reference whatever is made to papers on architecture, and contributions dealing strictly with engines, machinery, mining, electricity, etc. are very few in number compared with those in the traditional areas of civil engineering. It seems more accurate to assume that the major branches of engineering began to emerge as broad specialties during the Industrial Revolution after the Civil War. The real fragmentation of the profession would appear to have begun after 1900, when narrower fields of specialization began to appear.

Civil Engineering and Architecture

Although the profession of civil engineering is actually more closely related to the practice of architecture than it is to some branches of engineering, a regrettable barrier of provincialism has always separated the two professions. Interaction between them has been in the form of a sort of two-directional osmosis rather than true collaboration. This dichotomy has been detrimental to both professions, and to the beneficiaries of their services.

President Julius W. Adams, in his 1873 inaugural address, gave his explanation of the difference in aspect and attitude in architecture and engineering at that time. He saw the architect as the imitative perpetuator of art and design styles that characterized the monumental works of the ancients, which were built mainly for the luxurious enjoyment of the rich. On the other hand, he considered civil engineering to be a modern science that

> "has grown out of the wants induced by modern refinement and culture, and the more luxurious habits and comforts which, no longer confined to the governing classes, but extending throughout society, as the conditions of the masses were bettered, have become the necessities of life."

President Adams described the relationship between the civil engineer and the architect in these words:

"The science of construction in all its ramifications, however essential it may be to the Architect for embodying his ideas, yet, is entirely within the province of the Civil Engineer; whilst the decorative branch of art merely is all that properly belongs to the architect as his specialty."

Alonzo J. Hammond (President, 1933) held the view that, "In a comprehensive sense, Civil Engineering includes Architecture as a Mechanical Art, in distinction to Architecture as a Fine Art."

It should be noted here that qualified architects are eligible for membership in ASCE, and that they have been welcomed as such (Fig. 7 page 106). The 1891 Annual Report referred to "a recent occurrence" with the explanation that:

"Architects in good standing have always been eligible for membership; and especially at the present time, an architect competent to design and execute the complicated details of one of the large buildings now becoming so common in all large cities, is surely the peer of other engineers."

Through the years, a major source of irritation to the civil engineer has been the effort of the architectural profession to establish and insure its stature and jurisdiction by legislative edict. In March 1876 the Board of Direction took note of a "Bill to establish a Bureau of Architecture" in the Treasury Department. The legislation had been drafted by the AIA and introduced in the House of Representatives. Outcome of the legislation is not reported.

The issue of professional licensing arose in architecture several years before it appeared in engineering. The movement at the turn of the century was initially unwelcome to both AIA and ASCE, but both organizations lost no time in providing guidelines for registration laws as they were proposed in the state legislatures. From the beginning, the broadest possible definition of the practice of architecture was established in the architectural registration laws, with the result that barriers were raised to some areas of civil engineering practice, particularly in the design of buildings. These barriers have been vigorously defended by the architectural registration boards, and many engineers have been charged with encroachment into architectural practice although they were performing services for which they were fully qualified by education and experience.

Only occasionally has this reaction been reversed. A notable instance was exposed by the Pittsburgh Association of Members in 1920, when the Allegheny County Commissioners proposed to employ architects to design and supervise the construction of several important bridges in Pittsburgh. Naturally, ASCE protested such action as being "detrimental to the public interest to subordinate safety and economy, adequacy for future traffic, and cost of these structures to their appearance, although it is recognized that the embellishment and aesthetic features of bridges may properly be entrusted to those especially skilled in architecture." A similiar resolution had been adopted by the American Institute of Consulting Engineers (AICE). The Joint Committee of Engineering Council and AIA, however, disagreed with ASCE and AICE, taking the view that "whether the engineer is chief and the architect associate or vice versa is an administrative detail of relative unimportance." ASCE forcefully reminded the Engineering Council that it had no authority to represent the Society in a matter solely within the province of the civil engineer, and gave wide publicity to its resolutions on the subject. The issue remained moot until 1923, when ASCE reiterated its position in a communication to the Allegheny County Commissioners. This time, apparently, they paid heed.

While the Allegheny County controversy was going on, the Joint Committee of Engineering Council and AIA was adopting a resolution addressed to management of licensing problems in the two professions. At the same time AIA was urged to become a full member of Engineering Council. Apparently neither action was implemented.

Only rarely have architects been charged with transgression

into engineering practice under the licensing laws. Perhaps some engineers have been reluctant to file complaints against architects because they are occasionally engaged by the architect as the client.

The definition of the practice of architecture in architectural registration laws was the source of contention between the civil engineer and the architect in the late 1920's. It was temporarily resolved in 1931 by a Joint Committee on Registration of Architects and Engineers, whose work ". . . had a marked effect in clarifying the situation with regard to proposed legislation affecting the interest of engineers."

The use of the registration law as a device for staking out jurisdictional claims reached a peak in 1971 when the National Council of Architectural Registration Boards (NCARB) promulgated legislative guidelines defining architecture as covering "the design and construction of a structure or a group of structures which have as their principal purpose human habitation or use, and the utilization of space within and surrounding such structures." Where implemented into state registration laws, this definition in its entirety would have subordinated civil engineering and all other design professions to architecture!

ASCE and the other engineering societies protested loudly and alerted their state and local sections to be wary of proposed amendments to architectural registration laws. A possibly promising result was the creation in 1972 of a joint liaison committee of the NCARB and the National Council of Engineering Examiners (NCEE).

During World War II the term "Architect-Engineer" came into broad use as applied to those furnishing professional services on large Army projects. In 1945 President J. C. Stevens asked the Corps of Engineers to amend this practice, giving the consultant architect or engineer whichever title was appropriate. The Corps acceded readily to the request, stating that the term "Architect-Engineer" would continue to be used only in the declining number of instances where the services were representative of both professions. By this time, however, the term had found a place in civil practice, and was still in fairly general use in 1974.

On the positive side, an AIA historical document mentions that about 1902 ASCE was one of four bodies to be invited to send non-voting delegates to the AIA conventions. No reference to such an invitation was found, however, in the ASCE *Proceedings*. AIA and ASCE were both participants in a conference of eight engineering, architectural, and construction organizations held in December 1921 to produce a standard form of construction contract. The venture was highly successful in accomplishing its purpose.

In 1933 negotiations between the officers of AIA and ASCE produced the Engineer-Architect Joint Committee to develop means for ameliorating professional differences by furtherance of mutual attitudes, respect, and cooperation. An immediate objective was joint study and action on the then proposed federal Department of Public Works, which both organizations favored.

In 1938 a joint effort of ASCE, AIA, and the Engineering Council was directed toward improvement of procedures for employing engineers and architects on public works projects. This resulted in liberalization of policies governing the use of private consultants.

Another joint venture in 1941 by ASCE, AIA, ASLA, and ASME produced the first attempt to outline the specific areas of professional jurisdiction for civil and mechanical engineering and for architecture and landscape architecture on certain national defense projects. This significant statement on "Division of Responsibility and Work Among the Planning Professions of Architecture, Civil Engineering, Landscape Architecture and Mechanical Engineering on National Defense Housing Projects" was adopted by the four organizations, and proved to be an effective guideline in its narrow area.

After World War II the ASCE-AIA joint committee became known briefly as the Joint Committee of the Design Professions, and later as the ASCE-AIA Joint Cooperative Committee. In 1951 it served as the medium for united action in bringing about improvements in procedures for awarding professional service contracts on defense projects. During the ASCE Centennial year of 1952 AIA featured the "Reunion of Engineering and Architecture" in its convention, a gracious gesture, albeit more fanfare than fact. The cooperative committee sponsored a successful Public Works Conference in Washington in 1956 under the joint aegis of ASCE and AIA.

About 1958, in order to bring the point of view of mechanical and electrical engineers into the discussions, the ASCE representation in the joint committee was transferred to EJC. This was an unfortunate move; after a few years of lukewarm participation by EJC the liaison was dropped, leaving ASCE and AIA without a cooperative link.

A promising development was fostered in 1963 as a result of discussion initiated by the executive director of AIA with his counterpart in ASCE. Thus began the Interprofessional Commission (later changed to "Council") on Environmental Design (ICED). The original participant organizations included the American Institute of Planners (AIP) and the American Society of Landscape Architects (ASLA) with AIA and ASCE. Admitted soon after were the Consulting Engineers Council (later the American Consulting Engineers Council), the NSPE, and the American Society of Consulting Planners (ASCP).

ICED was intended to supplement any existing working liaison committees by providing "a top level mechanism for communication, long-range planning, the definition of major interprofessional problems and objectives and means for their solution and attainment." Especially significant was the requirement that the members of the Council be the President of each society at the time of his three-year appointment, and the Executive Director of each society. This insured a forum with authoritative representation at both the policy and executive levels of each society. The forthright and generally constructive discussions in ICED were immediately productive. The 1941 "Division of Responsibilities" policy was promptly updated to a new guide to "Professional Collaboration in Environmental Design." A mechanism for local mediation of interprofessional controversies was devised. A policy statement on professional criticism of public works projects was adopted, and problems in ethical standards, registration, and overlapping activities were confronted. A series of four ICED conferences — all managed by ASCE — held on professional collaboration, environmental design education, social aspects of design, and environmental impact were all successful.

Deliberations in ICED concerning the controversial definition of architecture proposed in 1971 resulted in formation of the Interprofessional Council on Registration, comprising the national organizations of registration boards in architecture, landscape architecture and engineering. A problem of great concern to ICED in 1974 was the widespread invitation of political contributions from engineers and architects seeking engagements on public projects.

ICED showed great potential as the "Key to Interprofessional Collaboration," but its good offices suffered several setbacks as the result of unilateral actions by AIA. A controversial film on urban transportation, a rule denying recognition of professional experience acquired by an architect in an engineering firm, and advertisements implying architectural expertise in traditional areas of engineering have caused irritation. For example, a national advertising campaign in 1973 invited industrial firms to call upon architects to solve pollution, traffic, noise, and similar environmental problems. It brought a bitter reaction.

Mutual trust and empathy between civil engineers and architects are vital to both professions and their clients. There is a long way to go.

Intersociety Relationships in Engineering

Interaction among American engineering societies has from their beginning generally been cordial and well-intentioned, but it has been highly variable in unanimity and effectiveness.

As the number of societies increased (to more than 150 national engineering and related organizations in 1972) the complexity and heterogeneity of the profession were compounded. Hence, the internal unification and organization of this nebulous system were infinitely more difficult to accomplish than was the case in the more compact disciplines of medicine and law.

In 1876, ASCE and AIME were the only national societies, and they joined in two ventures: A program for the Philadelphia Centennial Exposition and a Joint Committee of ASCE and AIME on Technical Education. In 1885, ASCE, AIME, ASME, and AIEE formed a Committee on Joint Library, which was unsuccessful in a three-year stint. The international expositions in Paris (1889), Chicago (1893), St. Louis (1904), and San Francisco (1915) afforded further opportunities for intersociety cooperation.

Late in 1889, an intriguing idea — which may have originated in ASME — was put before the Board of Direction by Henry R. Towne. It proposed an "Institute of Engineers," a super-society that would be composed of eminent engineers selected by the four basic societies. This body would elect from its membership a "Senate" that presumably would be the spokesman for the entire profession. After deliberation the Board determined that the plan would "be inexpedient," and dropped it. There was some confusion about this a few months later when Oberlin Smith, M. ASCE, believing that a federation of the four societies had been proposed by ASME, brought up the matter in the 1890 Annual Convention. Two motions aimed toward a joint committee study of some sort of federation were declared out of order.

For several years after the organization of AIEE in 1884, that organization met in the headquarters of ASCE. In 1890, AIEE expressed its gratitude for this hospitality and support by a gift of handsome fireplace fixtures.

The John Fritz Medal Board of Award was created in 1902 to administer this annual recognition for the four basic societies. It was still functioning in 1974.

In 1907 the Society for the Promotion of Engineering Education (later ASEE) brought the "Big Four" together in a Joint Committee on Engineering Education, which eventually was funded by the Carnegie Foundation. Its final report was made in 1919. Also in 1907 a Committee on Library was formed, with one member from each of the four societies, leading to establishment of the Engineering Societies Library in 1915.

In the meantime AIME, ASME, and AIEE had acted in 1904 to form the United Engineering Society as a corporation to administer a gift from Andrew Carnegie "to be used for advancement of the engineering arts and sciences, and all their branches, and for the maintenance of a free public engineering library." ASCE was not originally a participant in this Founders' Agreement — which was aimed toward erection of a headquarters building for the major societies — as ASCE had occupied its own new and palatial building for only a few years. In 1916, however, ASCE became the fourth Founder Society and joined the other three in the Engineering Societies Building. (AIChE was admitted as the fifth Founder Society in 1958, when the United Engineering Center project was undertaken.)

The proposed formation of an "Academy of Engineers" in 1918 elicited this response from the Board of Direction:

> ". . . that such an organization as is proposed under the title of 'Academy of Engineers,' if there is a field for its activities, should be organized under the initiative of the National Engineering Societies, and particularly that the membership of an Academy of Engineers should not be made up by self-appointment."

The Board of Direction reacted again in 1924 to the news that an American Society of Engineers was being formed in the Chicago area. The Society's legal counsel was requested to protest against the similarity of the proposed name and badge to those of ASCE. The matter appears to have been closed with this action.

Engineering Society Federations

Up to about 1912 the aforementioned exercises in intersociety collaboration, together with a few others dealing with technical studies, were strictly ad hoc in nature. It appears that the earliest continuing cooperative arrangement to be devised was a General Conference Committee of the National Engineering Societies, which reported in 1913-14 on several professional matters. In April 1917 this rather informal general joint committee originated the creation in the United Engineering Society of a new department called the Engineering Council, to be constituted initially of five representatives from each Founder Society:

> ". . . in order to provide for convenient cooperation between the Founder Societies, for the proper consideration of questions of general interest to engineers and to the public and to provide the means for united action upon questions of common concern to engineers."

Thus emerged the first formal federation of engineering societies in America. Provision was made for the admission of other approved engineering or technical organizations, and the American Society for Testing Materials and the American Railway Engineering Association joined the group.

The entrance of the United States into World War I gave the Engineering Council of National Technical Societies many opportunities to serve the public and the profession, and it did so with distinction. An office was set up in Washington in 1918, and activities ranged from legislation to cooperation with many federal agencies. A roster of "Civil Engineers for War Service" was produced as part of a comprehensive engineering manpower survey. In December 1918 the Engineering Societies Service Bureau, an employment service for members, was established under the direction of the Council, with the Founder Society secretaries as the managing committee.

Then came the first of the long series of dislocations that were to plague efforts to organize the engineering profession in the next half century. Three Founders, AIME, ASME, and AIEE, decided that an autonomous structure was desirable, and in 1920 formed the Federated American Engineering Societies (FAES), with the American Engineering Council as its executive arm. Despite the preference of ASCE and AREA for continuance of the Engineering Council, that body was dissolved in December 1920 when the proponents of the new federation withdrew their support.

A questionnaire submitted to the ASCE membership elicited an expression of almost 60% against affiliation with FAES.

The refusal of ASCE to become a Charter Member of FAES, which appeared partly to stem from legal questions and a reluctance to assume the $11,000 annual financial commitment, left the Society in something of a dilemma. On the one hand, the membership had endorsed the recommendation of the 1919 Committee on Development "to actively co-operate with other engineering and allied technical organizations in promoting the welfare of the Engineering Professions." On the other, there was no acceptable medium available for effecting such cooperation. A Committee on External Relations appointed to resolve the problem concluded that intersociety cooperation, "shall not involve the surrender of the name and standing of the Society into outside hands, or in any way permit the use of the Society's name in behalf of any cause of which it does not approve."

ASCE sought unsuccessfully, in 1922, to resolve the impasse when it asked Engineering Foundation to "make a study of Engineering Society Organizations in an endeavor to overcome the existing confusion and duplication of activities in Civil Engineering organizations." Engineering Foundation considered the request at its next meeting, but tabled the proposal when it appeared that it would not be supported outside the ASCE delegation.

After an ASCE membership referendum in 1923 affirmed the decision to forego affiliation with FAES, the Society enlisted the participation of the other three Founders in a Joint Committee on Co-operation in Public Matters, which promptly recommended that the Presidents and Secretaries of the four Founders establish a continuing forum for communication and interaction. The Joint Conference Committee was highly effective, and continued for the next seven years. This enabled ASCE to support independently such American Engineering Council causes as it chose. The Society also continued to support the Engineering Societies Service Bureau during the time that this function was under the direction of FAES.

In 1924 FAES took the name of its executive body, American Engineering Council. Three years later ASCE inquired as to the practicability and desirability of effecting certain amendments to the constitution, bylaws, and rules of the Council. This led to another referendum in the Society, this time overwhelmingly favorable to affiliation, which was consummated in 1928.

At this stage American Engineering Council comprised six national, five state, and fifteen local societies. Its interests were very strongly oriented to those of ASCE in such areas as reforestation, topographic surveys, national hydraulic laboratory, public health engineering, federal public works administration, water resources, highway safety, and public policy on power. Many other research studies and programs were directed toward industrial problems and appraisals.

The Council maintained its vigor through the early years of the Great Depression, but in 1934 deep curtailment of contributions from member organizations led to marked reduction in staff and programs. By July 1940 fiscal problems and general apathy had reached such levels that ASCE gave notice that it was withdrawing from the Council at the end of that year. The AEC leadership urged reconsideration, but to no avail; it was dissolved on January 1, 1941.

A few months later ASCE invited the other Founders to join in establishing a Washington office. While this did not develop, the Founders and AIChE did accept the suggestion that another Joint Conference Committee be reestablished, with the president and secretary of each society as delegates. It will be recalled that such a committee successfully provided inter-society communication and liaison from 1923 to 1930, when it was dissolved during the heyday of AEC.

Once again, the Joint Conference Committee, unhampered by administrative and hierarchial ballast, gave a fine account of itself. Among its concerns were an in-depth study of the organization of the engineering profession and of the economic status of the engineer, in addition to a range of matters relating to the role of the profession during and following World War II. Particularly significant were the reports on the industrial disarmament of Germany and Japan, produced by the five presidents in 1945 and 1946.

Engineers Joint Council

Continuing the pattern of some twenty years earlier, the Joint Conference Committee changed its name to Engineers Joint Council (EJC) in 1945, and added a program of international relations to its list of professional undertakings. The 1946 survey report, "The Engineering Profession in Transition," was the most comprehensive statistical study ever made of the engineering profession to that time. The Engineering Manpower Commission was initiated as an EJC function in 1950. By 1952 general preoccupation with unity in the engineering profession produced an Exploratory Group recommendation that EJC be reorganized, with delegates to be active members of the governing body of the societies they represented. The move was made, EJC membership was expanded to ten organizations, and programs were successfully developed and pursued.

The unity movement continued to bring many proposals and charts for restructuring the profession. Several EJC societies were nervous about their tax-exemption status in view of the non-educational professional nature of most EJC programs. Some leaders in EJC societies were also active in the NSPE (organized in 1934) and sought to establish that society as the professional spokesman for all engineering. In 1958, ASCE issued a policy statement:

> "In ASCE, unity means cooperative effort in the common interest. The Society welcomes the opportunity to work with other organizations in the attack on and solutions of common problems. At the same time, ASCE continues to apply itself to the same activities and problems in advancing the special interests of the civil engineer . . .
>
> "Recognizing the importance of the National Society of Professional Engineers as a vital and dynamic influence for the advancement of the engineering profession, and envisaging the great contribution to the profession that could result from the participation of that Society in both Engineers Joint Council and Engineers Council for Professional Development, [ASCE] would view with high favor the simultaneous affiliation of NSPE with both EJC and ECPD . . ."

NSPE would not accept membership in EJC, however, so this plan for "unity" was also unsuccessful.

EJC achieved its greatest visibility in 1960 when it began publication of a quarterly newsletter in tabloid format entitled *ENGINEER*, as the result of a proposal originating in ASCE. Productivity of the federation was at a high level at this time, but the restiveness of some of the constituent societies began to tell. The interests of the leadership were diverted, and this resulted in 1964 in the creation of the National Academy of Engineering (NAE) as a counterpart of the National Academy of Science. The objectives of NAE impinged somewhat upon those of EJC, and three years later EJC underwent another reorganization.

This one was all but disastrous, as it put the Council under the direction of a Board made up of engineers in responsible positions in industry but with almost no past or present relationship to the management of the constituent societies of the federation. Three societies — IEEE, AIChE, and ASHRAE — withdrew from EJC membership in 1968, even though some readjustments in organizational structure were hurriedly effected.

In 1970 ASCE, then the largest of all EJC societies, was concerned enough to review its own membership status. It was decided to continue, and to cooperate in rebuilding EJC as an effective forum and cooperative federation for the many segments of the engineering profession.

United Engineering Trustees, Inc.

The United Engineering Society, formed in 1904 by AIME, ASME, and AIEE and joined later by ASCE and AIChE (page 39), proved to be the most consistently viable of the older engineering federations. This is no doubt due to the specificity of its purpose — to administer as trustee the use of funds contributed for the general benefit of the engineering profession. In this capacity its primary function had been to build and operate the Engineering Societies Building from 1906 to 1961 and the United Engineering Center in New York City since 1961.

In 1914, the Engineering Foundation was created as a subsidiary of the United Engineering Society to administer substantial gifts made by Ambrose Swasey, a Past-President of ASME, for furtherance of research and advancement of the profession. The Engineering Societies Library also became a part of the United Engineering Society in 1915, although ASCE did not surrender its library until a year later.

The name, United Engineering Trustees, Inc. (UET), replaced the United Engineering Society in 1931 and has prevailed since then.

Engineering Societies Personnel Service (1918-1965)

Although it is no longer in existence, the Engineering Societies Personnel Service (ESPS) was an important intersociety activity. Events leading to its emergence in 1918 as the Engineering Societies Service Bureau have been recounted page 157). Engineering Council had reluctantly accepted the direction of the operation, even though it was being funded by the four Founder Societies and was being managed by a committee composed of the Founder Society secretaries.

There was no charge to members utilizing the service, and it was widely used by engineers seeking employment after World War I. In 1919 there were 1,256 placements of 5,377 registrants, and in 1920 placements totalled 1,606 from 2,256 registrants, at an average cost of $7.93 per placement. Local offices were opened in San Francisco and New York.

Direction of the Service was assumed by the FAES in 1921, but 18 months later it was turned back to the Founder Societies. ASCE continued to pay its share of the subsidy even though it was not a member of FAES.

In 1923 the Service was established as a cooperative undertaking of the Founders with management still under the committee of secretaries, and for the first time it was put on a fee basis with half the operating cost still being subsidized. A local office was opened in Chicago to supplement those in San Francisco and New York.

In this format the Service went into the depression decade of the 1930's, performing modestly but well. In 1932 there were 3,520 registrations and 573 placements. The usefulness of the Service increased as the depression waned. In 1941 it was incorporated as Engineering Societies Personnel Service, Inc., and local offices were opened temporarily in Detroit and Boston. With the Service under excellent management, income from fees exceeded expenses, and a substantial reserve was accumulated.

After World War II, however, a change in recruitment practices by employers of engineers, together with the expansion of government employment services brought disaster to ESPS. The reserve funds were gradually diminished by operating deficits, and an evaluation of the operation in 1965 led to a decision to terminate it in that year

Engineers Council for Professional Development

The organization of ECPD in 1930 has been noted previously (page 87). It has been on of the most successful of all the engineering federations, probably because its objectives are well defined, and shared by all segments of the profession.

ECPD has distinguished itself as the accreditation medium for engineering education, and has been wholeheartedly supported by ASCE since its formation.

The Continuing Quest for Unity

The endeavor to achieve a consolidated engineering profession assumed a new dimension in the latter 1950's, with the idea that certain major federations of societies should merge into a super-organization. This concept rose from the ashes of the attempt to bring NSPE into constituent membership in both EJC and ECPD. It took the form of a proposal to merge EJC and ECPD into an "American Engineering Association, Inc.," within which the functions of the two federations would be served by separate operating divisions. The plan was approved by ASCE in February 1959 but was never consummated.

From this background evolved, in 1960, the Board of Engineering Cooperation (BEC), an informal assembly of the Presidents, Past-Presidents, and Presidents-Elect of EJC, ECPD, and NSPE. Occasional meetings were held merely to review the goals and activities of the three organizations, thus to enhance cooperation and to minimize duplication. The American Society for Engineering Education (ASEE) became the fourth partner in the BEC in 1970.

Early in the 1960's it was decided, again informally, to invite the Presidents of the EJC constituent societies to join with the BEC in one of its meetings each year, and later the entire executive committee of each society was invited. An unusual feature of the "Joint Societies Forum," as it became known, was the fact that it was limited to elected officers and specifically excluded any staff representation.

The BEC and its yearly Joint Societies Forum served their primary purpose of communication among the elected officers of the national societies and their several federations. They also nurtured hopes that a more substantial form of engineering unity might be accomplished. Thus, these meetings became a launching pad for various ideas of an umbrella organization for the entire profession. In 1968 the oft-proposed concept of an individual-membership organization for all engineers was discussed for a few months. The certification concept, promoted by ECPD in 1934 (page 125), was resurrected in 1969 in a plan for "accrediting" engineers who did not choose or who might not be able to qualify for registration. The standards and evaluation of qualifications would have been made by a new "Institute for the Accreditation of Engineers." Joint Societies Forum referred the proposal to EJC in 1971, and three years later it was still under study in an EJC task force. ASCE was uncommitted on the matter in 1974.

Despite the opportunities for communication afforded by BEC and its Forum, another such mechanism found life in 1968 when the Founder Society Presidents held informal dinner meetings to discuss the problems then facing EJC. The scope of discussion and participation in these gatherings gradually broadened, until in 1973 the group called itself the "Coordinating Committee of Engineering Society Presidents."

Amidst this tangled skein the Joint Societies Forum fabricated in August 1973 a recommendation that an Ad Hoc Committee on Unification in Engineering be set up to devise a new organizational structure that would realize "a unified approach to many problems confronting the profession and the nation . . . beneficial to individual engineers and the public . . ." The BEC accepted the challenge, and in 1974 the movement to capture and contain the elusive spectre of engineering unity was again in full force, past failures notwithstanding.

Special Intersociety Relationships

It is impossible here to document all the external associations of ASCE. Those with the other Founder Societies are not detailed because they were so close and so cordial. Communication among the Founders was maximized after ASCE moved into the Engineering Societies Building in 1917, and informal cooperation was a matter of course. A joint committee had occasionally been set up with one of the Founders, such as the Joint ASCE-AIME Committee on Technical Education (1876) and the ASCE-AIChE Committee on Water Pollution (1932-38), but this was the exception. The spontaneous exchange and collaboration that occurred without official onus were far more productive. The Founders were very good neighbors, even though they did not always agree upon matters of professional import. This is not surprising, in view of the laborious evolution of professional policies in ASCE; the other societies underwent the same internal struggle.

Outside the Founder group ASCE had contact of some kind with all national organizations engaged with specialty fields of civil engineering, and a host of others as well. These relationships may have had to do with the origin, the goals, or the ultimate activities of such organizations. The following examples are illustrative of the nature and range of the Society's role within that uncharted complex known as the engineering profession.

Highway Research Board (HRB)

When, in 1920, the Engineering Division of the National Research Council undertook to create an organization to carry out a national program of highway research, ASCE was one of the twelve agencies invited to participate. The Society endorsed the proposal completely, pledging its moral support and assistance in raising funds and in publicity. In 1974 HRB changed its name to Transportation Research Board. ASCE'S interest and cooperation have continued through the Highway Division of the Society.

Chi Epsilon Fraternity

The scholastic honor society in civil engineering, Chi Epsilon, was formed in 1922 by students at the University of Illinois. One of the founders was Harold T. Larsen, who was later to devote most of his career to service on the publications staff of ASCE. He retired as Manager of Technical Publications in 1963.

When ASCE occupied its new offices in the United Engineering Center in 1961, Chi Epsilon conducted a special fund-raising drive that netted almost $15,000. These funds were contributed to the Society and used to furnish and equip its new conference room. The "Chi Epsilon Room" became one of the most attractive features of the Center.

Associated General Contractors of America (AGC)

ASCE collaboration with AGC began more than a half century ago in the Joint Conference of Engineers, Architects and Constructors, which undertook development of a standard form of contract for engineering construction. The venture was conceived in the ASCE Committee on Highway Engineering in 1921, and was consummated four years later by publication of the first edition of the "Standard Contract for Engineering Construction." By that time the Joint Conference included AASHO, AEC, AIA, AREA, AWWA, and the Western Society of Engineers, in addition to ASCE and AGC.

Again, in 1923 AGC and ASCE undertook discussion of the possibility of a buyers' strike in protest against inordinately rising costs of all elements of construction at that time. An AGC report on "The Seasonal Demand of Construction for Labor, Materials and Transportation" delineated the problem. The sections of ASCE were asked to use this document as the basis for local meetings on "Waste in the Building Industry," and also to express their views as to the desirability of forming a Construction Division of the Society. The Construction Division was activated in 1925.

Since 1948 the ASCE-AGC Joint Committee has provided an effective medium for liaison between the two bodies.

Water Pollution Control Federation (WPCF)

In 1927 the ASCE Sanitary Engineering Division directed a task committee to explore means of generating and disseminating knowledge and experience on the treatment of sewage and industrial wastes and on control of stream pollution. This committee, chaired by Harrison P. Eddy, M. ASCE, called a general meeting in Chicago on June 10, 1927, bringing together interested representatives of the AWWA, the APHA, the Conference of State Sanitary Engineers, and the American Society for Municipal Improvements. Subsequently a Committee of One Hundred was formed under the chairmanship of Charles A. Emerson, Jr., M. ASCE.

These efforts led to the organization of the Federation of Sewage Works Associations on January 19, 1928, during an ASCE Annual Meeting. Known since 1960 as the Water Pollution Control Federation (WPCF), this body has been significantly responsible for the advancement of scientific and technical knowledge in its area of concern.

Construction League of the United States

Formed in 1932 to mobilize and unify the far-flung construction industry during the Great Depression, the Construction League of the United States had a brief but useful life. ASCE was a Charter Member with AIA, AGC, ARBA, and AISC; some 15 other technical and trade associations concerned with production, installation, and erection in construction operations were also members.

The principal accomplishment of the League was the drafting of the Construction Industry Code of Fair Competition. This document provided procedures for coordinating the many facets of the construction industry. Chapter I, approved by President Roosevelt early in 1934, was said to include guidelines on bidding that "are a lasting norm of fair competition." The Code was administered by the National Recovery Administration (NRA).

Another chapter of the Fair Practice Code, drafted by the Engineering Division of the League, was the focus of interest on the part of ASCE. A draft of the so-called "Engineering Code" was published in *Civil Engineering,* June 1934, pp. 321-324, but was apparently never approved by NRA.

The League also provided input from its participant organizations toward decisions in public policy and legislation relating to construction during the post-depression years.

National Society of Professional Engineers (NSPE)

Several prominent ASCE members were among the organizers of the NSPE in 1934, but the movement was in no way sanctioned officially by ASCE. On the contrary, the ASCE Board of Direction directed the Secretary in January 1935 to issue a statement making it clear "that the Society has no official representative in the newly formed National Society of Professional Engineers."

Among the leaders in ASCE at this time were some who did not see the need for a new organization "devoted exclusively to the professional interests of all engineers." With the implementation of the Functional Expansion Plan in ASCE some years before, the Society had demonstrated its intent to serve the professional as well as the technical needs of the civil engineer, and to show that it was not just a "technical" society. The other Founder Societies were quite willing to accept NSPE, however, and the newcomer prospered in the post-depression demand for stronger economic-welfare programs for engineers.

NSPE membership was for 36 years limited to registered engineers, and as far more civil engineers were registered than in the other branches of the profession, there was substantial duplication of membership in ASCE and NSPE. Yet, the relationship between the two organizations was characterized by a restrained aloofness, with some aspect of competition in professional programs.

By 1955 enough change in leadership had taken place to reduce the barriers between the two societies, although ASCE was continuously strengthening its professional services on behalf of the civil engineer. Cooperation in reaching mutual goals actually began in the mid-forties when NSPE joined with ASCE and other bodies in supporting the professional employee provisions of the Taft-Hartley Act. The ASCE pronouncement on engineering unity in 1958 (page 313) evidenced the intent of the Society to work constructively with NSPE.

The relationship continued to improve through the sixties, with day-to-day interaction at the staff level and a growing number of joint representations in such bodies as ICED, COFPAES, ECPD, NCEE, and various task committees. In 1963 ASCE accepted an invitation to send representatives to the meetings of NSPE, but by this time such formal measures to demonstrate comity were no longer necessary. In 1974 the mutual attitudes of the two societies were cordial and collaborative in every respect.

The 1934 activities directed at that time toward organization of NSPE and other new engineering bodies may have had considerable impact on the course of ASCE. The Board set up a task force to analyze the objectives of these movements and:

> "To study the present activities of the Society with respect to human relations, welfare, and public relations . . . and to make recommendations for changes or amplification of such activities."

This could readily explain the sharply increased emphasis in ASCE upon improvement of civil engineering salaries and other economic welfare services in the late 1930's and through the 1940's (page 158 et seq.)

American Academy of Environmental Engineers (AAEE)

The AAEE is particularly noteworthy here because (1) it is the first (and in 1974 the only) roster of engineers certified in a specialty field in the manner done in medicine, and (2) its roots go back originally to ASCE.

In the early 1950's the Committee on Advancement of Sanitary Engineering of the ASCE Sanitary Engineering Division noted the need for sanitary engineers engaged in the public health field to identify themselves at a high level of competency in their specialty, in order to compete with medical public health personnel for ranking administrative positions. Because of the interdisciplinary nature of sanitary engineering, the ASCE committee was expanded in 1952 to the Joint Committee for the Advancement of Sanitary Engineering, encompassing ASCE, APHA, ASEE, AWWA, and the

Federation of Sewage and Industrial Wastes Associations (later WPCF).

From the deliberations of this body emerged the American Sanitary Engineering Intersociety Board, Inc., founded in 1955 with its constitution patterned after that of the American College of Surgeons. This Board set up criteria and procedures for certifying registered engineers on the basis of their education, experience, and an examination, for proficiency in sanitary engineering. Candidates thus certified were recognized as diplomates of the American Academy of Sanitary Engineers.

ASCE support of the Intersociety Board was demonstrated in tangible form by a loan of $7,500 to cover start-up costs and the provision of complimentary office facilities from 1955 to 1958 at the headquarters of the Society.

In the mid-1960's the names of the Board and Academy were changed to the American Environmental Engineering Intersociety Board, Inc., and the American Academy of Environmental Engineers. In a move to simplify the operation in 1973 the Board was merged into the Academy. This made the Academy of Environmental Engineers essentially autonomous, although its 18 trustees included representatives of ASCE, AIChE, APHA, ASEE, AWWA, WPCF, APCA, and APWA.

American Consulting Engineers Council (ACEC)

The ACEC is the product of a 1973 merger of the American Institute of Consulting Engineers (AICE) and the Consulting Engineers Council of the U.S.A. (CEC).

AICE, formed in 1910, was a small but prestigious group in which membership was held by prominent consulting engineers as individuals. Membership was largely limited to the New York area at first, but ultimately reached a nationwide total of about 450.

Most AICE members were also members of ASCE. The monthly luncheons and annual dinner meeting of AICE were its main activities, but there was a mutuality of interest with

ASCE in ethical matters and in public issues impinging upon private practice. The relationship was always cordial and highly professional.

CEC was founded in 1959 to fill the need for an association to further the promotional and business techniques of the consulting engineer as is done by AIA for the architect. Membership was limited to firms, either partnership or corporation.

When CEC suggested to ASCE in 1960 that a joint liaison committee be set up, it was recognized that CEC filled a need that could not properly be served by ASCE. A three-way liaison was established comprising CEC, AICE, and ASCE, all having special interest in the private practice of engineering. The "Tripartite Committee" continued until 1965, when its purpose was fulfilled within the Coordinating Committee on Relations of Engineers in Private Practice and Government (later COFPAES). In the meantime, CEC met its goal most successfully and grew to a membership of 2,500 firms.

In the late 1960's a movement was launched to combine AICE, CEC, and the Professional Engineers in Private Practice (PEPP) Division of NSPE. The latter abstained from the 1973 merger action, however, which brought AICE and CEC together as the American Consulting Engineers Council.

Committee on Federal Procurement of Architect-Engineer Services (COFPAES)

Frustration with the insistence of some federal construction agencies on requesting competitive prices from professional consultants led to formation, in 1964, of the Coordinating Committee on Relations of Engineers in Private Practice with Government, with participation by ASCE, AICE, CEC, NSPE, and ARBA. With the addition of AIA in 1968, the name was changed to the Committee on Federal Procurement of Architect-Engineer Services (COFPAES).

This energetic joint venture gave material assistance to a congressional task force appointed in 1967 to study A-E contract negotiation procedures. It afforded a forum for communication during the anti-trust actions by the Department of

Justice against ASCE and AIA in 1971-72. Its greatest accomplishment, however, was the original drafting and successful support of the Brooks Bill, enacted in 1972 to require that all professional services incident to construction projects involving federal funds be contracted by professional negotiation.

The 1973 disclosures resulting in the resignation of the Vice President of the United States because of political contributions by consulting engineers in Maryland, however, confronted COFPAES with considerable difficulty in maintaining this progress. Some public officials and newspapers took the erroneous view that the selection of professional consultants by competitive bidding would preclude such corruption.

American Society of Certified Engineering Technicians (ASCET)

In 1967, the Board of Direction resolved that:

> "The Society should provide an appropriate recognition of the civil engineering technician, in order to foster his growth and development as 'a member of the team,' to stimulate the contributions he can make to the profession, and to encourage a viable, fruitful relationship between the professional civil engineer and the technician as his subprofessional supporter."

This policy was implemented under the principle that the engineer and technician should carefully preserve their independent identities, high standards of competency, and organizational autonomy, while establishing positive means through their national organizations for communication, cooperation, and maximum service to members.

The ASCE-ASCET Joint Committee was formed in 1968 on this foundation. An additional innovation, however, was the "Civil Engineering Technician Service Roster," open, upon payment of a modest fee, to any ASCET member who wished to receive ASCE publications and meeting notices. The plan emphasized ASCE-ASCET interaction at the local level, and was still progressing in 1974.

Joint Cooperative Committees

Such mutuality of interest prevails between some organizations and ASCE that standing "joint cooperative" committees have been established from time to time. These liaisons with AGC, ASCET, and with AIA from 1933-68 have previously been mentioned. More recent joint committees have been established with the American Institute of Planners (AIP) from 1959-63, the American Railway Engineering Association (AREA) in 1966, and the American Public Works Association (APWA) in 1968.

Close working relationships with other engineering and related bodies have prevailed, of course, without formal organizational structure. Among such societies are the American Road Builders Association, the American Water Works Association, American Concrete Institute, Prestressed Concrete Institute, American Institute of Steel Construction, American Iron and Steel Institute, Western Society of Engineers, American Association of State Highway Officials, American Arbitration Association, American Geophysical Union, American Institute of Aeronautics and Astronautics, American Nuclear Society, Institute of Traffic Engineers and many, many others.

ASCE "cooperation in the common interest" was also implemented with many other organizations through the technical divisions and the local sections.

International Engineering Relationships

As the first national engineering entity in the United States it is not surprising that ASCE has had a basic role in the development of the engineering profession in the nation. The Society has been equally influential, however, in the furtherance of engineering relationships throughout the world. The record is expansive both in time and in geography.

The Corresponding Member grade authorized in the original Constitution gave ASCE an international dimension from its beginning. Such members were non-residents of the United States, eminent or knowledgeable in an engineering specialty. They paid no dues, but were required to correspond with the Society at least once a year.

The first international interaction occurred as a communication in 1874 from the Association of Engineers and Architects of Austria, which proposed an exchange of publications and invited ASCE members to visit its headquarters in Vienna. The Society accepted, and took initiative itself in extending courtesy to foreign engineers attending the Philadelphia Centennial Exposition in 1876. A general card of introduction and "List of Members" were authorized for issuance to duly accredited members of foreign societies, to assist them in their American contacts.

A series of exchanges of visits were a consequence of the International Engineering Congresses held in Paris in 1889, Chicago in 1893, Paris in 1900, St. Louis in 1904, and San Francisco in 1915. Fifty-two ASCE members and 33 wives were part of the delegation that visited France, England, and Germany incident to participation in the 1889 celebration of the Centennial of the French Republic in Paris. Highlights of the tour were two events in England: An Institution of Civil Engineers dinner in the famous London Guildhall and "the special permission given by Her Most Gracious Majesty, the Queen, to visit and inspect her Royal Palaces and Domains at Windsor and in the Metropolis."

Noting the possibility of ASCE participation in the Paris Exposition of 1900, the Institution of Civil Engineers invited the Society to hold a full-scale convention in the London headquarters of the Institution. Thus, after enjoying the hospitality of the Société des Ingenieurs Civiles de France in Paris, 68 ASCE members moved on to London for the 32nd Annual Convention of the Society. The business meeting was limited, but a fine technical program was featured. This time the party was received by Her Majesty, the Queen, at Windsor Castle and by the Earl and Duchess of Warwick at Warwick Castle. Again, the ICE held a reception at the City of London Guildhall.

ASCE went all out to reciprocate for foreign guests attending the International Engineering Congress arranged and financed solely by ASCE in connection with the 1904 Louisiana Purchase Exposition. Each visitor was given a letter of introduction, and host committees were set up in 16 major cities to accommodate the guests in the course of their travels. A delegation of 104 members and guests from the ICE accepted the invitation of the Society to take part in a four-day program of technical visits and excursions in and near New York City, climaxed by a dinner on September 16, 1904, at Delmonico's Restaurant. The range and eloquence of the oratory on this occasion, all documented in *Proceedings*, November 1904, are believed to stand unsurpassed, before or since!

A lasting result of these international experiences was the Board of Direction action in 1905 authorizing a communication ". . . to all engineering societies, both in this country and abroad, stating that this Society would be glad to welcome any of their members to the use of this House and Library, and to any of its regular meetings." Sixteen foreign societies promptly accepted and reciprocated these courtesies. The Annual Yearbook carried a listing of them until 1949, by which time 30 foreign societies were included.

This informal relationship was replaced in 1950 by a Bylaw providing for a reciprocal membership agreement, for the Secretary or equivalent officer, with the Founder Societies and a list of 20 engineering societies in Europe and Latin America. This arrangement was expected to establish a continuing working relationship between the executive officers of the societies, with a view toward greater cooperation as well as broader service to visiting members. The idea was sound, but the purpose was better served through the personal contacts made possible later by international organizations.

An open policy on the establishment of ASCE sections in foreign countries prevailed until 1958. Sections were formed in Panama (1931), Venezuela (1947), Brazil (1948), Mexico (1949) and Colombia (1957). A review of policy in 1958 elicited the conclusion that every nation needs a strong engineering profession of its own in order to best develop its natural and human resources. To comply with this premise, it was further concluded that ASCE should encourage its members abroad to become a part of the engineering community of the country of their residence, and that ASCE should not compete in any way with a national engineering society in its own country. This led to two important developments: A decision to suspend any further authorization of ASCE sections outside the United States, and a requirement that a candidate for membership from a foreign country must hold the equivalent grade of membership in a recognized national engineering society in his own country. These measures were well received abroad.

An informal "Overseas Unit" plan was introduced on a trial basis in Australia and Pakistan only for the identification and communication of ASCE members in those countries. This was not too successful. The Sections in Brazil and Venezuela were liquidated in the early 1960's.

Institution of Civil Engineers (Great Britain)

From the frequent and respectful references to the ICE on the part of early ASCE leadership, it is evident that there is a first-order kinship between the two organizations. ICE was founded in 1818, receiving its royal charter ten years later. It was a virile body of some 700 members when ASCE appeared on the scene, and a stimulating model for its American counterpart.

An early gesture of courtesy and friendship was the contribution of the first twenty volumes of the ICE *Proceedings* to the ASCE library in 1882. The exchanges of visiting delegations in connection with the several international congresses held between 1889 and 1915 (page 327) established a spirit of friendship and rapport that has prevailed without interruption ever since.

In the early 1900's the Institution held periodic General Engineering Conferences, to which ASCE members were always welcomed. The first of these held after World War I in 1921 became something of a special occasion when it was attended by a delegation from the four American Founder Societies. At this time the John Fritz Medal was conferred upon Sir Robert A. Hadfield in London and upon Charles P. E. Schneider in Paris.

The New York World's Fair in 1939 was to have been the setting for a British-American Engineering Congress, as a joint venture of ASCE, ICE and the Engineering Institute of Canada. After an outstanding technical and social program had been arranged, unfortunately it became necessary to cancel the event almost on the eve of its opening. The precarious state of international relationships in Europe in September 1939 made it impossible for the ICE to participate.

After 1949 ASCE and ICE found opportunities to interact in the course of their participation in international federations such as EUSEC (page 335) and WFEO (page 336). One result was the 1969 Conference on World Airports in London, in which the ASCE Aero-Space Transport Division took part. Another was a plan to hold biennial joint conferences on topics of timely mutual concern.

The first ICE-ASCE Joint Conference was held in Bermuda in 1970, on the theme "The Engineer in the Community." It was highly successful in meeting its objectives in every way. The second, at Disney World, Lake Buena Vista, Florida, in 1972 featured the theme "Public Works and Society." Eminent panelists with outstanding discussions in both conferences produced published proceedings of high reference value. Continuation of such conferences will insure close contact between ICE and ASCE in the future.

Société des Ingenieurs Civiles de France (SICF)

ASCE was the beneficiary of the hospitality of the SICF on the occasions of the Paris Expositions in 1889 and 1900, and the French "Congress General du Genie Civil" in 1919. The latter event gave rise to a strong interest in formation of a permanent Franco-American Engineering Committee. The French representatives were appointed on the spot, and the members of the American delegation to the Congress agreed to serve temporarily until their sponsors (the four Founder Societies) could designate permanent members.

The ASCE Board of Direction acted promptly to endorse the new committee and to name members to serve with similar appointees from the other Founder Societies as the American side. Similar enthusiasm was not shown by the other Founders, however, and there is no indication that the joint committee was ever fully organized. The Annual Yearbook of ASCE listed representatives on the committee until 1929, but there is no record of activity.

An American Section of the SICF was established in New York City some time prior to 1923. The periodic dinner meetings of the American Section did not conflict in any way with the operations of ASCE, and courtesy invitations were always sent to the Secretary of the Society. The 75th anniversary of the SCIF in 1923 was celebrated by the American Section in a joint meeting with the Founder Societies.

The American Branch of SCIF was still active in 1974.

Engineering Institute of Canada (EIC)

Only about a dozen Canadian civil engineers were included in the ASCE roster in 1881 when the Society held its 13th Annual Convention in Montreal. It was at this meeting that ASCE became involved with the uniform standard time effort (page 228 et seq.). There was no national engineering society in Canada at that time.

The Canadian Society of Civil Engineers (CSCE) was founded in 1887 as "a vehicle of communication between engineers." Although reciprocity in courtesies to visiting members was established with ASCE in 1908, there was no proposal of joint activities until 1917. ASCE held Annual Conventions at Quebec in 1897 and at Ottawa in 1913, in which participation by the CSCE appeared to have been limited to the manning of local-arrangements committees.

The onset of World War I gave rise to a 1917 resolution by the ASCE Board of Direction expressing " . . . its desire that the two Societies should cooperate for mutual advancement to the greatest extent possible . . ." and further recording " . . . its approval of a plan of holding joint meetings of the two Societies at such times and places as may be found convenient . . ." This gesture was acknowledged in most cordial fashion by the Canadian Society, but the effort made to carry out its provisions was apparently superseded by the exigencies of World War I.

In 1918 the CSCE became the Engineering Institute of Canada (EIC), under the provisions of a new federal charter. When ASCE held a convention at Montreal in 1925 the Montreal Branch of the Institute extended hospitality, but there was no other EIC participation. The first full-scale joint convention was held at Victoria in 1934, and this was the pattern for highly successful meetings at Niagara Falls in 1942, Toronto in 1950, and Montreal in 1974. EIC was also a participant in the planned 1939 British-American Engineering Congress, which was cancelled because of events in Europe leading to World War II.

In 1968 a Conference on Great Lakes Water Resources was staged jointly by the EIC and ASCE at Toronto, to fill a need for delineation of international problems.

For more than 50 years EIC and ASCE have exchanged representation by their top officers at their annual meetings, and a high order of mutual respect and understanding has derived from these contacts. For example, although there was no formal agreement, it was always understood that ASCE would not establish local sections or student chapters in Canada. It will be noted that this is in full accord with the international relations policy adopted by ASCE in 1966.

The absence of the restraints of a formal agreement made it possible in 1960 to accommodate an enterprising group of Toronto civil engineers who wished to associate for professional and social purposes. ASCE joined with EIC and the British ICE to subsidize modestly the Toronto Area Joint Civil Engineering Group, which was highly successful in serving the Toronto members of the participating societies.

In 1969 EIC underwent extensive reorganization, which

provided for the creation within the Institute of "Societies" in major branches of engineering. Such constituent societies were formed for mechanical engineering in 1970, civil engineering and geotechnology in 1972, and electrical engineering in 1973. At the first meeting of the new Canadian Society for Civil Engineering a spirit of complete rapport was established with ASCE. Almost 1,300 ASCE members were residents of Canada at that time.

Latin-American Relationships

The importance of interaction between the national engineering societies of the Western Hemisphere was realized in the leadership of the Founder Societies as early as 1916. At that time a Pan-American Joint Committee was appointed "... to try and form some closer relationship between engineers of North and South America." The movement was not well-timed, however, because of the imminence of World War I. Overtures to various societies in South America were not encouraging, and the committee was disbanded.

Six years later, Fred Lavis, M. ASCE, who had served as secretary of the 1916 joint committee, urged that a delegation of North American engineers endeavor to make the 1922 Brazilian Centennial the occasion for an organizational effort. Secretary Dunlap was instructed to confer with Mr. Lavis and representatives of the other Founder Societies, and their efforts brought about significant participation by North Americans in the Rio de Janeiro Congress. Although there were discussions during the Congress of a plan for structuring cooperation, there was no action toward that end.

Responsive to the encouragement of an Inter-American Development Commission in 1942, the Founders again formed a Committee on Inter-American Engineering Cooperation. A year later ASCE pledged its cooperation with the engineering societies of South America in the conduct of biennial or triennial hemispherical engineering congresses. These discussions came to realization at a convention in São Paulo, Brazil, in 1949, although the North American representatives favored a somewhat simpler organization than that provided for in the tentative constitution.

Finally, on April 19-22, 1951, the Union Panamericana des Associaciones des Ingenieros (UPADI) held its first convention in Havana, Cuba. The constituent societies of EJC with the EIC represented North America, and twelve national societies represented Latin America. The constitutional aims of UPADI were:

> "To encourage, promote, expand and guide the work of the engineers of the Western Hemisphere; to encourage the holding of periodical Pan-American engineering conventions; to promote individual and collective visits, interchange of teachers, lecturers, engineering literature, engineers and students; and to organize connections between engineering associations at administrative level for the purpose of advancing their common purposes."

UPADI held twelve biennial conventions in the period 1951-74, always with strong participation by ASCE. From 1965-72 the Executive Director was a member of the UPADI Board of Directors, and the Society unfailingly supported every proposal toward engineering comity in the Western Hemisphere through the medium of UPADI.

With official representation in UPADI limited to only two North American entities (EJC and EIC), as compared to as many as 23 Latin American organizations, the programs and meetings reflected the same imbalance. The majority of the operating budget was borne, however, by the North American bodies. A proposal initiated by ASCE in 1969 urged a review of structure to enable greater technological and professional input from North America.

European Relationships

Following the close of World War II, there developed a feeling in Europe that there was need for a mechanism to engender communication and cooperation among the national engineering societies. To this end the three chartered engineering institutions in Great Britain, ICE, IME, and IEE, issued an invitation to fourteen societies to convene in London in October 1948. This assembly laid the organizational foundation for the Conference of Engineering Societies of Western Europe and the U.S.A., which emerged from a reconvened meeting in London in September 1949. The acronym "EUSEC" soon came to identify a dynamic and very effective agency for international rapprochement in engineering.

ASCE, ASME, AIEE, AIME, and AIChE were charter members of EUSEC, together with the three British institutions and national societies in Austria, Belgium, Denmark, Finland (two societies), France, West Germany, Italy, the Netherlands, Norway, Sweden, and Switzerland. Greece, Ireland, Portugal and Spain affiliated later. The EUSEC bylaws specified the president and secretary of each society as official delegates, thus bringing together the policy leaders and the continuing executive officers.

The accomplishments of EUSEC in the 22 years of its existence covered the gamut of such functions as engineering education, standards of professional practice, information documentation and exchange, manpower studies, training of technicians, exchange of membership services, international registration practices, etc. In addition, of course, participation in EUSEC affairs created personal acquaintanceship and day-to-day cooperation among the executive officers of the societies to an extent never known before.

One professional contribution of EUSEC, however, stood out above all others. This was the 1961 *Report on Education and Training of Professional Engineers,* a comparative study of systems of engineering education in 19 countries. It was by far the most comprehensive international study of engineering education undertaken up to 1974. Chairman of the committee that developed the report was Dr. Thorndike Saville, Hon. M. ASCE, and its secretary was the Executive Secretary of ASCE. The latter negotiated the $30,000 Ford Foundation grant that was matched by grants from the Organization for Economic Cooperation and Development and the EUSEC societies to finance the project.

EUSEC held biennial General Assemblies, the 1958 assembly being held in New York. In 1956 the Executive Secretary of ASCE was designated General Secretary of EUSEC. He filled this post until 1963, when he was elected chairman of the federation. He was still serving in this capacity in 1968, when UNESCO proposed that the governing council of EUSEC be made the nucleus of a commission to create a world federation of engineering organizations along the EUSEC pattern. The challenge was accepted, with the understanding that EUSEC would be disestablished when the viability of a world federation was clearly demonstrated. This took place in 1971.

Worldwide Relationships

The 1921 visit to England and France by a delegation of the American Founder Societies (page 329) aroused spirited discussion of the desirability of an infrastructure to bring together engineers throughout the world. The spark was kept alive by the Federated American Engineering Societies, and a plan for a world engineering federation was put before the 1929 World Engineering Congress in Tokyo. The proposal resurfaced after World War II in the formation of the World Engineering Conference in Paris in 1946. ASCE, ASME, and AIChE acted in 1947 to set up a U.S. National Committee of the World Engineering Conference, under the auspices of EJC. At this point, unfortunately, the venture subsided, for reasons unknown.

This was the background for the movement toward a world organization undertaken by the EUSEC Advisory Committee, with the support of the United Nations Educational, Scientific and Cultural Organization (UNESCO). The first meeting of the Organizing Commission was held in Paris in April 1966. Eastern Europe and Latin America were represented together with the EUSEC countries. As chairman of the EUSEC group, the Executive Secretary of ASCE was designated chairman of the Organizing Commission.

The Commission met twice in developing a draft constitution, the second session being required to overcome objections raised by the British institutions. In March 1968 the

World Federation of Engineering Organizations (WFEO) held its inaugural General Assembly, with 61 nations represented. The presence of ASCE was manifest through its Executive Secretary, who gave the keynote address and was elected to the WFEO Executive Committee for a four-year tenure.

With its secretariat established in the Institution of Electrical Engineers in London, WFEO has assumed the role filled so well by EUSEC for its constituent societies. Successful General Assemblies were held in Paris in 1969, Varna, Bulgaria, in 1971, and New York in 1973. At the latter meeting there was still some uncertainty as to the relationship between WFEO and the regional international bodies such as UPADI and the Commonwealth Engineering Council of the U.K. Barring a serious lapse in leadership, however, WFEO appeared in 1974 to be soundly founded.

One aim of the 1959 policy on international relations sought to encourage engineers professionally engaged in a foreign country to become a part of the engineering community in that country. To implement this, ASCE initiated in 1963 an Intersociety Agreement under which a national society, for a modest fee, would extend certain privileges to visiting engineers from other countries. These privileges included: (1) Registration in a Guest Mailing List and invitation to all national and local meetings of the host society, (2) an identification card or "Professional Passport" for introduction purposes, (3) a subscription to the regular periodical magazine of the host society, etc. The plan was reciprocal between the signatories.

In 1974 ASCE had eighteen Intersociety Agreements in force with national societies around the world. The principle of this ASCE innovation was endorsed by EUSEC, and has been adopted by a number of societies abroad.

International Technical Federations

ASCE has welcomed every opportunity to collaborate with the several international technical federations engaged in areas of civil engineering concern. Some of these liaisons go back to the early 1900's, as those with the International Navigation Conference and the International Association for Testing Materials. In 1923 the Society assigned a task force "to formulate a basis of cooperation between International Technical Bodies."

More recently, a number of U.S.A. national committees of such international associations have been spawned in the Technical Divisions of the Society. Among the international bodies in which ASCE has been actively represented are the World Energy (formerly Power) Conference, Commission on Large Dams, International Association for Theoretical and Applied Mechanics, International Association for Bridge and Structural Engineers, International Association for Soil Mechanics and Foundation Engineering, International Association for Water Pollution Research, International Association for Earthquake Engineering, International Association for Shell Structures, and the International Commission on Irrigation and Drainage.

The Bridge of International Cooperation

A young member of the Society, after a highly productive series of technical visits in Europe, reported on the "bridge of cooperation" that has been built by the international activities of ASCE (*Civil Engineering*, March 1973, p.61). He stated that the introduction provided by the Society "was to have remarkable significance; it would have a magical effect across Europe, opening doors and leading to the right people."

His closing admonition to the younger members of the Society is quoteworthy:

> "Consider the bridge of international cooperation. Consider the broadened horizon it gives. Consider that maintenance is also necessary."

* * *

There was a vast spread between the reaction of a group of Rochester members to the proposed Joint Engineering Building in 1903, in part as follows:

> ". . . Our A.S.C.E. is an organization that is recognized by the whole civilized world as a power. We have our library and our property and are self-supporting, and it does not strike [us] favorably that we should lower our standard of qualifications and throw all this into a common fund to be participated in by all societies, however ephemeral or social or scientific, that may from time to time come into existence . . ."

and the 1958 policy of the Society which began:

> "In ASCE, unity means cooperative effort in the common interest. The Society welcomes the opportunity to work with other organizations in the attack on and solution to common problems . . ."

The record shows conclusively, however, that the gregarious nature of ASCE in its intersociety relationships far transcended its occasional aloofness. The original reluctance of the Society to affiliate with United Engineering Society (1903) and the Federated American Engineering Societies (1920) was more than countered by its strong involvement in the successors to those organizations, and in a host of others.

A number of technical specialty organizations trace their origin to ASCE support in one form or another, as does the multidisciplinary body, ICED. No other engineering organization in the world surpassed the ASCE contribution to cooperation between engineering societies internationally.

These activities, however, were structurally organized, from joint committees to federations. Just as important, perhaps more so, was the informal relationship between ASCE and other organizations on a day-to-day, current-problem basis. This close, impromptu interaction was especially fruitful among the Founder Societies, but it also extended to others including AICE, NSPE, CEC (later ACEC), and many more. Communication and interchange at this level may have constituted real "working unity," all the more effective because it was unhampered by any kind of bureaucratic regimen.

Considering the character of the engineering profession, ASCE has committed itself creditably — and at times in distinguished fashion — in relating constructively with other organizations. In 1974 the extroverted attitude of the Society was firmly established, both at home and abroad, and "cooperative effort in the common interest" was a way of life.

CHAPTER VIII

FROM THE PAST TO THE FUTURE

As he stepped down from the presidency of ASCE in 1921, Dr. Arthur P. Davis quoted the following:

> "Over the past not Heaven itself has power.
> What has been has been,
> And I have had my hour."

The foregoing chapters document "what has been" in the American Society of Civil Engineers for almost a century and a quarter, to provide a factual historical reference. Detail and repetition have been suffered in order to make the record accurate and complete. If the story at times is heavy and tedious, it is because ASCE was sometimes a rather dull organization, just as it was vital and dynamic on other occasions.

Beginning as a handful of earnest veteran engineers, ASCE was very much a struggling New York City coalition until the nationalization process began with the Louisville Convention in 1873. The geographical expansion did not, however, extend to the attitudes of the Society, which remained conservatively introverted through the 19th century. Lack of compromise became almost traditional in this period. Paradoxically, the democratic requirement of membership ballots on important measures was coexistent with an autocratic control exercised by the Board of Direction in determining which issues were put to referendum and the manner of their presentation. This ambivalence prevailed until 1921.

The accelerated urbanization of America at the turn of the century brought government into the sponsorship and financing of major public works in addition to the powerful canal and railroad companies that served these functions earlier. This was accompanied by the entrance of a growing number of young civil engineers into the profession and the Society. More progressive interpretations of the aims of ASCE were not long in coming, and a gradual transformation began. World War I, the Great Depression, and World War II placed demands upon the Society that accelerated the liberalization of concept and action, and maturation was brought to completion by the post-Sputnik explosion in technology. Thus, there was little resemblance in ASCE between the pluralism of 1900 and the stability of 1974.

The productive era of ASCE began about 1870, and thenceforth the Society was without question importantly instrumental in the advancement of the industrial revolution in America. By ". . . the establishment of a central point of reference and union for its members," and especially through the medium of its technical publications, the Society provided much of the leverage that elevated American civil engineering technology from its empirical beginning to its present scientific sophistication. ASCE contributions to the advancement of theory and technique in the fields of engineering mechanics, structures, soil mechanics, hydraulics, and water resource development have been unsurpassed throughout the life of the Society. Equally significant contributions in sanitary engineering and transportation engineering made in the early years of the Society have more recently been shared with other specialty organizations in these fields.

It is axiomatic that a nation must have a strong engineering profession if it is to be able to develop fully its natural and human resources, and here ASCE has played an unique and vitally important role. Not only was the engineering profession in America first identified as a national entity by the Society, but it also provided leadership in establishing the standards of education, competence, and mode of practice that characterize a true profession. Having no predecessor, ASCE had to initiate and develop these basic elements by trial and error. At the same time it was creating the economic, social, and political climate appropriate for the practitioner of civil engineering. That it was a good model is amply demonstrated by the strength and effectiveness of the many national engineering organizations that have found guidance and inspiration in the Society.

Despite its pioneering role, ASCE found the means on many occasions to represent the public interest in affairs of national concern. Such public-service functions in the past were necessarily constrained by limited resources and preoccupation with internal problems. There is a vast opportunity here for the Society to distinguish itself in the years ahead.

Behind every significant progression or accomplishment of ASCE there were one or more individuals imbued with perspective, initiative, persistence, and other personal qualities that distinguished themselves as leaders. Many of these, although not all, ascended to the presidency or other high office in the Society. Some were visionaries, who created ideas; others were "exciters," who stimulated action; still others were the "doers," who turned ideas into realities.

The founder of ASCE, James Laurie, personified all of these attributes, as did Sandford Fleming, champion of the standard time venture (1881-1900), and John P. Hogan, who led the movement culminating in the federal public works programs of the 1930's.

The first concern for engineering education in ASCE was sparked by Professor Estavan A. Fuertes of Cornell in 1874, but it was not until 1907 that a real program was launched and carried through by Desmond FitzGerald's education committee. Samuel P. Whinery in 1893 and again in 1901 strove to initiate action toward a code of ethics, but it took another touch of the spurs by Percival M. Churchill in 1912 to accomplish that end. Later, Professor Daniel W. Mead gave impetus to the implementation of ethical standards and policies.

ASCE concern with engineering registration was ignited by S. C. Thompson in 1897-99, and pursued by Samuel P. Whinery in 1901. Mr. Whinery was also an outspoken proponent of the "democratization" of the Society brought about by the new Constitution in 1921.

The same Percival M. Churchill who supported the ethical code in 1912 was also successful, after five years of effort (1912-17), in generating action by the Society in providing employment services and salary studies. The employment-conditions programs were first carried to fruition, however, by E. P. Goodrich in the period 1927-39.

John C. Hoyt appears to have originated the concept of the technical divisions in the Society, and this move was planned and promptly executed under the direction of Professor Arthur N. Talbot in 1922. Similarly, the local section program formalized in 1921 had actually been envisioned in 1885 by Arthur M. Wellington.

In the realm of intersociety relations, the 1916 proposal by Elmer L. Corthell of the concept of the joint conference committee as a device for achieving "solidarity of the engineering profession" was significant. It was Secretary Charles Warren Hunt who saw this idea through to realization.

Up to this point, interpretive comment has been limited to that considered necessary to further the understanding of important events as they occurred. Care has been exercised to avoid bias in the interest of preserving credibility. In this chapter, however, indulgence in some personal evaluation is undertaken with a view toward constructive utilization of 122 years of valuable experience.

Most civil engineers will be gratified with the success that has marked much of the work of ASCE. Some may be surprised by the not infrequent failures, and the shortcomings of the Society from time to time. It must be remembered that any society is made up of people, and that people — even though well-intentioned — will sometimes procrastinate, obfuscate, err, and otherwise fail to utilize fully their collective capabilities. In ASCE it must also be noted that its leaders were men of substance who were obliged to give priority to their professional responsibilities in projects of great magnitude and importance, to which the interests of ASCE had to defer. This point is not made in apology; it is a simple fact.

There are lessons to be learned both from the successes and the failures. Those noted here are believed to be significant.

Society Objectives

As the need for professional leadership arose from time to time in ASCE, there was often difficulty in deciding upon the role that the Society might assume. The overall record shows that the greatest success accompanied those efforts in which the Society served as a catalyst in bringing about interaction on the part of outside interests and forces. The standard time venture (1882), the economic recovery legislation (1932), the Taft-Hartley law amendments (1947), and the research and public relations programs of the 1960's are examples of such catalytic effect.

The lesson here may be the programming of a role for the Society that is realistic and within the capacity of its resources. This may sometimes call for early decision as to whether ASCE should proceed unilaterally to attack a problem or whether it should enlist the interest and participation of other entities in a joint effort.

Management Observations

The Committee of Policy of the Society, in a report accepted by the Board on November 3, 1875, made a most perceptive observation:

> "While admitting the importance of revision of the laws governing the Society, and of some additions to them, the Committee cannot refrain from expressing a belief, that the danger in associations like this, as well as civil affairs, is that there be too much government and too many laws, rather than the contrary. It is undoubtedly better to postpone legislation until evils become so imminent as to unmistakenly threaten danger, than to attempt to provide for contingencies that do not yet exist."

This admonition would have been very much in order before 1951, when regimentation in ASCE reached such a state that the Constitution and Bylaws required supplementation by a set of Rules of Policy and Procedure. It was still in order in 1974.

Countless times in the past 122 years ASCE has met its problems by the formula, "appoint a committee." From the record it is clear that there must have been a better way. Not only was the percentage of success something less than might have been desired, but also the cost in membership effort, staff input, and in money was excessive.

Bureaucracy is the worst enemy of efficiency. In government a new problem means a new agency and here, also, there is usually mediocre performance at high cost. The approach in industry, however, in attacking a problem or initiating a new enterprise is "to find the right man." The oft-demonstrated success of this management concept merits its consideration in ASCE.

In execution, the Board of Direction would, upon recognizing a problem, request its executive officer to propose a course of procedure. He would explore the various alternatives that might include the use of existing staff, the employment of staff specialists on an ad hoc basis, the engagement of outside consultants and services, or even the appointment of a committee of members. His recommendation to the Board would be accompanied by budget requirements and a time-table, so that action might be authorized. Responsibility for implementation would rest upon the executive officer and those involved in his plan of action. Upon completion, all extra commitments made to the venture would be terminated. Otherwise, a staff bureaucracy would result.

This process requires strong management capability in the staff, with delegation of authority and responsibility, and adequate resources to get the job done.

Another management imbalance of long standing is the disproportionate under-representation of the civil engineer in government in the leadership and direction of the affairs of the Society. This is actually a side effect of a much deeper concern.

A high order of professional orientation and motivation on the part of all employees of government is very much in the public interest, particularly in view of the increasing influence and power of the bureaucracy in interpreting and enforcing legislated policy. It is vitally important that such professional attitudes prevail among the civil engineers who are so strongly involved in the implementation of significant programs at all government levels. This is a purview of ASCE.

Elected and politically appointed officials must somehow be impressed with the essentiality of professionalism as a staff quality. A long and painstaking process of education directed to political leadership and the public will be required. Only in this way will professional attitude and stature become recognized as valuable personnel qualifications for those in government service.

In the meantime, civil engineers in public practice must be brought into the mainstream of ASCE to considerably greater extent than in the past. Agencies must be induced to allow time and travel expense for professional staff to attend meetings and to participate in activities that are relevant to their professional development. Nominating committees at all levels of the Society will do well to give consideration to the proportionality of the various major practice areas of the membership (public, private, industry, education, etc.) in reviewing qualifications of candidates for office.

Using the Membership Resource

The inadequacy of the committee approach to problem-solving in ASCE is not necessarily a reflection upon the member called upon to participate in the system. In many cases the charge given a committee was too much to expect of a group of busy men as an extracurricular duty. For members of the Board of Direction such demands were often compounded by time-consuming appointments to several committees.

Too often, also, members were expected to resolve problems of which they themselves were a part. Objective analysis of a situation in which one has a deep self-interest is, at best, difficult if not impossible. The ad hoc approach employing independent professional expertise would circumvent this dilemma, although the solution of problems in this manner could not always be expected to satisfy completely all divergent interests.

The greatest value of the tremendous brainpower potential of the membership resource has not been utilized effectively in ASCE, and this avenue for professional service deserves study. This problem-solving capability could best be applied to the problems of society rather than those of "the Society." New means for mobilizing this brainpower for the guidance of political and public policy decision-makers must be developed.

Doing this will not be easy. The gathering of a body of expertise to attack a problem area involves a form of selective osmosis that will accept those who are knowledgeable in the area and capable of articulating a rational policy and action plan, while excluding those who are equally knowledgeable and articulate but who may be otherwise unqualified by reason of bias or self-interest. Such a process will entail the highest order of altruism and discipline — but are these not the prime attributes of a profession?

Engineering Interaction

The futility of the long quest for engineering unity as a structured entity is starkly clear from past experience. Is the engineering profession truly a "unit" that can be regimented to think and act in unison? It is actually more realistic to think of it as a dynamic, pluralistic complex, with a wide spectrum of diverse aims, interests and capabilities. As stated by Secretary Emeritus William N. Carey, "The various branches of engineering, when mixed, always have formed an emulsion, never a true solution."

It seems logical that effective interaction among the various branches of engineering can hardly be expected until each of the branches has crystallized its own professional objectives and policies. Also, in order for joint ventures to be productive, the separate branches must each have found the ways and means to mobilize and utilize the expertise represented by its own membership. Because of these underlying deficiencies, past unity movements have brought on frustration, which in turn has led to the bureaucratic treadmill in which the spinning of organizational wheels gives the illusion of progress.

Past experience has indicated repeatedly that the effectiveness of inter-engineering relationships is proportional to the degree of communication, and inversely proportional to the degree of formal regimentation. Best results have been realized with the "Joint Conference Committee" approach, in which policy and action are personified at the levels of the presidency and the chief executive officer. The JCC serves only to define the problem and to decide upon strategy, including the identification of participating organizations. The problem-solving operation is ad hoc using whatever personnel and resources are best suited to the task. Permanent organizational structure is held to a minimum.

This general plan does not preclude a joint action by all members of the coalition in the occasional case where all may have a common interest and may be able to agree upon a position. It must provide flexibility, however, in permitting some societies to join in a venture without involving others having no real concern with the matter at issue.

Looking ahead from 1974, it seems likely that "engineering unity" will be much less important than "engineering community." It is no longer enough that the branches of engineering communicate and interact only among themselves; they must now be willing and able to establish working rapport with a host of non-engineering disciplines. The ad hoc approach to specific, well-defined problems will accommodate the participation and input that can be provided advantageously from such fields as anthropology, sociology, economics, political science, law, medicine, and fine arts.

Nothing less than the integrated effective collaboration of our best problem-solvers will suffice in overcoming the massive dilemma confronting America and the world in the management of population, food, energy, and of land, water, and air resources. The need is urgent, and there is an opportunity and a challenge here for ASCE leadership at a crucial time.

Ethical Policy

The founders and leaders of ASCE always have been strongly imbued with the service ethic as individuals. However, a collective approach to ethical standards as a function of the Society was long in developing. When it did come, it was focused inwardly upon internal relationships rather than the responsibilities of the civil engineer to the public. In recent years the growing acknowledgment within the Society of societal responsibility has been salutary, but there is need for greater specificity of purpose here. This might be accomplished in a formal manner by broadening of the Code of Ethics.

The original "short form" code of ASCE was adopted and has been maintained in that format because of its amendability to enforcement action. This is commendable, to be sure, but does this advantage necessarily exclude standards of more general import that might never actually be enforced as the letter of the law? The ethical standards of other professions are not similarly restricted.

Expansion of the Code of Ethics might give greater emphasis to "professional community" responsibility, to recognition of societal impacts in the course of professional practice, and to guidance of the individual practitioner who may be called upon to perform services that he may hold to be questionable to the public welfare.

It is timely that the ethical standards of ASCE should delineate the humanistic and social considerations that are inherent to the practice of civil engineering in this fast-changing world.

Civil Engineering Education

The uniqueness of the civil engineer among the representatives of other branches of the engineering profession has been highlighted by this historical study. "The Civils are different" was a comment often heard in intersociety discussions, yet there was little thought given to the reconciling of the variance in professional attitude and climate that set the civil engineer apart from other engineering disciplines.

In the realm of education, it appears inevitable that civil engineering must adopt some aims and parameters that better fit the areas of technology and the mode of practice that are specific to that field. It is unrealistic to carry on the fallacy that the engineering of public works and services must be fabricated from the same template as the engineering of industrial processes and products.

It is not proposed that civil engineering education should be segregated altogether. ASCE should, however, assume a stronger role in the curriculum and accreditation functions presently being served for all engineering by ECPD. The strong moves in engineering education being made by the Society in 1974 should be carried to fruition.

There may also be a message in the 1916 survey made by Professor C. R. Mann, which indicated that the preponderant quality sought in the engineer was "character," encompassing the elements of integrity, responsibility, resourcefulness, and initiative. Professor Mann properly responded that development of these personal values was not a responsibility of the educator. But this does not mean that the engineering education experience must be limited to the cold world of mathematics and science alone. It is obvious in 1974 that the engineering student must be given a basic understanding of the society in which he is expected to provide technological solutions — an evaluation of his profession in terms of prevailing and emerging cultural values.

Engineering education should prepare the student for professional practice — for "the pursuit of a learned art in the spirit of service." Professional attitude and enthusiasm cannot be taught, but they can be instilled in the student by dedicated professional teachers, or by academic programs that bring the student into contact with highly-principled professional practitioners.

ASCE should be able to do something about this.

Historical Perspective

Maturity comes with time, but it is not necessarily automatic. Maturation is a combination of growth and development; in people and organizations it comes through learning by experience — through the beneficial utilization of the successes and mistakes of the past.

ASCE has long since come of age, and it has always been proud of its stature and traditions in American engineering. Members of the Society have also cherished their professional identification and association with the great civil engineers of bygone days who had so much to do with the ascendence of the United States of America to world leadership.

Detailed review of the history of ASCE reveals, however, a surprising number of cases where ineffective measures were repeated time and again, and where lessons might have been learned from fruitful endeavors. The Society is rich in experience, but it has not taken full advantage of the substantial investment in membership energy, staff input, and financial resource that is represented in its history. Historical perspective of its purposes and goals is essential to the professional maturity of ASCE.

The establishment in 1965 of the Committee on History and Heritage of American Civil Engineering was one of many examples of ASCE leadership in the engineering profession. Sponsorship of this book is another manifestation of the realization by the Society of the rich heritage of its past. Much is yet to be done, however, in the dissemination of this message to the public, the student, and the practicing civil engineer. Historical perspective must be integrated into the professional education experience, as well as into the public information and continuing education programs of the Society. Architecture has provided a fine example in its insistence upon a knowledge of the history of that profession as one requisite in the licensing process.

An awareness of historical origins and models of behavior is an exceedingly important element in professionalization. Every civil engineer will enhance his potential by cultivating such awareness in his own attitudes and conscience.

APPENDIX I

ROSTER OF PRESIDENTS

As terms of presidents extend a short period in a prior or subsequent year they appear in this table as the year, or years, covering the major part of the term.

(The names of deceased past Presidents are printed in capital and lower case letters.)

Laurie, James	1853-1867
Kirkwood, James Pugh	1868
McAlpine, William Jarvis	1868-1869
Craven, Alfred Wingate	1870-1871
Allen, Horatio	1872-1873
Adams, Julius Walker	1874-1875
Greene, George Sears	1876-1877
Chesbrough, Ellis Sylvester	1878
Roberts, William Milnor	1879
Fink, Albert	1880
Francis, James Bicheno	1881
Welch, Ashbel	1882
Paine, Charles	1883
Whittemore, Don Juan	1884
Graff, Frederic	1885
Flad, Henry	1886
Worthen, William Ezra	1887
Keefer, Thomas Coltrin	1888
Becker, Max Joseph	1889
Shinn, William Powell	1890
Chanute, Octave	1891
Cohen, Mendes	1892
Metcalf, William	1893
Craighill, William Price	1894
Morison, George Shattuck	1895
Clarke, Thomas Curtis	1896
Harrod, Benjamin Morgan	1897
Fteley, Alphonse	1898
FitzGerald, Desmond	1899
Wallace, John Findley	1900
Croes, John James Robertson	1901
Moore, Robert	1902
Noble, Alfred	1903
Hermany, Charles	1904
Schneider, Charles Conrad	1905
Stearns, Frederic Pike	1906
Benzenberg, George Henry	1907
Macdonald, Charles	1908
Bates, Onward	1909
Bensel, John Anderson	1910
Endicott, Mordecai Thomas	1911
Ockerson, John Augustus	1912
Swain, George Fillmore	1913
McDonald, Hunter	1914
Marx, Charles David	1915
Corthell, Elmer Lawrence	1916
Herschel, Clemens	1916
Pegram, George Herndon	1917
Talbot, Arthur Newell	1918
Curtis, Fayette Samuel	1919
Davis, Arthur Powell	1920
Webster, George Smedley	1921
Freeman, John Ripley	1922
Loweth, Charles Frederick	1923
Grunsky, Carl Ewald	1924
Ridgway, Robert	1925
Davison, George Stewart	1926
Stevens, John Frank	1927
Bush, Lincoln	1928
Martson, Anson	1929
Coleman, John Francis	1930
Stuart, Francis Lee	1931
Crocker, Herbert Samuel	1932
Hammond, Alonzo John	1933
Eddy, Harrison Prescott	1934
Tuttle, Arthur Smith	1935
Mead, Daniel Webster	1936
Hill, Louis Clarence	1937
Riggs, Henry Earle	1938
Sawyer, Donald Hubbard	1939

Hogan, John Philip 1940
Fowler, Frederick Hall 1941
Black, Ernest Bateman 1942
Whitman, Ezra Bailey 1943
Pirnie, Malcolm 1944
Stevens, John Cyprian 1945
Horner, Wesley Winans 1946
Hastings, Edgar Morton 1947
Dougherty, Richard Erwin 1948
Thomas, Franklin 1949
Howard, Ernest Emmanuel 1950
HATHAWAY, GAIL ABNER 1951
Proctor, Carlton Springer 1952
Huber, Walter Leroy 1953
TERRELL, DANIEL VOIERS 1954
Glidden, William Roy 1955
Needles, Enoch Ray 1956
Lockwood, Mason Graves 1957
HOWSON, LOUIS RICHARD 1958
Friel, Francis de Sales 1959
Marston, Frank Alwyn 1960
HOLCOMB, GLENN WILLIS 1961
EARNEST, GEORGE BROOKS 1962
FRIEDMAN, EDMUND 1963
BOWMAN, WALDO GLEASON 1965
CHADWICK, WALLACE LACY 1965
HEDLEY, WILLIAM JOSEPH 1966
ANDREWS, EARLE TOPLEY 1967
TATLOW, RICHARD HENRY, III ... 1968
NEWNAM, FRANK HASTINGS, JR. . 1969
NILES, THOMAS McMASTER 1970
BAXTER, SAMUEL SERSON 1971
BRAY, OSCAR SIMON 1972
RINNE, JOHN ELMER 1973
YODER, CHARLES WILLIAM 1974

APPENDIX II

ROSTER OF OFFICERS

The figures of the year during which office was held are printed without the first two digits for the 1900's. Terms overlapping into prior or subsequent years are listed simply as the year, or years, of the principal service. For example, the Directors' terms from January 15, 1947 to January 18, 1950 appear as 47-49. Past Presidents who have served as Directors by virtue of their office are not listed in this table as Directors.

Following the adoption of amendments to the Constitution, in April 1950, a change occurred in the districting, and in the date of annual meetings. As a result, some terms of office ended in January (old system), at the same time that others ended in October (new system). In either case, this tabulation lists only the years of coverage. To accomplish the necessary transition in the succeeding few years, some terms were less than normal while others were extended.

(The names of deceased officers are printed in capital and lower case letters.)

Name	Pres.*	Vice Pres.	Secy.	Treas.	Director
ACKERMANN, WILLIAM CARL					72-74
Adams, Arthur Lincoln					07-09
Adams, Julius Walker	1874-75	1868-73			1853, 1876
Agg, Thomas Radford		43-44			38-40
Alexander, Randle Burette					57-59
Allen, Horatio	1872-73				
ALTER, AMOS JOSEPH					74-76
Alvord, John Watson					17-18, 19-21
Ammann, Othmar Hermann					34-36
Anderson, George Gray					21-23
Anderson, James Aylor					38-40
ANDREWS, EARLE TOPLEY	67				61-63
Andrews, Horace					08-10
Archbald, James					1887

*All Presidents who assumed office after October 1968 also served a prior one-year term as President-Elect.

a Assistant Secretary.

b Assistant Treasurer.

Name	Pres.*	Vice Pres.	Secy.	Treas.	Director
Arneson, Edwin Percival					36-38
Ayres, Louis Evans					38-40
Baillie, David Gemmell, Jr.					62-64
Bakenhus, Reuben Edwin					43-45
BAKER, WOODROW WILSON					60-62
BARBER, CHARLES MERRILL					61-63
Barbour, Frank Alexander					34-36
BARGE, DANIEL BLYTHEWOOD, JR.					67-69
BARRY, BENJAMIN AUSTIN					72-74
Bates, Onward	09	06-07			
BAXTER, SAMUEL SERSON	71				60-62
BAY, ROBERT DEWEY					74-76
Beahan, Willard					19-21
Beam, Carl Eugene			23-41a		
Becker, Max Joseph	1889				
Begien, Ralph Norman					23-24
Belknap, William Ethelbert					10-12
Bell, Gilbert James					27-29
Bellinger, Lyle Frederick		37-38			
BELT, ROBERT McCOLL					64-66
Benham, Webster Lance					48-50
Bensel, John Anderson	10	07-08			1899-01
Benzenberg, George Henry	07	01-02			1895-97
BESPALOW, EUGENE FREDERICK					63-65
BINGER, WALTER DAVID					52-54
Bissell, Hezekiah					05-07
Black, Ernest Bateman	42				32-34
Blair, Clarence Moore					40-42
BLAKESLEE, HAROLD LAW					49-51
BLESSEY, WALTER EMANUEL					72-74
Bogart, John			1877-91	1876-77, 1891-94	1873-75
Bogue, Virgil Gay					16
Boller, Alfred Pancoast		11-12	1870-71		1872
Bontecou, Daniel		15-16			1896-98
BOOTH, ARCHIBALD ALLAN KIRSCHNER					53-55
Boucher, William James				26-29b	
Boughton, Van Tuyl					42-44
Bouscaren, Louis Gustave Frederic		03-04			1881
Bowen, Oscar Sidney		24-25			
Bowman, Austin Lord					05-07
BOWMAN, WALDO GLEASON	64	58-59			49-51
Brackett, Dexter					08-10
BRANDOW, GEORGE EVERETT					63-65
Braune, Gustave Maurice					25-27
BRAY, OSCAR SIMON	72				64-66
Breed, Charles Blaney					43-45
Bres, Edward Sedley					39-41
Briggs, Josiah Ackerman					01-03
Brillhart, Jacob Herbst					27-29
BRITZIUS, CHARLES WESLEY					59-61
Brooks, Robert Blemker		49-50			39-41
Brown, Baxter Lamont					21-23
BROWN, BEVAN WOOD, JR.					73-75
BROWN, CAREY HERBERT		62-63			56-58
Brown, Paul Goodwin					24-26
Browne, George Hamilton					1895-97
Brush, Charles Benjamin		1892-93			1888-91
Bryan, Charles Walter, Jr.					45-47
Buchholz, Carl Waldemar					1899-01
Buck, Con Morrison					27
Buck, Henry Robinson					31-33
Buck, Henry Wolcott					61-63
Buck, Leffert Lefferts					1892-94
Buck, Richard Sutton					02-04
Budd, Ralph					29-31

Name	Pres.*	Vice Pres.	Secy.	Treas.	Director
Burdick, Charles Baker		41-42			35-37
Burpee, George William		52-53		42-54b	42-44
Burr, William Hubert					1894-96
Bush, Lincoln	28	24-25		15-16	12-14
Butler, Gordon Hubert					49-52
Cain, William					12-14
Cappel, Curry Glenn					48-50
CAREY, WILLIAM NELSON			45-55		41-43
Carlton, Ernest Wilson					54-56
CARTER, ARCHIE NEWTON					74-76
Carter, Edward Carlos					01-03
Cartwright, Robert		1899-00			1895-97
Casey, Thomas L.					1882
Cattell, William Ashburner					12-14
CHADWICK, WALLACE LACY	65				51-53
CHANDLER, ELBERT MILAM			21-22		
Chandler, Emerson Lawrence			48-60a	60-65b	
Chanute, Octave	1891	1880-81			1874-76, 85
CHENEY, LLOYD THHEODORE					65-67
Chesbrough, Ellis Sylvester	1878				1870
Chesbrough, I.C.					1855-67
Chester, John Needles		31-32			22-24
Chevalier, Willard Townshend					25-27
Christie, James					07-09
CHRISTENSEN, NEPHI ALBERT					62-64
Churchill, Charles Samuel		12-13			08-10
Clapp, Otis Francis					16-17
Clark, George Hallett					19-21
Clark, Jacob Merrill		1872-73			1870-71
Clarke, Eliot Channing					1889
Clarke, George Calbraith					11-13
Clarke, Thomas Curtis	1896	1871			1879
Cohen, Mendes	1892	1890			1888
Coleman, John Francis	30	18-19			15-17
Collingwood, Francis			1891-95		1873-76
Condron, Theodore Lincoln					23-25
Connor, Edward Hanson					13-15
Cooley, Mortimer Elwyn					14-16
Cooper, Theodore					1884-85
Copeland, Charles W.		1853-69			
CORBETT, DON MELVIN					55-57
Corthell, Elmer Lawrence	16	1889, 93-94			
Courtenay, William Howard					11-13
Cowper, John Whitfield					41-43
Craighill, William Price	1894				1892-93
CRANDALL, LIONEL LeROY					71-73
Craven, Alfred		16-17			03-05
Craven, Alfred Wingate	1870-71	1854-67			1853, 68-69, 1872-73
Crawford, Ivan Charles					35-37
Critchlow, Howard Thompson					45-47
Crocker, Herbert Samuel	32	19-20	20-21		15-17
Croes, John James Robertson	01	1888		1878-87	1877
Crosby, Benjamin Lincoln					1897-99
Crowell, Foster					1893-95
Crum, Roy Winchester					46-49
Cummings, Robert Augustus		20-21			14
Cunningham, John Wilbur		48-49			41-43
Curtis, Fayette Samuel	19	04-05			1895-97
Curtis, William Giddings					1890
DAMES, TRENT RAYSBROOK		71-72			60-62
Darling, William Lafayette					17-19
Darrow, Frank Tenney					21-23
DAVIES, CLARENCE W.E.					74-76
Davies, John Vipond					15-17
Davis, Arthur Powell	20				17-19

Name	Pres.*	Vice Pres.	Secy.	Treas.	Director
DAVIS, HARMER ELMER					61-63
Davis, Joseph Phineas		1884			1878,81-83
DAVIS, ROLAND PARKER		40			37-39
Davison, George Stewart	26	23-24			03-05
Dawson, Francis Murray					53-55
DAWSON, RAYMOND FILLMORE					54-56
Dean, Arthur Warren					37-39
DeBerard, Wilford Willis					38-40
Dennis, Harry Whiting		36-37			27-29
Dewell, Henry Dievendorf		34-35			25-27
Deyo, Solomon LeFevre		04-05			1898-00
Dickinson, William Dewoody					42-44
DOBBS, ELWOOD DeWITT					71-73
DODDS, ROBERT HUNGERFORD				73-b	65-67
Dohm, Edward Clarence					52
DORNBLATT, BERNHARD					60-62
DOUGHERTY, NATHAN WASHINGTON					43-45
Dougherty, Richard Erwin	48	44-45			28-30
Dresser, George Warren					1882
Dufour, Frank Oliver		33-34			27-29
Dunlap, John Hoffman			22-24		
Dunnells, Clifford George					40-42
Durkee, E. Leland		65-66			57-59
Duryea, Edwin					16-18
Dusenbury, Allan Theodore					30-32
DWYRE, BURTON GOLDING		62-63			51-53
Dyer, Arthur James		29-30			22-24
DYSON, GERALD ROLAND					71
Eads, James Buchanan		1882			
EARNEST, GEORGE BROOKS	62	53-54			50-51
Eckel, Clarence Lewis					56-58
Eddy, Harrison Prescott	34				28-30
Edwards, Dean Gray					43-45
Edwards, James Harvey					13-15
Ellis, John Waldo					04-06
Ellis, Theodore Grenville		1874-77			1879
ELSENER, LAWRENCE ALOIS		60-61			55-57
Elwell, Charles Clement					19-21
Ely, Theodore Newel					1892-93
Endicott, Mordecai Thomas	11	08-09			01-03
Enger, Melvin Lorenius					32-34
ERICKSON, DAVID LEONARD					47-49
Etcheverry, Bernard Alfred					34-36
EVANS, WESTON SUMNER, SR.					58-60
Fanning, John Thomas		10-11			1893-95
Farnham, Robert					24-26
Fay, Frederic Harold					17-18
Fenkell, George Harrison					23-25
Ferebee, James Lumsden		39-40			35-37
Finch, James Kip					36-38
Fink, Albert	1880	1878-79			
Fisher, Edwin Augustus					05-07
FitzGerald, Desmond	1899	1895-96			1892-94
Flad, Edward					11
Flad, Henry	1886	1883			
Flinn, Alfred Douglas					17-19
FOGG, ROBERT KNOWLTON					69-71
Foote, Arthur DeWint					10-12
Ford, James K.					1868-69
Forney, Matthias N.					1877
Forsyth, Robert					1887
Fort, Edwin John					18-20
Fowler, Frederick Hall	41				28-30
FOWLER, LLOYD CHARLES					71-73
FOX, ARTHUR JOSEPH, JR.					69-71
Francis, James Bicheno	1881	1870, 79-80			

Name	Pres.	Vice Pres.	Secy.	Treas.	Director
FRATAR, THOMAS JOSEPH					59-61
Frazier, James Lewis					02-04
FRAZIER, JOHN WARREN		72-73			68-70
Freeman, John Ripley	22	02-03			1896-98
FRENCH, WILLIAM DANIEL			72-a		
FRIEDMAN, EDMUND	63	53-54			49-51
Friel, Francis de Sales	59	57-58			50-51
Fteley, Alphonse	1898	1889-91			1888
FUCIK, EDWARD MONTFORD					69-71
Fuertes, Estevan Antonio					1892
Fuller, Georgee Warren		28-29			14-16
Gamble, Raleigh Welch					44-46
Gardiner, Edward		1853	1853-54	1853-54	
Gardiner, John Haines					45-47
Gardner, William Montgomery					09-11
GARRELTS, JEWELL MILAN		66-67		64-65b	55-57
Gerber, Emil					12-14
GESSEL, CLYDE DAVID					65-67
Gibbs, George					06-08
GIBBS, WILLIAM READ		74-75			71-73
GILLIS, LYMAN RIDGWAY					74-76
Gillmore, Quincy Adams					1876
GILMAN, CHARLES					25-27
GILMAN, ROGER HOWE					61-63
Glidden, William Roy	55	51-52			46-48
Goethals, George Washington					18
Goodkind, Morris					49-51
Goodrich, Charles Francis					43-45
Goodrich, Ernest Payson					40-42
Gorsuch, Robert Bennett			1852-53	1852-53	
Gottlieb, Abraham					1892-93
Gowdy, Roy Cotsworth		37-38			30-31
Gowen, Charles Sewall					04-06
Graff, Frederic	1885				1884
Gram, Lewis Merritt					47-49
GRANT, ALBERT ABRAHAM					74-76
Gray, Samuel Merrill		1892			
Greeley, Samuel Arnold					47-49
Green, Bernard Richardson		06-07			1894-96
GREEN, ROY MELVIN					62-64
Greene, Carleton					20-22
Greene, George Sears	1876-77				1868-69, 71, 73, 75, 77-79
Greene, George Sears, Jr.		1885		1888-90	1882-84, 86
Gregory, John Herbert					32-34
Greiner, John Edwin					15-17
Griffin, William McKenna					48-50
Grunsky, Carl Ewald	24	22-23			19-21
Gzowski, Sir Casimir Stanislaus					1894
Haertlein, Albert		50-51			46-48
Haines, Henry Stevens		01-02			1897-99
Hamilton, William Gaston					1883-87
Hammond, Alonzo John	33	29-30			26-28
Hanover, Clinton DeWitt Jr.					57-59
Hardesty, Shortridge					46-48
Harding, Sidney Twichell					49-51
Harrington, Arthur William		46-47			38-40
Harrison, Charles Lewis					08-10
Harrod, Benjamin Morgan	1897	1895-96			1892-94
HARTWELL, OLIVER WHITCOMB					54-56
Harwood, George Alec					15-17
Haskell, Eugene Elwin					14-16
HASSLER, PAUL CLIFFORD					72-74
Hastings, Edgar Morton	47	43-44			
HATHAWAY, GAIL ABNER	51	46-48			44-46
Hatton, Thomas Chalkley					26-28

Name	Pres.	Vice Pres.	Secy.	Treas.	Director
Hawgood, Harry					18-20
Hawley, John Blackstock					15-17
HAYES, JOHN MARION		69-70			64-66
HAZELET, CRAIG POTTER					58-60
Hazen, Allen		26-27			07-09
HAZEN, RICHARD					67-69
Hedges, Samuel Hamilton					13-15
HEDLEY, WILLIAM JOSEPH	66	61-62			57-59
HEMPEL, HUGH WILLIAM					73-75
Henny, David Christiaan		32-33			20-22
Hering, Rudolph		00-01			1891, 97-99
Hermann, Frederick Charles					31-33
Hermany, Charles	04	1891			1880
Herschel, Clemens	16	15-16			1891
Hidinger, Leroy Lemaynne					36-38
Higgs, James Allan					52-54
Hill, Albert Banks					1892
Hill, Louis Clarence	37	28-29			
Hill, Raymond Alva					36-38
Hill, William Ryan					17-18
HINDS, JULIAN					48-50
Hodgdon, Frank Wellington					07-09
Hodge, Henry Wilson					13-15
Hoffman, Robert					32-34
Hogan, John Philip	40	34-35			21-23
HOLCOMB, GLENN WILLIS	61	56-57			53-55
Holden, Otto					50-52
Holland, Clifford Milburn					22-24
HOLLAND, PAUL LEACH		59-60			49-51
Holleran, Leslie Gilbert					31-33
Holley, Alexander Lyman		1877			1876
HOLLISTER, SOLOMON CADY					44-46
Holman, Minard Lafever		05-06			
Holmes, Glenn Dickinson					23-25
Horner, Wesley Winans	46				33-35
Horton, Horace Ebenezer					07-09
Hovey, Otis Ellis				21-41	
Howard, Ernest Emmanuel	50	45-46			41-43
Howe, Joseph Milton		30-31			24-26
HOWSON, LOUIS RICHARD	58	55-56		61-63b	50-51
Hoyt, John Clayton		27-28			20-22
Huber, Walter Leroy	53	26-27			22-24
Hudson, Clarence Walter					20-22
Hudson, Harold Walton					39-41
Huie, Irving Van Arnam					46-48
Humphrey, Richard L.		25-26			21-23
Humphreys, Alexander Crombie					16-18
Hunt, Andrew Murray		21-22			18-20
Hunt, Charles Warren			1895-1920		
HUTCHINSON, RALPH WHITE					68-70
Hutton, William Rich		1896-97			1884-86
Hyde, Charles Gilman					40-42
ISAAK, ELMER BRAMWELL					73-75
Jackson, William					02-04
Jacobs, Joseph		40-41			29-31
Jagger, James Edwin			45-47a		
JANSEN, CARL B.					66-68
JESSUP, WALTER EDGAR			41-42a		
JOHNSON, EMORY EMMANUEL					68-70
Johnson, James Moreland					06-08
Johnson, Lorenzo Medici					1897-99
Johnson, Thomas Humrickhouse					00-02
Johnston, James Houstoun					28-30
Jonah, Frank Gilbert		33-34			16-17
JONES, RUSSEL CAMERON					70-71, 73-75
Just, George Alexander					1896-98

Name	Pres.	Vice Pres.	Secy.	Treas.	Director
Justin, Joel DeWitt					48-49
Katte, Walter					1885,1889
KAVANAGH, THOMAS CHRISTIAN					71-73
Keefer, Charles Henry					14-16
Keefer, Thomas Coltrin	1888	1886-87			1882
KENNEDY, DANIEL					68
Kennedy, Sir John					1898-00
Kennedy, William Harlin					02
Ketchum, Milo Smith		25-26			18-20
Khuen, Richard, Jr.					16-18
Kiersted, Wynkoop					06-08
Kimball, George Albert					10-12
KIMBALL, WILLIAM PHELPS			66-69a		
KINDSVATER, CARL EDWARD					64-66
King, Donald D.			66-67a		
Kirkwood, James Pugh	1868				1853-67
Kittredge, George Walson		17-18			08-10
Knap, Joseph Moss				00-12	1894-96
KNAPP, LLOYD DUNAWAY		59-60			53-55
KNOOP, FREDERICK RIPPEL, JR.					71-73
Knowles, Morris					28-30
Koch, Oscar Henry					45-47
KONSKI, JAMES LOUIS		72-73			67-70
KRAL, GEORGE JOE					70-72
Kuichling, Emil		05-06			01-03
LA LONDE, WILLIAM SALEM, JR.		64-65		66-	54-56
Lamb, George William					50-52
Landreth, Olin Henry					1893-95
Landreth, William Barker					05-07
LANG, EDMUND HERMAN			70-a		
Langthorn, Jacob Stinman					19-21
Laurie, James	1853-67				
LAVERTY, FINLEY BURNAP					57-59
Leeds, Charles Tileston					39-41
Legare, Thomas Keith					37-39
Leisen, Theodore Alfred					35-37
LENZ, ARNO THOMAS					71-73
Leonard, Henry Read					13-15
Leverich, Gabriel			1871-77		
Lewis, Eugene Castner					03-05, 12-14
Lewis, Harold MacLean					39-41
Lewis, Nelson Peter		18-19			04-06
Lilly, Scott Barrett					42-44
Lockwood, Mason Graves	57	54-55			
Loomis, Horace		12			10-12
Loweth, Charles Frederick	23	14-15			10-12
Lucas, George Latimore		40-41			27-29
Ludlow, William					1890
Lupfer, Edward Payson		36-37			30-31,32-34
LYON, OSCAR T.					72-74
MacCrea, Don Alexander					30-32
Macdonald Charles	08	1893-94			1871, 74-75
MacLeod, John		1892			
Maitland, Alexander, Jr.					24-26
Malony, Walden LeRoy					47-49
Manley, Henry					1898-00
Marston, Anson	29	23-24			20-22
Marston, Frank Alwyn	60	56-57			52-54
Martin, Charles Cyril		1894-95			
Marx, Charles David	15	12-13			04-06
Mason, George Cotner					23-25
Massey, George Bragg					41-43
MATTERN, DONALD HECKMAN		61-62			58-60
McAlpin, George Washington					52-54
McAlpine, William Jarvis	1868-69				1854-67, 1870
McCABE, JOSEPH			69-a		

Name	Pres.	Vice Pres.	Secy.	Treas.	Director
McCALL JOHN EVAN					72-74
McConnell, Ira Welch					21-23
McDonald, Frederick Honour					34-36
McDonald, Hunter	14	10-11			03-05
McGEE, HENDERSON ENGLAND					62-64
McKEE, JACK EDWARD					66-68
McMath, Robert Emmet					1889
McMINN, JACK HUDSON					73-75
McNew, John Thomas Lamar		46			42-44
McNulty, George Washington					1892-93
McVean, John Jay					1898-00
Mead, Charles Adriance					31-33
Mead, Daniel Webster	36				
Mead, Elwood					03-05
Mendell, George Henry		1897-98			
Mendenhall, Herbert Drummond					31-33
MENDENHALL, IRVAN FRANK					74-76
MERRIFIELD, JOHN TURNER					71-73
Merrill, William Emery					1883
Merriman, Thaddeus					24-26
Metcalf, Leonard		19-20			13-15
Metcalf, William	1893				1883-84
Meyer, Thomas C.			1869-70		
MILHOLLIN, AUSTIN BARLOW					68-70
Modjeski, Ralph					04-06
MOLINEAUX, CHARLES BORROMEO		60-61			53-55
Montfort, Richard		16			14-16
MOORE, NORMAN ROBERT		57-58			51-53
Moore, Robert	02	1888, 99-00			1892-93
MOORE, WILLIAM WALLACE					67-69
Mordecai, Augustus					1895-97
Morell, W.H.					1853
MORGAN, ARTHUR ERNEST		27-28			
Morison, George Shattuck	1895				
Morris, Clyde Tucker					29-31
Morris, Henry Gurney					1886
Morris, Samuel Brooks		58-59			54-56
Morse, Edwin Kirtland					31-33
Morse, Henry Grant					1897-99
Morse, Howard Scott					35-37
Morse, James Otis			1854-69	1855-75	1854-67, 77
Myers, Charles Hayward					1892
Myers, Clarence Eugene					36-38
Myers, Edmund Trowbridge Dana					1892
Needles, Enoch Ray	56	54-55		59-60b	37-39
NEWNAM, FRANK HASTINGS, JR.	69				63-65
Nichols, Othniel Foster					1892-93
Nicholson, Frank Lee					29-31
NILES, THOMAS McMASTER	70	63-64			60-62
Noble, Alfred	03	00-01			1895
Noble, Frederick Charles					17-18
Norcross, Paul Howes					25
North, Edward Payson		1898-99			1891
Norton, George Harvey					29-30
Noyes, Edward Newton		38-39			33-35
O'BRIEN, EARL FRANCIS					59-61
Ockerson, John Augustus	12	07-08			
O'Connor, John Adam					18-19, 20-22
O'HARRA, WAYNE GILDER					59-61
O'Rourke, John Francis					00-02
Osborn, Frank Chittenden					01-03
Osgood, Joseph Otis					03-05
Owen, James					1897-99
Paine, Charles	1883				
Paine, William H.		1882-84			1877-81
PANHORST, FREDERICK WILLIAM					46-48

Name	Pres.	Vice Pres.	Secy.	Treas.	Director
PARISI, PAUL ALBERT			72-a		
Parker, Glenn Lane					39-40
PARKS, WARREN WRIGHT					53-54
Parsons, William Barclay					1896-98
Paul, Charles Howard					24-28
Paulsen, Carl Gustav					53-55
Paulson, Frederick Holroyd					55-57
PECK, RALPH BRAZELTON					63-65
PECKWORTH, HOWARD FARON					57-59
Pegram, George Herndon	17	09-10			02-04
Perry, John Prince Hazen					33-35
PERSON, HJALMAR THORVAL					65-67
PETERSON, DEAN FREEMAN		73-74			
Peterson, Peter Alexander		1896-97			1892-93
Piatt, William McKinney					46-48
Pierson, George Spencer					05-07
Pirnie, Malcolm	44	38-39			28-30
PIRNIE, MALCOLM, JR.					68-70
PLETTA, DAN HENRY					70-72
Polk, Armour Cantrell		45-46			40-42
Poole, Charles Arthur					35-37
Pope, Willard Smith					1893-95
Pratt, T. W.					1854
Prentis, Edmund Ashley					48
PRICHARD, MASON CARTER					56-58
Proctor, Carlton Springer	52	48-49		55-58b	36-38
Prosser, Thomas					1870
Prout, Henry Gosiee					1893-95
Quinlan, George Austin					00-01
Ramsey, Joseph					00-02
Randolph, Isham					16
Rawn, A M		52-53			42-44
Ray, George Joseph					27-29
Raymer, Albert Reesor					25-27
Read, Robert Leland					1892-93
Reed, Ralph John					33-35
Reichmann, Albert Ferdinand					29-31
Reppert, Charles Miller		39			
REQUARDT, GUSTAV JAEGER					41-43
REYNOLDS, DON POTTER			60-a		
REYNOLDS, GARDNER MEAD					66-68
RHODES, FRED HAROLD, JR.					59-61
RICHARDSON, GEORGE SHERWOOD					55-57
Richardson, Henry Brown					00-02
Ricketts, Palmer Chamberlaine		16-17			1899-01
Ridgway, Arthur Osbourne					24-26
Ridgway, Robert	25	22-23			11-13
Riggs, Henry Earle	38	35-36			32-34
Rights, Lewis Daniel					17-19
RILEY, JOHN PHILIP					56-58
RINNE, JOHN ELMER	73	68-69			58-60
Roberts, Percival, Jr.					10-12
Roberts, William Minor	1879	1874-76,78			1877
ROGERS, CRANSTON ROBINSON					72-75
Root, Joseph Eugene					38-40
ROWE RAPHAEL ROBINSON					56-58
Rowland, Thomas Fitch		1886-87			1871-73
Rumery, Ralph Rollins				30-41	
Rundlett, Leonard Warren					11-13
Rust, Charles Henry		13-14			
RUTLEDGE, PHILIP CASTEEN					58-60
Ryan, Walter Jackman					50-51
RYDELL, LOUIS ERNEST					56-58
SALGO, MICHAEL NICHOLAS		70-71			64-66
SANDERS, AZEL LABON RALPH					66-68
Sanborn, James Forrest					33-35

Name	Pres.	Vice Pres.	Secy.	Treas.	Director
SANGSTER, WILLIAM McCOY	75				66-67
Saville, Thorndike					45-47
Sawin, Sanford Wales					39-41
Sawyer, Donald Hubbard	39	35-36			26-28
SAWYER, JAMES ERNEST					70-72
SCALZI, JOHN BAPTIST					61
Schneider, Charles Conrad	05	02-03			1887, 98-00
Schuyler, James Dix		03-04			1892, 99-01
Scobey, Frederick Charles		50-51			43-45
Seabury, George Tilley			25-45		
Seaman Henry Bowman					00-02
See, Horace					1896-98
SESSUMS, FRANCIS BLAKE					69-71
SHANNON, WILLIAM DAY		47			44-46
Shaver, Robert Ezekial					67-69
Shea, William James					37-39
Shedd, Thomas Clark					54-56
Shelburne, Tilton Edward					58-60
SHELTON, MERCEL JOSEPH		64-65			53-55
Sherlock, Robert Henry					56-58
Sherman, Henry John		49-50			33-35
Sherrerd, Morris Robeson					05-07
Shinn, William Powell	1890				1889
Sidel, W. H.					1853
SIMS, JAMES REDDING		70-71			66-68
Singstad, Ole					30-32
Slattery, John Rodolph					30-32
Smith, Charles Shaler					1878
Smith, Charles Vandervoort					1878-81
Smith, J. Waldo		13-14			06-08
SMITH, KIRBY					49-51
Smith, T, Guilford					1894-96
SMITH, WALDO EDWARD					62-64
Snow, J. Parker					11-13
Sooysmith, Charles					1895-97
Spofford, Charles Milton		42-43			25-27
Stabler, Herman					35-37
Staniford, Charles Wilkinson					12-13
Stanton, Robert Brewster					1894-96
Stanton, Thomas Elwood		42-43			37-39
Stearns, Frederic Pike	06	1893-99			1893-95
Steele, Isaac Cleveland					52-54
Stevens, Charles Henry		41-42			30-32
Stevens, John Cyprian	45				32-34
Stevens, John Frank	27				
Storey, William B.					06-08
Stott, Henry Gordon					11
Strobel, Charles Louis		11-12			1886, 94
Stuart, Francis Lee	31	20-21			09-11
Sumner, Horace Augustus					09-11
Swain, George Fillmore	13	08-09			01-03
Svensson, Emil		09-10			06-08
Symons, Thomas William					1896-98
Taber, Edward Gray					26-28
Talbot, Arthur Newell	18	11			09-10
Talcott, William Hubbard					1854-68
TATLOW, RICHARD HENRY, III	68			66b	63-65
TERRELL, DANIEL VOIERS	54	51-52			47-49
Thomas, Franklin	49	44-45			30-32
Thomas, Henry Freeman					45-47
Thompson, Samuel Clarence					09-11
Thomas, Harry Freeman					45-47
Thomson, John				1895-99	1892-94
Thomson, Reginald Heber					17-19
Thomson, Thomas Kennard					12-14
Tiffany, Ross Kerr					38-39
Tillson, George William				17-18	07-09

Name	Pres.	Vice Pres.	Secy.	Treas.	Director
TIMBY, ELMER KNOWLES				67-73b	60-62
Tipton, Royce Jay		65-66			44-46
Tolles, Frank Clifton					44-46
Towle, Stevenson					1887-88
Trout, Charles Eliphalet				41-59	34-36
TURCHICK, ALBERT WILLIAM			72-a		
Turner, Edmund Kimball					1899-01
Tuttle, Arthur Smith	35	32-33		19-20	14-16
TYSON, CHRISTOPHER GILBERT					73-75
Van Buren, Robert					1890
Van Hoesen, Edmund French					02-04
Van Horne, John Garret					1892
Van Winkle, Edgar Beach					1880
Vaughan, Frederic Willis					1885
VEATCH, NATHAN THOMAS					59-61
Ventres, Daniel Brainerd		63-64			59-61
VIEST, IVAN MIROSLAV		74-75			69-71
Voorhees, Theodore					1890
VREELAND, ROBERT PAUL, JR.		68-69			65-66
Wagner, Samuel Tobias					18-20
Waite, Henry Matson		31-32			
WALKER, LELAND JASPER		66-67			62-64
Wall, Edward Everett		21-22			18-20
Wallace, John Findley	00	1897-98		13-14	
WALLACE, RALPH HOWES					71-73
Ward, John Frothingham					1868-72
WARD, JOSEPH SIMEON					70-72
Warren, Gouverneur K.					1880
WATSON, JOHN DARGAN					61-63
Weaver, Frank Lloyd		55-56			50-51
WEBER, EUGENE WILLIAM					68-70
Webster, George Smedley	21	17-18			04-06
Welch, Ashbel	1882	1881			
Whinery, Samuel		1892-93			1891, 99-01
WHITACRE, HORACE J.					65-67
Whitcomb, Henry Donald					1894-96
White, Lazarus					40-42
White, William Howard					1886
Whitman, Ezra Bailey	43				23-25
Whitman, Thomas Jefferson		1885			
Whittemore, Don Juan	1884				1881
Wiley, Ralph Benjamin		47-48			41-43
Wilgus, William John					02-04
Wilkerson, Thomas Jefferson					34-36
Wilkins, William Glyde					09-11
WILLIAMS, CLIFFORD DAVID		67-68			64-66
Williams, Frank Martin					26-28
Williams, Gardner Stewart		14-15			08-10
Wilson, Joseph Miller		1894-95			1885
WILLOUGHBY, GRAHAM PAUL		69-70			55-57
WILSON, JAMES DAVID		71-72			65-67
Wilson, Milton Theurer					51-52
Wilson, Wilbur M.					44-46
Winsor, Frank Edward		30-31			22-24
WISE, LAURESS LEE					70-72
WISELY, WILLIAM HOMER			55-72		
Wisner, George Y.					1893-00
Wood, De Volson					1874
WOODMAN, LONGINO ALPHONSE		73-74			67-69
Woods, Harland Clark					47-49
Worthen, William Ezra	1887				1872
Yates, Joseph Johnson					22-24
YODER, CHARLES WILLIAM	74	67-68			62-64
ZACK, SAMUEL ISADOR					63-65
ZOLLER, JAMES HAROLD					67-69
ZWOYER, EUGENE M.			72-		69-71

APPENDIX III

ROSTER OF EXECUTIVE OFFICERS

Name	*Term of Office*	*Title*
Robert B. Gorsuch	1852-53	Secretary
Edward Gardiner	1853-54	Secretary
James O. Morse	1854-69	Secretary
Thomas C. Meyer	1869-70	Secretary
Alfred P. Boller	1870-71	Secretary
Gabriel Leverich	1871-77	Secretary and Librarian
John Bogart	1877-91	Secretary and Librarian
Francis Collingwood	1891-95	Secretary
Charles W. Hunt	1895-1920	Secretary
Herbert S. Crocker	1920-21	Secretary
Elbert M. Chandler	1921-22	Secretary
John H. Dunlap*	1922-24	Secretary
George T. Seabury†	1925-45	Secretary
William N. Carey	1945-55	Secretary and Executive Secretary
William H. Wisely	1955-72	Secretary and Executive Director
Eugene M. Zwoyer	1972—	Secretary and Executive Director

**Deceased July 29, 1924, while in office*
†Deceased May 25, 1945, while in office

APPENDIX IV

ROSTER OF HONORARY MEMBERS

An Honorary Member shall have attained acknowledged eminence in some branch of engineering or in the arts and sciences related thereto.

(The names of deceased Honorary Members are printed in capital and lower case letters)

Historic Roster

Name	Date of Election to Honorary Membership
Abert, John James	March 2, 1853
Bache, Alexander Dallas	March 2, 1853
Long, Stephen H.	March 2, 1853
Mahan, Dennis Hart	March 2, 1853
Totten, Joseph Gilbert	March 2, 1853
Burden, Henry	March 28, 1853
Robinson, Moncure	July 6, 1853
Whipple, Squire	May 6, 1868
Jervis, John Bloomfield	December 2, 1868
Barnard, John Gross	April 7, 1873
Humphreys, Andrew Atkinson	May 7, 1873
Allen, Horatio	March 4, 1874
Ericsson, John	October 2, 1879
Wright, Horatio Gouverneur	March 3, 1880
Dirks, Justus	June 2, 1880
Weber, Baron Christian Phillipp Max Maria von	June 2, 1880
Hawkshaw, Sir John	November 3, 1880
Malezieux, Emile	November 3, 1880
Newton, John	April 30, 1884
Duane, James Chatham	November 20, 1886
Adams, Julius Walker	October 26, 1888
Greene, George Sears	October 26, 1888
McAlpine, William Jarvis	October 26, 1888
Campbell, Allan	March 1, 1892
Young, William Clark	March 1, 1892
Francis, James Bicheno	April 5, 1892
Wilson, William Hasell	August 2, 1892
Storrow, Charles Storer	January 24, 1893
Worthan, William Ezra	April 4, 1893
Gray, George Edward	June 5, 1894
Craighill, William Price	March 23, 1896
Baker, Sir Benjamin	May 5, 1897
Davidson, George	May 5, 1897
Fritz, John	September 5, 1899
Melville, George Wallace	December 20, 1899
Rowland, Thomas Fitch	December 20, 1899
Fox, Sir Douglas	March 5, 1901
White, Sir William Henry	December 16, 1904
Gorsuch, Robert Bennett	January 9, 1905
Haswell, Charles Haynes	May 12, 1905
Mackenzie, Alexander	May 12, 1905
Mills, Hiram Francis	November 30, 1909
Whittemore, Don Juan	January 6, 1911
Keefer, Thomas Coltrin	October 1, 1913
Pickett, William Douglas	April 1, 1914
Dodge, Grenville Mellen	March 2, 1915
Rea, Samuel	June 6, 1921
Swazey, Ambrose	June 6, 1921
Carson, Howard Adams	October 10, 1921
Luiggi, Luigi	October 10, 1921
Schneider, Charles Prosper Eugene	October 10, 1921
Foch, Ferdinand	December 6, 1921
Chagnaud, Leon-Jean	June 19, 1922
Fitzmaurice, Sir Maurice	June 19, 1922
Herschel, Clemens	June 19, 1922
Stevens, John Frank	June 19, 1922
Unwin, William Cawthorne	June 19, 1922
Bates, Onward	October 15, 1923
FitzGerald, Desmond	October 15, 1923
Hoover, Herbert	December 1, 1924

Name	Date of Election to Honorary Membership
Parsons, William Barclay	November 16, 1925
Talbot, Arthur Newell	November 16, 1925
Marx, Charles David	October 10, 1927
Smith, Jonas Waldo	October 1, 1928
Furuichi, Baron Koi	July 8, 1929
Fisher, Edwin Augustus	October 7, 1929
Lindenthal, Gustav	October 7, 1929
Swain, George Fillmore	October 7, 1929
Freeman, John Ripley	September 29, 1930
Kittredge, George Watson	October 5, 1931
Mead, Daniel Webster	October 5, 1931
Pegram, George Herndon	October 5, 1931
Ricketts, Palmer Chamberlaine	October 5, 1931
Bush, Lincoln	October 3, 1932
Greiner, John Edwin	October 3, 1932
Strobel, Charles Louis	October 3, 1932
Turneaure, Frederick Eugene	September 26, 1933
Ketchum, Milo Smith	October 2, 1934
Ridgway, Robert	October 2, 1934
Darling, William Lafayette	November 15, 1934
Alvord, John Watson	October 14, 1935
Coleman, John Francis	October 14, 1935
Cooley, Mortimer Elwyn	October 14, 1935
Wilgus, William John	October 14, 1935
Dow, Alex	October 11, 1936
Duggan, George Herrick	October 11, 1936
Hoffmann, Robert	October 11, 1936
Lippincott, Joseph Barlow	October 11, 1936
Waddell, John Alexander Low	October 11, 1936
Davison, George Stewart	October 4, 1937
Hovey, Otis Ellis	October 4, 1937
McDonald, Hunter	October 4, 1937
Worcester, Joseph Ruggles	October 4, 1937
Allen, C. Frank	October 10, 1938
Marston, Anson	October 10, 1938
Tuttle, Arthur Smith	October 10, 1938
Wall, Edward Everett	October 10, 1938
Weymouth, Frank Elwin	October 10, 1938
Binnie, William James Eames	July 24, 1939
Fairbairn, John Morrice Roger	July 24, 1939
Crocker, Herbert Samuel	October 6, 1939
Jacoby, Henry Sylvester	October 6, 1939
Taylor, Thomas Ulvan	October 6, 1939
Berkey, Charles Peter	October 14, 1940
Fenkell, George Harrison	October 14, 1940
Galloway, John Debo	October 14, 1940
Jonah, Frank Gilbert	October 14, 1940
Thomson, Reginald Heber	October 14, 1940

Name	Date of Election to Honorary Membership
Budd, Ralph	October 13, 1941
KELLY, WILLIAM	October 13, 1941
Riggs, Henry Earle	October 13, 1941
Savage, John Lucian	October 13, 1941
Waite, Henry Matson	October 13, 1941
Hammond, Alonzo John	October 12, 1942
Lawson, Lawrence Milton	October 12, 1942
MOREELL, BEN	October 12, 1942
Somervell, Brehon Burke	October 12, 1942
Woodward, Sherman Melville	October 12, 1942
Connor, Edward Hanson	October 11, 1943
Crowe, Francis Trenholm	October 11, 1943
MacDonald, Thomas Harris	October 11, 1943
Matthes, Gerard Hendrik	October 11, 1943
Dyer, Arthur James	October 9, 1944
GROSS, CHARLES PHILIP	October 9, 1944
King, Horace Williams	October 9, 1944
Sabin, Louis Carlton	October 9, 1944
Surveyer, Arthur	October 9, 1944
Bakhmeteff, Boris Alexandrowitch	July 17, 1945
Purcell, Charles Henry	July 17, 1945
Kettering, Charles Franklin	October 15, 1945
Billings, Asa White Kenney	October 14, 1946
Burdick, Charles Baker	October 14, 1946
Greensfelder, Albert Preston	October 14, 1946
Sherman, Leroy Kempton	October 14, 1946
Challies, John Bow	October 13, 1947
Cross, Hardy	October 13, 1947
McAlpine, William Horatio	October 13, 1947
Terzaghi, Karl	October 13, 1947
Buford, Charles Homer	October 11, 1948
CLAY, LUCIUS DuBIGNON	October 11, 1948
Derickson, Donald	October 11, 1948
Dunn, Gano	October 11, 1948
Weiss, Andrew	October 11, 1948
Breed, Charles Blaney	November 1, 1949
Creager, William Pitcher	November 1, 1949
DeBerard, Wilford Willis	November 1, 1949
von Karman, Theodor	November 1, 1949
Wilson, Wilbur M.	November 1, 1949
Greeley, Samuel Arnold	June 12, 1951
Hyde, Charles Gilman	June 12, 1951
Jones, Jonathan	June 12, 1951
Ohrt, Frederick	June 12, 1951
BAILEY, ERVIN GEORGE	June 16, 1952
Coyne, Andre Leon Jules	June 16, 1952

Name	Date of Election to Honorary Membership
MACKENZIE, CHALMERS JACK	June 16, 1952
Mudd, Harvey Seeley	June 16, 1952
Torres, Ary Frederico	June 16, 1952
Ammann, Othmar Herman	June 16, 1953
MORGAN, ARTHUR ERNEST	June 16, 1953
Page, John Chatfield	June 16, 1953
Spofford, Charles Milton	June 16, 1953
Cummins, Robert James	June 15, 1954
Hardesty, Shortridge	June 15, 1954
Lupfer, Edward Payson	June 15, 1954
Banks, Frank Arthur	June 14, 1955
Colbert, Leo Otis	June 14, 1955
Martin, Park Hussey	June 14, 1955
Parcel, John Ira	June 14, 1955
Burpee, George William	June 5, 1956
Haertlein, Albert	June 5, 1956
Robins, Thomas Matthews	June 5, 1956
Singstad, Ole	June 5, 1956
Wiley, Ralph Benjamin	June 5, 1956
Huntington, Whitney Clark	June 4, 1957
Imhoff, Karl	June 4, 1957
Morse, Howard Scott	June 4, 1957
PEREZ CASTRO, LORENZO	June 4, 1957
Brunnier, Henry John	June 24, 1958
DOUGHERTY, NATHAN WASHINGTON	June 24, 1958
Rawn, AM	June 24, 1958
Austin, Herbert Ashford Robertson	May 5, 1959
EDGERTON, GLEN EDGAR	May 5, 1959
HINDS, JULIAN	May 5, 1959
Masters, Frank Milton	May 5, 1959
REQUARDT, GUSTAV JAEGER	May 5, 1959
Atkinson, Guy Frederick	June 20, 1960
HOLLISTER, SOLOMON CADY	June 20, 1960
Kerekes, Frank	June 20, 1960
Scobey, Fred C.	June 20, 1960
Morris, Samuel B.	April 10, 1961
RICHARDSON, GEORGE SHERWOOD	April 10, 1961
Saville, Thorndike	April 10, 1961
WOLMAN, ABEL	April 10, 1961
BARTHOLOMEW, HARLAND	May 14, 1962
Eckel, Clarence Lewis	May 14, 1962
Fair, Gordon Maskew	May 14, 1962
FRY, ALBERT STEVENS	May 14, 1962
Hill, Raymond Alva	May 14, 1962
Holden, Otto	May 14, 1962

Name	Date of Election to Honorary Membership
Clyde, George Dewey	May 13, 1963
Diemer, Robert Bernard	May 13, 1963
Lane, Emory Wilson	May 13, 1963
VEATCH, NATHAN THOMAS	May 13, 1963
AMIRIKIAN, ARSHAM	May 13, 1964
MADIGAN, MICHAEL JOHN	May 13, 1964
MAYER, ARMAND	May 13, 1964
Camp, Thomas Ringgold	May 18, 1965
CASAGRANDE, ARTHUR	May 18, 1965
Gilchrist, Gibb	May 18, 1965
Growdon, James	May 16, 1966
NEWMARK, NATHAN M.	May 16, 1966
Praeger, Emil H.	May 16, 1966
Steele, I. Cleveland	May 16, 1966
SVERDRUP, LEIF J.	May 16, 1966
HEALD, HENRY T.	May 8, 1967
HIGGINS, THEODORE R.	May 8, 1967
Wynne-Edwards, Robert	May 8, 1967
DAVIS, ROLAND P.	May 13, 1968
Forrest, T. Carr	May 13, 1968
Kyle, John M., Jr.	May 13, 1968
MATTERN, DONALD H.	May 13, 1968
TURNBULL, WILLARD J.	May 13, 1968
WILBUR, LYMAN D.	May 13, 1968
BAITY, HERMAN G.	April 14, 1969
JOHNSTON, BRUCE G.	April 14, 1969
KEULEGAN, GARBIS H.	April 14, 1969
LABOON, JOHN F.	April 14, 1969
Land, James L.	April 14, 1969
RADER, F. LLOYD	April 14, 1969
REESE, RAYMOND C.	April 14, 1969
GOTAAS, HAROLD B.	April 6, 1970
MOSES, ROBERT	April 6, 1970
BROWN, GEORGE RUFUS	April 19, 1971
FERGUSON, PHIL MOSS	April 19, 1971
GREER, DEWITT CARLOCK	April 19, 1971
GRINTER, LINTON ELI	April 19, 1971
JANSEN, CARL B.	April 19, 1971
LAVERTY, FINLEY BURNAP	April 19, 1971
McCARTHY, GERALD TIMOTHY	April 19, 1971
PALO, GEORGE PAYNE	April 19, 1971
BUGGE, WILLIAM ADAIR	April 26, 1972
CLEARY, EDWARD JOHN	April 26, 1972
HAZELET, CRAIG POTTER	April 26, 1972
HEDGER, HAROLD EVERETT	April 26, 1972
KNAPP, LLOYD DUNAWAY	April 26, 1972
RADY, JOSEPH J.	April 26, 1972
TIMBY, ELMER KNOWLES	April 26, 1972
WINKELMAN, DWIGHT WILLIAM	April 26, 1972

Name	Date of Election to Honorary Membership
ARMSTRONG, ELLIS L.	April 8, 1973
ROUSE, HUNTER	April 8, 1973
TURNER, FRANCIS C.	April 8, 1973
WHITTON, REX M.	April 8, 1973
WINTER, GEORGE	April 8, 1973
WISELY, WILLIAM H.	April 8, 1973
NORRIS, CHARLES H.	April 20, 1974
RICH, GEORGE R.	April 20, 1974
SHUMATE, CHARLES E.	April 20, 1974
TURNER, NATHANIEL P.	April 20, 1974
VANCE, JAMES A.	April 20, 1974

Alphabetical Roster

Name	Date of Election to Honorary Membership
Abert, John James	March 2, 1853
Adams, Julius Walker	October 26, 1888
Allen, C. Frank	October 10, 1938
Allen, Horatio	March 4, 1874
Alvord, John Watson	October 14, 1935
AMIRIKIAN, ARSHAM	May 13, 1964
Ammann, Othmar Herman	June 16, 1953
ARMSTRONG, ELLIS L.	April 8, 1973
Atkinson, Guy Frederick	June 20, 1960
Austin, Herbert Ashford Robertson	May 5, 1959
Bach, Alexander Dallas	March 2, 1853
BAILEY, ERVIN GEORGE	June 16, 1952
BAITY, HERMAN G.	April 14, 1969
Baker, Sir Benjamin	May 5, 1897
Bakhmeteff, Boris Alexandrowitch	July 17, 1945
Banks, Frank Arthur	June 14, 1955
Barnard, John Gross	April 7, 1873
BARTHOLOMEW, HARLAND	May 14, 1962
Bates, Onward	October 15, 1923
Berkey, Charles Peter	October 14, 1940
Billings, Asa White Kenney	October 14, 1946
Binnie, William James Eames	July 24, 1939
Breed, Charles Blaney	November 1, 1949
BROWN, GEORGE RUFUS	April 19, 1971
Brunnier, Henry John	June 24, 1958
Budd, Ralph	October 13, 1941
Buford, Charles Homer	October 11, 1948
BUGGE, WILLIAM ADAIR	April 26, 1972
Burden, Henry	March 28, 1853
Burdick, Charles Baker	October 14, 1946
Burpee, George William	June 5, 1956
Bush, Lincoln	October 3, 1932
Camp, Thomas Ringgold	May 18, 1965
Campbell, Allan	March 1, 1892
Carson, Howard Adams	October 10, 1921
CASAGRANDE, ARTHUR	May 18, 1965
Chagnaud, Leon-Jean	June 19, 1922
Challies, John Bow	October 13, 1947
CLAY, LUCIUS DuBIGNON	October 11, 1948
CLEARY, EDWARD JOHN	April 26, 1972
Clyde, George Dewey	May 13, 1963
Colbert, Leo Otis	June 14, 1955
Coleman, John Francis	October 14, 1935
Connor, Edward Hanson	October, 11, 1943
Cooley, Mortimer Elwyn	October 14, 1935
Coyne, Andre Leon Jules	June 16, 1952
Craighill, William Price	March 23, 1896
Creager, William Pitcher	November 1, 1949
Crocker, Herbert Samuel	October 6, 1939
Cross, Hardy	October 13, 1947
Crowe, Francis Trenholm	October 11, 1943
Cummins, Robert James	June 15, 1954
Darling, William Lafayette	November 15, 1934
Davidson, George	May 5, 1897
DAVIS, ROLAND P.	May 13, 1968
Davison, George Steward	October 4, 1937
DeBerard, Wilford Willis	November 1, 1949
Derickson, Donald	October 11, 1948
Diemer, Robert Bernard	May 13, 1963
Dirks, Justus	June 2, 1880
Dodge, Grenville Mellen	March 2, 1915
DOUGHERTY, NATHAN WASHINGTON	June 24, 1958
Dow, Alex	October 11, 1936
Duane, James Chatham	November 20, 1886
Duggan, George Herrick	October 11, 1936
Dunn, Gano	October 11, 1948
Dyer, Arthur James	October 9, 1944
EDGERTON, GLEN EDGAR	May 5, 1959
Eckel, Clarence Lewis	May 14, 1962
Ericsson, John	October 2, 1879
Fair, Gordon Maskew	May 14, 1962
Fairbairn, John Maurice Roger	July 24, 1939

Name	Date of Election to Honorary membership
Fenkell, George Harrison	October 14, 1940
FERGUSON, PHIL MOSS	April 19, 1971
Fisher, Edwin Augustus	October 7, 1929
FitzGerald, Desmond	October 15, 1923
Fitzmaurice, Sir Maurice	June 19, 1922
Foch, Ferdinand	December 6, 1921
Forrest, T. Carr	May 13, 1968
Fox, Sir Douglas	March 5, 1901
Francis, James Bicheno	April 5, 1892
Freeman, John Ripley	September 29, 1930
Fritz, John	September 5, 1899
FRY, ALBERT STEVENS	May 14, 1962
Furuichi, Baron Koi	July 8, 1929
Galloway, John Debo	October 14, 1940
Gilchrist, Gibb	May 18, 1965
Gorsuch, Robert Bennett	January 9, 1905
GOTAAS, HAROLD B.	April 6, 1970
Gray, George Edward	June 5, 1894
Greeley, Samuel Arnold	June 12, 1951
Greene, George Sears	October 26, 1888
Greensfelder, Albert Preston	October 14, 1946
GREER, DEWITT CARLOCK	April 19, 1971
Greiner, John Edwin	October 3, 1932
GRINTER, LINTON ELI	April 19, 1971
GROSS, CHARLES PHILIP	October 9, 1944
Growdon, James	May 16, 1966
Haertlein, Albert	June 5, 1956
Hammond, Alonzo John	October 12, 1942
Hardesty, Shortridge	June 16, 1954
Haswell, Charles Haynes	May 12, 1905
Hawkshaw, Sir John	November 3, 1880
HAZELET, CRAIG POTTER	April 26, 1972
HEALD, HENRY T.	May 8, 1967
HEDGER, HAROLD EVERETT	April 26, 1972
Herschel, Clemens	June 19, 1922
HIGGINS, THEODORE R.	May 8, 1967
Hill, Raymond Alva	May 14, 1962
HINDS, JULIAN	May 5, 1959
Hoffmann, Robert	October 11, 1936
Holden, Otto	May 14, 1962
HOLLISTER, SOLOMON CADY	June 20, 1960
Hoover, Herbert	December 1, 1924
Hovey, Otis Ellis	October 4, 1937
Humphreys, Andrew Atkinson	May 7, 1873
Huntington, Whitney Clark	June 4, 1957
Hyde, Charles Gilman	June 12, 1951

Name	Date of Election to Honorary Membership
Imhoff, Karl	June 4, 1957
Jacoby, Henry Sylvester	October 6, 1939
JANSEN, CARL B.	April 19, 1971
Jervis, John Bloomfield	December 2, 1868
JOHNSTON, BRUCE G.	April 14, 1969
Jonah, Frank Gilbert	October 14, 1940
Jones, Jonathan	June 12, 1951
Keefer, Thomas Coltrin	October 1, 1913
KELLY, WILLIAM	October 13, 1941
Kerekes, Frank	June 20, 1960
Ketchum, Milo Smith	October 2, 1934
Kettering, Charles Franklin	October 15, 1945
KEULEGAN, GARBIS H.	April 14, 1969
King, Horace Williams	October 9, 1944
Kittredge, George Watson	October 5, 1931
KNAPP, LLOYD DUNAWAY	April 26, 1972
Kyle, John M., Jr.	May 13, 1968
LABOON, JOHN F.	April 14, 1969
Land, James L.	April 14, 1969
Lane, Emory Wilson	May 13, 1963
LAVERTY, FINLEY BURNAP	April 19, 1971
Lawson, Lawrence Milton	October 12, 1942
Lindenthal, Gustav	October 7, 1929
Lippincott, Joseph Barlow	October 11, 1936
Long, Stephen H.	March 2, 1853
Luiggi, Luigi	October 10, 1921
Lupfer, Edward Payson	June 15, 1954
MacDonald Harris Thomas	October 11, 1943
Mackenzie, Alexander	May 12, 1905
MACKENZIE, CHALMERS JACK	June 16, 1952
MADIGAN, MICHAEL JOHN	May 13, 1964
Mahan, Dennis Hart	March 2, 1853
Malezieux, Emile	November 3, 1880
Marston, Anson	October 10, 1938
Martin, Park Hussey	June 14, 1955
Marx, Charles David	October 10, 1927
Masters, Frank Milton	May 5, 1959
MATTERN, DONALD H.	May 13, 1968
Matthes, Gerard Hendrik	October 11, 1943
MAYER, ARMAND	May 13, 1964
McAlpine, William Horatio	October 13, 1947
McAlpine, William Jarvis	October 26, 1888
McCARTHY, GERALD TIMOTHY	April 19, 1971
McDonald, Hunter	October 4, 1937
Mead, Daniel Webster	October 5, 1931
Melville, George Wallace	December 20, 1899

Name	Date of Election to Honorary Membership
Mills, Hiram Francis	November 30, 1909
MOREELL, BEN	October 12, 1942
MORGAN, ARTHUR ERNEST	June 16, 1953
Morris, Samuel B.	April 10 1961
Morse, Howard Scott	June 4, 1957
MOSES, ROBERT	April 6, 1970
Mudd, Harvey Seely	June 16, 1952
NEWMARK, NATHAN M.	May 16, 1966
Newton, John	April 30, 1884
NORRIS, CHARLES H.	April 20, 1974
Ohrt, Frederick	June 12, 1951
Page, John Chatfield	June 16, 1953
PALO, GEORGE PAYNE	April 19, 1971
Parcel, John Ira	June 14, 1955
Parsons, William Barclay	November 16, 1925
Pegram, George Herndon	October 5, 1931
PEREZ CASTRO, LORENZO	June 4, 1957
Pickett, William Douglas	April 1, 1914
Praeger, Emil H.	May 16, 1966
Purcell, Charles Henry	July 17, 1945
RADER, LLOYD F.	April 14, 1969
RADY, JOSEPH J.	April 26, 1972
Rawn, A M	June 24, 1958
Rea, Samuel	June 6, 1921
REESE, RAYMOND C.	April 14, 1969
REQUARDT, GUSTAV JAEGER	May 5, 1959
RICH, GEORGE R.	April 20, 1974
RICHARDSON, GEORGE SHERWOOD	April 10, 1961
Ricketts, Palmer Chamberlaine	October 5, 1931
Ridgway, Robert	October 2, 1934
Riggs, Henry Earle	October 13, 1941
Robins, Thomas Matthews	June 5, 1956
Robinson, Moncure	July 6, 1853
ROUSE, HUNTER	April 8, 1973
Rowland, Thomas Fitch	December 20, 1899
Sabin, Louis Carlton	October 9, 1944
Savage, John Lucian	October 13, 1941
Saville, Thorndike	April 10, 1961
Schneider, Charles Prosper Eugene	October 10, 1921
Scobey, Fred C.	June 20, 1960
Sherman, Leroy Kempton	October 14, 1946
SHUMATE, CHARLES E.	April 20, 1974
Singstad, Ole	June 5, 1956
Smith, Jonas Waldon	October 1, 1928
Somervell, Brehon Burke	October 12, 1942
Spofford, Charles Milton	June 16, 1953
Steele, I. Cleveland	May 16, 1966
Stevens, John Frank	June 19, 1922
Storrow, Charles Storer	January 24, 1893
Strobel, Charles Louis	October, 3, 1932
Surveyer, Arthur	October 9, 1944
SVERDRUP, LIEF J.	May 16, 1966
Swain, George Fillmore	October 7, 1929
Swazey, Ambrose	June 6, 1921
Talbot, Arthur Newell	November 16, 1925
Taylor, Thomas Ulvan	October 6, 1939
Terzaghi, Karl	October 13, 1947
Thomson, Reginald Heber	October 14, 1940
TIMBY, ELMER KNOWLES	April 26, 1972
Torres, Ary Frederico	June 16, 1952
Totten, Joseph Gilbert	March 2, 1853
TURNBULL, WILLARD J.	May 13, 1968
Turneaure, Frederick Eugene	September 26, 1933
TURNER, FRANCIS C.	April 8, 1973
TURNER, NATHANIEL P.	April 20, 1974
Tuttle, Arthur Smith	October 10, 1938
Unwin, William Cawthorne	June 19, 1922
VANCE, JAMES A.	April 20, 1974
VEATCH, NATHAN THOMAS	May 13, 1963
von Karman, Theodor	November 1, 1949
Waddell, John Alexander Low	October 11, 1936
Waite, Henry Matson	October 13, 1941
Wall, Edward Everett	October 10, 1938
Weber, Baron Christian Phillipp Max Maria von	June 2, 1880
Weiss, Andrew	October 11, 1948
Weymouth, Frank Elwin	October 10, 1938
Whipple, Squire	May 6, 1868
White, Sir William Henry	December 16, 1904
Whittemore, Don Juan	January 6, 1911
WHITTON, REX M.	April 8, 1973
WILBUR, LYMAN D.	May 13, 1968
Wilgus, William John	October 14, 1935
Wilson, Wilbur, M.	November 1, 1949
Wilson, William Hasell	August 2, 1892
Wiley, Ralph Benjamin	June 5, 1956
WINKELMAN, DWIGHT WILLIAM	April 26, 1972
WINTER, GEORGE	April 8, 1973
WISEL' WILLIAM H.	April 8, 1973
WOLMAN, ABEL	April 10, 1961

Name	Date of Election to Honorary Membership
Woodward, Sherman Melville	October 12, 1942
Worcester, Joseph Ruggles	October 4, 1937
Worthan, William Ezra	April 4, 1893
Wright, Horatio Gouveneur	March 3, 1880
Wynne-Edwards, Robert	May 8, 1967
Young, William Clark	March 1, 1892

APPENDIX V

ASCE AWARDS AND RECIPIENTS

The honors program of the Society has as its basic objective the advancement of the engineering profession by emphasizing exceptionally meritorious achievement. Traditionally, such accomplishments have been in the form of technical papers, although some awards are based on other contributions. The awards are made by the Board of Direction, in the name of the entire Society, on the recommendation of Society agencies designated in each particular case.

Most of these honors may be bestowed yearly. Details, including eligibility and presentation, are given in the following pages. The ceremony of conferring awards is normally held at the Annual Meeting of the Society in October.

In addition to the following awards, prizes, and fellowships —which are listed in the order in which they were established — ASCE participates in the following joint-society awards: John Fritz Medal, Hoover Medal, Alfred Noble Prize, and Washington Award.

The Norman Medal

The Norman Medal of the American Society of Civil Engineers was instituted and endowed in 1872 by George H. Norman, M. ASCE. With the assent and the approval of the donors, on June 1, 1897, the Society assumed responsibility for the payment in perpetuity of The Norman Medal.

I. Competition for the Norman Medal is restricted to members of the Society.

II. A gold medal, bronze duplicate, and certificate are awarded as hereinafter provided.

III. All original papers presented to the Society by members in any grade, are open to the award, provided that such papers have not been previously contributed in whole or in part to any other association, nor have appeared in print prior to their publication by the Society. Papers written jointly by members and nonmembers are not eligible.

IV. The medal, duplicate, and certificate are awarded to the author, or authors, of a paper which shall be judged worthy of special commendation for its merit as a contribution to engineering science.

1874. J. James R. Croes
1875. Theodore G. Ellis
1877. William W. Maclay.
    Book Prize to Julius H. Striedonger.
1879. Edward P. North.
    Book Prize to Max E. Schmidt.
1880. Theodore Cooper
1881. L.L. Buck
1882. A. Fteley and F.P. Stearns
1883. William P. Shinn
1884. James Christie
1885. Eliot C. Clarke
1886. Edward Bates Dorsey
1887. Desmond FitzGerald
1888. E.E. Russel Tratman
1889. Theodore Cooper
1890. John R. Freeman
1891. John R. Freeman
1892. William Starling
1893. Desmond FitzGerald
1894. Alfred E. Hunt
1895. William Ham. Hall
1896. John E. Greiner
1897. Julius Baier
1898. B.F. Thomas
1899. E. Herbert Stone
1900. James A. Seddon
1902. Gardner S. Williams, Clarence W. Hubbell, and George H. Fenkell
1904. Emile Low
1905. C.C. Schneider
1906. John S. Sewell
1907. Leonard M. Cox
1908. C.C. Schneider
1909. J.A.L. Waddell
1910. C.E. Grunsky
1911. George Gibbs
1912. Wilson Sherman Kinnear
1913. J.V. Davies
1914. Caleb Mills Saville
1915. Allen Hazen
1916. J.A.L. Waddell
1917. Benjamin F. Groat
1918. L.R. Jorgensen
1919. William Barclay Parsons
1920. J.A.L. Waddell
1922. Charles H. Paul
1923. D.B. Steinman
1924. B.F. Jakobsen
1925. Harrison P. Eddy
1926. Julian Hinds
1927. B.F. Jakobsen
1928. Charles E. Sudler
1929. G.T. Rude
1930. Charles Terzaghi
1931. Floyd A. Nagler and Albion Davis
1933. Hardy Cross
1934. Leon S. Moisseiff
1935. D.C. Henny
1936. Daniel W. Mead
1937. J.C. Stevens
1938. Hunter Rouse
1939. Charles H. Lee
1940. Shortridge Hardesty and Harold E. Wessman
1941. J.A. Van den Broek
1942. Karl Terzaghi
1943. Thomas E. Stanton
1944. Ralph B. Peck
1945. Merrill Bernard
1946. Karl Terzaghi
1947. Boris A. Bakhmeteff and William Allan
1948. Alfred M. Freudenthal
1949. Gerard H. Matthes
1950. Friedrich Bleich
1951. D.B. Steinman
1953. Friedrich Bleich and L.W. Teller
1954. Robert H. Sherlock
1955. Karl Terzaghi
1956. Carl E. Kindsvater and Rolland W. Carter
1957. Alfred M. Freudenthal
1958. Anestis S. Veletsos and Nathan M. Newmark
1959. Willard J. Turnbull and Charles R. Foster
1960. Carl E. Kindsvater and Rolland W. Carter
1961. Lorenz G. Straub and Alvin G. Anderson
1962. William McGuire and Gordon P. Fisher
1963. Bruno Thurlimann
1964. T. William Lambe
1965. Gerald A. Leonards and Jagdish Narain
1966. Charles H. Lawrance
1967. Daniel Dicker
1968. H. Bolton Seed and Kenneth L. Lee
1969. Basil W. Wilson
1970. Cyril J. Galvin, Jr.
1971. John H. Schmertmann
1972. Nicholas C. Costes, W. David Carrier III, James K. Mitchell, and Ronald F. Scott
1973. Bobby O. Hardin and Vincent P. Drenivich
1974. James R. Cofer

The Thomas Fitch Rowland Prize

The Thomas Fitch Rowland Prize was originally instituted by the Society at the Annual Meeting of 1882. It was endowed in 1884 by Thomas Fitch Rowland, past Vice President and Hon. M. ASCE.

I. The award is not restricted to members of the Society.

II. The prize consists of a wall plaque and a certificate.

III. In the award of this prize preference is given to papers describing in detail accomplished works of construction, their cost, and errors in design and execution. Only papers published by the Society in the twelve-month period ending with June of the year preceding the year of award are eligible.

IV. The award will be made on the basis of a report of the Executive Committee of the Construction Division, made in writing to the Executive Director and subsequently ratified by the Board of Direction of the Society.

1883. G. Lindenthal
1884. Hamilton Smith, Jr.
1885. A.M. Wellington
1886. Charles C. Schneider
1887. William Metcalf
1888. Clemens Herschel
1889. James D. Schuyler
1890. O. Chanute, John F. Wallace, and William H. Breithaupt
1891. William H. Burr
1892. Samuel M. Rowe, Stillman W. Robinson, and Henry H. Quimby
1893. William Murray Black
1894. David L. Barnes
1895. William R. Hill
1896. H. St. L. Coppee
1897. Arthur L. Adams
1898. Henry Goldmark
1899. R.S. Buck
1900. Allen Hazen
1901. L.G. Montony
1902. William W. Harts
1903. George W. Fuller
1904. George Cecil Kenyon
1905. Charles L. Harrison and Silas H. Woodard
1906. George B. Francis and W.F. Dennis
1907. James D. Schuyler
1908. Edward E. Wall
1909. William J. Wilgus
1910. John H. Gregory
1911. B.H.M. Hewett and W.L. Brown
1912. Eugene Klapp and W.J. Douglas
1913. Burgis G. Coy
1914. H.T. Cory
1915. Charles W. Staniford
1916. E.L. Sayers and A.C. Polk
1917. John Vipond Davies
1918. F.W. Scheidenhelm
1919. O.H. Ammann
1920. Charles Evan Fowler
1921. Ernest E. Howard
1922. Gustav Lindenthal
1923. F.W. Peek, Jr.
1925. H. de B. Parsons
1926. Nicholas S. Hill, Jr.
1927. L.S. Stiles
1928. Roderick B. Young
1929. D.B. Steinman and William G. Grove
1930. R. McC. Beanfield
1931. Samuel A. Greeley and William D. Hatfield
1932. Clifford Allen Betts
1933. J.C. Baxter
1934. Miles I. Killmer
1935. W.H. Kirkbride
1936. A.V. Karpov and R.L. Templin
1937. Eugene A. Hardin
1940. J.D. Galloway
1941. O.J. Todd and S. Eliassen
1942. Shortridge Hardesty and Alfred Hedefine
1943. Paul Baumann
1944. Eugene L. Grant
1945. Donald N. Becker
1946. James B. Hays
1947. R.F. Blanks and H.S. Meissner
1948. M.M. Fitzhugh, J.S. Miller, and Karl Terzaghi
1949. H.M. Westergaard
1950. R.N. Bergendoff and Josef Sorkin
1951. William K. Boyd and Charles R. Foster
1952. Clarence E. Keefer
1953. E. Montford Fucik
1954. A. Warren Simonds

1955. Maurice N. Quade
1956. Jonathan Jones
1957. Harry N. Hill, Ernest C. Hartmann, and John Wood Clark
1958. Charles I. Mansur and Robert I. Kaufman
1959. J. George Thon and Gordon L. Coltrin
1960. Jack W. Hilf
1961. H. Bolton Seed, Robert L. McNeill, and Jacques de Guenin
1962. James D. Parsons
1963. Marvin J. Kudroff
1964. Francis E. Mullen
1965. Kent S. Ehrman
1966. L. Earl Tabler, Jr.
1967. John W. Kinney, Herman Rothman, and Frank Stahl
1968. C.Y. Lin
1969. James Douglas
1970. Norman L. Liver
1971. Richard E. Whitaker
1972. Hans Sacrison
1973. John V. Bartlett, Tadeusz M. Noskiewicz, and James A. Ramsay
1974. Russell C. Borden and Carl E. Selander

The Collingwood Prize

The Collingwood Prize was instituted and endowed in 1849 by Francis Collingwood, Past Secretary and M. ASCE.

I. The competition for the prize is restricted to Associate Members of the Society who hold that grade at the time the paper is submitted in essentially its final form.

II. The prize consists of a wall plaque and a certificate.

III. The prize is awarded to the author or authors of a paper (1) describing an engineering work with which he or they have been directly connected, or (2) recording investigations contributing to engineering knowledge to which he or they have contributed some essential part, and (3) containing a rational digest of results. Any mathematical treatment must show immediate adaptability to professional practice. Accuracy of language and excellence of style are factors in the award.

IV. Papers published by the Society during the twelve-month period ending with June of the year preceding the year of award are eligible.

V. These rules may be modified by the Board of Direction.

1895. Morton L. Byers
1896. Herbert Waldo York
1899. Julius Kahn
1900. Robert P. Woods
1901. F.A. Kummer
1903. Isaac Harby
1904. Herbert J. Wild
1905. E.P. Goodrich
1908. D.W. Krellwitz
1909. H.L. Wiley
1911. A. Kempkey, Jr.
1912. W.W. Clifford
1914. J.S. Longwell
1915. George Schobinger
1916. Harold Perrine and George E. Strehan
1917. Clement E. Chase
1918. James B. Hays
1919. Floyd A. Nagler
1920. Floyd A. Nagler
1921. L. Standish Hall
1923. Jacob Feld
1926. Cecil Vivian von Abo
1927. William Breuer
1928. Franklin Hudson, Jr.
1929. William J. Cox
1932. A.R.C. Markl
1933. Bernard L. Weiner
1934. G.H. Hickox and G.O. Wessenauer
1935. C. Maxwell Stanley
1936. Clinton Morse
1937. Victor L. Streeter
1938. Douglas M. Stewart
1939. B.K. Hough, Jr.

1940. Kenneth D. Nichols
1941. Elmer Rock
1942. John F. Curtin
1943. Roy K. Linsley, Jr. and William C. Ackermann
1944. Walter L. Moore
1945. Carl E. Kindsvater
1946. C.O. Clark
1947. F.L. Ehasz
1948. John K. Vennard
1949. Alfred Machis
1950. Charles A. Lee and Charles E. Bowers
1951. John W. Forster and Raymond A. Skrinde
1952. T. William Lambe
1953. Kuang-Han Chu
1954. Vaughn E. Hansen
1955. William J. Bauer
1956. John H. Schmertmann
1957. George E. Mac Donald
1958. Turgut Sarpkaya
1959. Norman H. Brooks
1960. Ralph L. Barnett
1961. Sidney A. Guralnick
1962. Ronald T. McLaughlin
1963. Robert L. Kondner
1964. Melvin T. Davisson and Henry L. Gill
1965. George G. Goble
1966. Russel C. Jones and John A. Hribar
1967. Richard N. White and Pen Jeng Fang
1968. Maurice L. Sharp
1969. Richard D. Woods
1970. Kenneth L. Lee and C.K. Shen
1971. P. Aarne Vesilind
1972. James M. Duncan and G. Wayne Clough
1973. Thomas M. Lee
1974. James C. Anderson and Raj P. Gupta

The J. James R. Croes Medal

This prize was established by the Society on October 1, 1912, and is named in honor of the first recipient of the Norman Medal, John James Robertson Croes, Past President, ASCE.

I. The award consists of a gold medal, bronze duplicate, and certificate.

II. The medal is awarded annually, under the rules governing the award of the Norman Medal.

III. The medal, duplicate, and certificate are awarded to the author, or authors, of such paper as may be judged worthy of the award and be next in order of merit to the paper to which the Norman Medal is awarded; or, if the Norman Medal is not awarded, then to the author, or authors, of a paper, if any, which is judged worthy of the award of this prize for its merit as a contribution to engineering science.

1913. B.F. Cresson, Jr.
1914. J.B. Lippincott
1915. Richard R. Lyman
1916. C.E. Smith
1917. Henry S. Prichard
1918. Israel V. Werbin
1919. D.B. Steinman
1920. B.A. Smith
1921. Fred A Noetzli
1922. William Cain
1923. James F. Sanborn
1924. Joel D. Justin
1925. Charles S. Whitney
1926. Clarence S. Jarvis
1927. Henry Clay Ripley
1930. H. de B. Parsons
1931. John R. Freeman
1932. David L. Yarnell and Floyd A. Nagler
1933. Earl I. Brown
1934. H.M. Westergaard
1935. A.T. Larned and W.S. Merrill
1936. Wilbur M. Wilson
1937. Inge Lyse and Bruce G. Johnston
1938. E.C. Hartmann
1939. C.A. Mockmore
1940. Edward J. Rutter, Quintin B. Graves, and Franklin F. Snyder
1941. Earl I. Brown
1942. Charles F. Ruff
1943. C.H. Gronquist
1944. N.M. Newmark
1945. G.H. Hickox
1946. Gail A. Hathaway
1947. Thomas R. Camp
1948. Karl DeVries
1949. F.H. Kellogg
1950. L.F. Harza
1951. M.E. Von Seggern

1953. Paul Rogers
1954. Fu-Kuei Chang and Bruce G. Johnston
1955. John S. McNown
1956. Jack W. Carter, Kenneth H. Lenzen, and Lawrence T. Wyly
1957. William E. Wagner
1958. William J. Oswald and Harold B. Gotaas
1959. Charles I. Mansur and Robert I. Kaufman
1960. H. Bolton Seed and Clarence K. Chan
1961. George Winter
1962. H. Bolton Seed and C.K. Chan
1963. Melvin L. Baron, Hans Bleich and Paul Weidlinger
1964. Daryl B. Simons and Maurice L. Albertson
1965. Jack G. Bouwkamp
1966. John W. Clark and Richard L. Rolf
1967. Carl H. Plumlee
1968. William R. Hudson and Hudson Matlock
1969. Hugo B. Fischer
1970. John M. Henderson
1971. Erich J. Plate and John H. Nath
1972. H. Bolton Seed and Izzat M. Idriss
1973. David J. D'Appolonia, Harry G. Poulos, and Charles C. Ladd
1974. Subrata K. Chakrabarti

The James Laurie Prize

The James Laurie Prize was established by the Society on October 1, 1912, and is named in honor of the first President of the Society. In the period from 1912 through 1965, the prize was awarded under the rules for the Thomas Fitch Rowland Prize for the paper judged to be next in order of merit. By action of the Board of Direction on October 19-20, 1964, beginning in 1966 the prize has been awarded on the basis of contributions to the advancement of transportation engineering.

I. The award is made annually to a member of the American Society of Civil Engineers who has made a definite contribution to the advancement of transportation engineering, either in research, planning, design, or construction, these contributions being made either in the form of papers or other written presentations, or through notable performance or specific actions which have served to advance transportation engineering.

II. Not more than one award of the prize is made each year unless the achievement upon which the award is based is considered to be the joint contribution by more than one man, such as the joint authorship of a paper. In such cases duplicate or multiple prizes will be awarded. No one shall receive the honor more than once.

III. Each Section and Branch and Technical Division Executive Committee shall be entitled to enter the name of one nominee annually. These names, and supporting documents, will be assembled by the Executive Director and referred to the judging committee.

IV. The judging committee shall recommend the recipient, if any, to the Board of Direction for final action.

V. The award consists of a plaque and certificate, suitably inscribed with the name of the recipient and the circumstances of the award.

Previous criteria:

1913. M.M. O'Shaughnessy
1914. Samuel Tobias Wagner
1015. J.E. Greiner
1916. William G. Grove and Henry Taylor
1917. H.R. Stanford
1918. Charles W. Staniford
1919. F.W. Gardiner and S. Johannesson
1920. Dabney H. Maury
1921. W.C. Curd
1922. Arthur T. Safford and Edward Pierce Hamilton
1923. R.W. Gausmann and C.M. Madden
1924. C.M. Allen and I.A. Winter
1925. William Kelly
1926. Lewis A. Perry
1927 John R. Baylis

1928. James F. Case
1929. R.H. Keays
1930. John H. Gregory, C.B. Hoover, and C.B. Cornell
1932. Earl I. Brown
1933 W.W. Saunders
1934. E.W. Bowden and H.R. Seely
1935. Wilson T. Ballard
1936. Paul Baumann
1937. Boris A. Bakhmeteff and Arthur E. Matzke
1938. Leon S. Moisseiff
1939. Stanley M. Dore
1940. AM Rawn, A. Perry Banta, and Richard Pomeroy
1941. Samuel A. Greeley
1942. W. Watters Pagon
1943. T.A. Middlebrooks
1944. Gordon R. Williams
1945. Ole Singstad
1946. L.A. Schmidt, Jr.
1947. Ross M. Riegel
1948. Hyde Forbes
1949. G.H. Hickox, A.J. Peterka, and R.A. Elder
1950. Harris G. Epstein
1951. Hans H. Bleich
1952. John M. Kyle
1953. Harold N. Fisk
1954. Thomas A. Middlebrooks
1955. Joseph N. Bradley
1956. Samuel I. Zack
1957. Walter L. Dickey and Glenn B. Woodruff
1958. Chesley J. Posey
1959. Bramlette McClelland and John A. Focht, Jr.
1960. Ernest F. Masur
1961. J. Barry Cooke
1962. Marvin Gates
1963. Norman J. Magneson
1964. Jean J. Martin
1965. Gail B. Knight

Current criteria:
1966. Kenneth B. Woods
1967. Harmer E. Davis
1968. Robert Horonjeff
1969. Walter S. Douglas
1970. Charles E. Shumate
1971. Francis C. Turner
1972. Donald S. Berry
1973. Robert W. Brannan
1974. William A. Bugge

The Arthur M. Wellington Prize

This prize was instituted and endowed by *The Engineering News-Record* in 1921.

I. The award is not restricted to member of the Society.

II. The prize consists of a wall plaque and a certificate.

III. The prize is awarded for papers on transportation, on land, on the water, or in the air. At the request of *The Engineering News-Record,* the scope of the award was broadened by action of the Board, in July 1946, to include papers on foundations and closely related subjects, but not including contributions in the form of reports and manuals.

1923. J.P. Newell
1924. Rufus W. Putnam
1925. William M. Black
1926. Charles W. Kutz
1927. John A. Miller, Jr.
1928. Roy G. Finch
1929. F.C. Carstarphen
1930. George Gibbs
1931. G.F. Schlesinger
1932. Fred Lavis
1933. D.J. Kerr
1934. J.C. Evans
1935. Hawley S. Simpson
1937. E.C. Harwood
1938. Charles M. Noble
1939. Rufus W. Putnam
1941. Fred Lavis
1942. William J. Wilgus
1943. Milton Harris
1946. James H. Stratton
1948. Joseph Barnett
1949. Arthur Casagrande
1950. A.A. Anderson
1951. Alexander Hrennikoff
1952. Basil Wrigley Wilson
1953. L.A. Nees
1954. Claude H. Chorpening
1955. R.J. Ivy, T.Y. Lin, Stewart Mitchell, N.C. Raab, V.J. Richey, and C.F. Scheffey

1956. John Hugh Jones and Robert Horonjeff
1957. Wesley G. Holtz and Harold J. Gibbs
1958. Hamilton Gray
1959. Frank H. Newnam, Jr.
1960. John M. Biggs, Herbert S. Suer, and Jacobus M. Louw
1961. T. William Lambe
1962. James D. Parsons
1963. F.E. Richart, Jr.
1964. H.W. Reeves
1965. Ralph B. Peck and Tonis Raamot
1966. William J. Oswald, Clarence G. Golueke, and Donald O. Horning
1967. Leopold H. Just, Ira J. Levy, and Vladimir F. Obrician
1968. H. Bolton Seed and Stanley D. Wilson
1969. Melvin L. Baron, Hans H. Bleich, and Joseph P. Wright
1970. Edwin W. Eden, Jr.
1971. Jean J. Janin and Guy F. LeSciellour
1972. Daniel Dicker
1973. James L. Sherard, Rey S. Decker, and Norman L. Ryker
1974. William K. Mackay

The Freeman Fellowship

The Freeman Fund was established in 1924 by John R. Freeman, Past President and Honorary Member, ASCE. The income from the fund is to be used in the aid and encouragement of young engineers, especially in research work for objectives as indicated in the rules.

In addition to the Society's Freeman Fund, Mr. Freeman established similar trusts with the Boston Society of Civil Engineers and the American Society of Mechanical Engineers. In 1949 ASME and ASCE entered an informal agreement by which each would award a fellowship every other year, in alternating years.

I. The ASCE grant is offered in odd numbered years (e.g., the Society's fiscal year beginning October 1973). In even numbered years a similar grant is offered by the American Society of Mechanical Engineers, all applications being judged on their merits regardless of Society affiliation. There is no established application form as such. The recipient will be selected on the basis of the quality of his application. Six copies of the application are required.

II. Grants are made toward expenses for experiments, observations, and compilations to discover new and accurate data that will be useful in engineering.

III. Each application shall include a statement, in general terms, of the purposes for which the funds are expected to be used.

IV. The fellowship also recognizes underwriting fully or in part some of the loss that may be sustained in the publication of meritorious books, papers, or translations pertaining to hydraulic science and art which might, except for some such assistance, remain mostly inaccessible.

V. The grant can be in the form of a prize for the most useful paper relating to the science or art of hydraulic construction.

VI. Traveling scholarships are available. These are open to members younger than 45 years, in any grade of membership, in recognition of achievement, or promise. These are also for the purpose of aiding the candidate to visit engineering works in the United States, or in any other part of the world, where there is good prospect of obtaining information useful to engineers.

VII. The grant may be for assisting in the translation, or publication in English, of papers or books in foreign languages pertaining to hydraulics.

1927. Lorenz G. Straub; F. Theodore Mavis; Morrough P. O'Brien
1928. Clarence E. Bardsley
1929. James G. Woodburn.
1930. Hans Kramer
1931. Hans Kramer
1932. Herbert H. Wheaton; Donald P. Barnes
1935. Paul W. Thompson
1936. John Hedberg
1938. Douglas C. Davis
1939. Miles M. Dawson

1940. Haywood G. Dewey, Jr.
1947. George F. Dixon
1953. Ira A. Hunt, Jr.
1955. Walter J. Tudor
1957. Norbert L. Ackerman
1959. Willard E. Fraize
1961. Jacques William Delleur
1963. Roger J.M. DeWeist
1965. Hsieh W. Shen
1967. Frederick A. Locher
1969. F.G. Alden Burrows
1971. V.W. Goldschmidt
1973. Walter H. Graf

The Rudolph Hering Medal

The Rudolph Hering Medal was instituted and endowed in 1924 by the Sanitary Engineering Division (now the Environmental Engineering Division) of the American Society of Civil Engineers, in honor of Rudolph Hering, past Vice President of the Society. The medal is awarded in accordance with the rules of the Society, except that the paper for which the medal is awarded should be recommended by a committee of three members of the division appointed for the purpose by the executive committee of the division, which committee reviews the recommendation and acts upon it not later than February 15.

I. A bronze medal and certificate may be awarded annually as hereinafter provided.

II. All original papers authored or co-authored by ASCE members dealing with water works, sewerage works, drainage, refuse collection and disposal, or any branch of environmental engineering which are presented to the Society, in finished form, whether presented to the Environmental Engineering Division or otherwise, are open to the award, provided that such papers have not been previously contributed or published elsewhere. Papers published by the Society in the twelve-month period ending with June of the year preceding the year of award are eligible.

III. The medal is awarded to the author, or authors, of the paper which contains the most valuable contribution to the increase of knowledge in, and to the advancement of, the environmental branch of the engineering profession.

IV. In any year when the excellence of more than one paper justifies it, the prize committee designates a "second order of merit." A paper so recognized is considered eligible to compete in the award for the next succeeding year.

1927. Harrison P. Eddy
1931. Samuel A. Greeley and W.D. Hatfield
1935. John H. Gregory, R.H. Simpson, Orris Bonney, and Robert A. Allton
1937. W.W. Horner and F.L. Flynt
1939. A.J. Schafmayer and B.E. Grant
1940. J.W. Ellms
1941. Thomas H. Wiggin
1942. Robert T. Regester
1943. George J. Schroepfer
1945. Langdon Pearse
1947. A.L. Genter
1948. C.E. Jacob
1953. Ralph Stone and William F. Garber
1954. Harold A. Thomas, Jr. and Ralph S. Archibald
1955. W.F. Langelier, Harvey F. Ludwig, and Russell G. Ludwig
1956. Thomas R. Camp
1957. Alfred C. Ingersoll, Jack E. McKee, and Norman H. Brooks
1958. William J. Oswald and Harold B. Gotaas
1959 Donald J. O'Connor and William E. Dobbins
1960. Charles G. Gunnerson
1961. A.L. Tholin and Clint J. Keifer
1962. AM Rawn, F.R. Bowerman, and Norman H. Brooks
1963. Peter A. Krenkel and Gerald T. Orlob
1964. Ross E. McKinney
1965. William E. Dobbins
1966. Donald J. O'Connor
1967. Charles G. Gunnerson
1968. Robert L. Johnson and John L. Cleasby
1969. Raymond C. Loehr and Robert W. Agnew

1970. Kou-Ying Hsung and John L. Cleasby
1971. Committee on Environmental Quality Management of the Sanitary Engineering Division
1972. Harvey F. Collins, Robert E. Selleck, and George C. White
1973. John D. Parkhurst and Richard D. Pomeroy
1974. Kenneth S. Price and Richard A. Conway

The J. Waldo Smith Hydraulic Fellowship

This fellowship was established by the Board of Direction of the Society in 1938 and was made possible by J. Waldo Smith, past Vice President and Hon. M. ASCE, who bequeathed funds to the Society.

I. The fellowship is offered every third year, beginning with the fall term of the academic year 1973-1974. It runs for one full academic year and provides $2,000, plus as much more up to $1,000 as may be required for physical equipment connected with the research. Such equipment becomes the property of the institution upon completion of the work.

II. Applications for the fellowship come through the various institutions applying. There is no established application form as such. The recipient will be selected on the basis of the quality of his application. Six copies of the application are required.

III. Each application shall include a statement, in general terms, of the purposes for which the funds are expected to be used.

IV. Administration is in part through the institution which invites cooperation, through its engineering faculty.

V. The award is made to that graduate student, preferably an Associate Member of the Society, who gives promise of best fulfilling the ideals of the fellowship.

VI. Under the provisions adopted by the Board, the scope of the fellowship should be restricted to research in the field of experimental hydraulics as distinguished from that of purely "theoretical hydraulics." To this end, emphasis is to be placed on practical experiments designed and executed for the purpose of advancing knowledge with respect to the laws of hydraulic flow, rather than to the type of research which proceeds on the theory of mathematical analysis based on assumptions of unknown validity. The essence of the purpose of the research is to test the assumptions which are currently made, and also to develop a better understanding of fluid flow.

1939. Walter J. Meditz
1940. James M. Robertson
1941. Percy H. Bliss
1948. Russell J. Kennedy
1949. Joseph C. Kent
1950. Arthur W. Van't Hul
1952. James R. Barton
1953. Robert E. Templeton
1954. Gunnar Sigurdsson
1955. Peter L. Monkmeyer
1958. William W. Troutman
1961. John A. Hoopes
1964. Carl R. Goodwin
1967. Rolf Kellerhals
1970. George D. Ashton
1973. Qais N. Fattah

The A.P. Greensfelder Construction Prize

In 1939, an ASCE Construction Engineering Prize was provided for by A.P. Greensfelder, Hon. M. ASCE, as part of the St. Louis Regional Planning and Construction Foundation Trust. By Board action in October 1961, the name of the award was changed to "The A.P. Greensfelder Construction Prize."

I. Competition for the A.P. Greensfelder Construction Prize is open to all authors, members and nonmembers alike, of articles as defined in Rule III.

II. The prize consists of $150 in cash, a plaque, and an appropriate certificate. In case of more than one author, the cash will be divided, and each will receive a plaque and a certificate.

III. The prize may be awarded annually to the author, or authors, of the best original scientific or educational article on construction, printed in *Civil Engineering*, during the specified annual period, but not previously published. The article should deal primarily with the construction phase of an engineering project, which shall have been performed by a construction organization, corporation, company, partnership, or individual having its headquarters in the United States of America, using methods applicable in the United States. The presentation should accentuate one or more of the following: (a) Ingenuity, novelty or efficiency of field methods, (b) unit cost-time data, (c) engineering basis for advanced practice, (d) engineering analyses or discussions in which the problems inherent in a given situation are recognized and developed.

IV. In judging articles as published in abridged form, due weight may be given to the original complete manuscripts.

V. Articles printed in the twelve issues of *Civil Engineering* ending with June of the year preceding the year of award are eligible for consideration.

VI. The award will be made on the basis of favorable report of the Executive Committee of the Construction Division made in writing to the Executive Director not later than February 15, and subsequently ratified by the Board of Direction of the Society.

VII. In any year when the excellence of more than one article justifies it, the prize committee will designate in addition to the prize article a "second order of merit." Such additional recommendation is not to be publicly announced but the article so recognized is considered eligible to compete in the award for the next succeeding year.

VIII. The public presentation of the prize will be made at the Annual Meeting next following submission of the report.

1939. Howard L. King
1940. Russell G. Cone
1941. E. Leland Durkee
1942. Frederic R. Harris
1943. Carlton B. Jansen
1944. Fred W. Stiefel
1945. C. Glenn Cappel
1946. George K. Leonard
1948. C. Glenn Cappel
1950. W.C. Mason
1951. James G. Tripp
1952. John N. Newell
1953. Ben C, Gerwick, Jr.
1954. Samuel D. Sturgis, Jr.
1955. John A. Dominy, Charles C. Zollman, and Henon Pearce
1956. Arve S. Wikstrom
1957. John N. Newell
1958. Myers Van Buren
1959. Edward E. White
1960. Joseph Peraino
1961. Gail Knight
1962. John W. Fowler
1963. Bertold E. Weinberg
1964. Robert E. White
1965. Gabriel A. Reti
1966. Ernest Graves, Jr.
1967. Louis W. Riggs, Jr.
1968. J.A. Sterner
1969. W.W. Wolcott and J. Birkmyer
1970. Martin S. Kapp
1971. Paul J. Varello
1972. Carlton T. Wise and D.D. Gillespie
1973. Edward Peterson and Peter Frobenius
1974. Ernest C. Harris and John A. Talbott

The Karl Emil Hilgard Hydraulic Prize

This prize was instituted in 1939 as a result of an endowment left to the Society for the purpose by Karl Emil Hilgard, M. ASCE.

I. The prize consists of a wall plaque and a certificate.

II. Papers published by the Society during the twelve-month period ending with June of the year preceding the year of award are eligible.

III. The award is given to the author, or authors, of that paper which is judged to be of superior merit, dealing with a problem of flowing water, either in theory or practice. Preferably, the award is given to a paper that is not otherwise recognized by receiving another Society prize. The value of the paper is judged both on the basis of the subject matter and the method of presentation.

IV. In any year when the excellence of more than one paper justifies it, the prize committee designates a "second order of merit." A paper so recognized is considered eligible to compete in the next award.

V. The Executive Committee of the Hydraulics Division each year considers the possibilities for the Hilgard prize, whether or not it is to be awarded that year and makes recommendation accordingly.

VI. The rules of this award and the honorarium are subject to the administration of the Board of Direction.

1941. Thomas R. Camp
1943. Harold A Thomas and Emil P. Schuleen
1945. L. Standish Hall
1947. A.A. Kalinske
1949. Vito A. Vanoni
1951. Maurice L. Albertson, Y.B. Dai, Randolph A. Jensen, and Hunter Rouse
1953. Arthur T. Ippen
1955. James M. Robertson and Donald Ross
1957. Donald Ross
1959. Emmett M. Laursen
1961. Hunter Rouse, Tien To Siao, and S. Nagaratnam
1963. Gerald T. Orlob
1965. Francis F. Escoffier and Marden B. Boyd
1967. Hubert J. Tracy
1968. Samuel O. Russell and James W. Ball
1969. Donald Van Sickle
1970. Norman H. Brooks and Robert C.Y. Koh
1971. Edward R. Holley, Donald R.F. Harleman, and Hugo B. Fischer
1972. Robert W. Zeller, John A. Hoopes, and Gerard A. Rohlich
1973. Wayne C. Huber, Donald R.F. Harleman, and Patrick J. Ryan
1974. George D. Ashton and John F. Kennedy

The Daniel W. Mead Prizes

These prizes were established and endowed in 1939 by Daniel W. Mead, Past President and Honorary Member, ASCE.

I. The Daniel W. Mead Prizes for Associate Members and for Students are awarded annually on the basis of papers on professional ethics. Each year the specific topics (if any) of the contests for the forthcoming year shall be selected by the Committee on Younger Members in the case of the Associate Member award and by the Committee on Student Chapters for the Student award.

II. Associate Members and members of the Student Chapters of the Society who are in good standing at the time their papers are presented (see item III) are eligible for the respective contests, unless they have previously received the national award for which they are competing.

III. To be entered in the national contests a paper must have been presented before a local, regional, or national meeting or conference of the Society, or its Divisions, Sections, or Student Chapters. If the paper is presented as an entry in a local paper contest, only the winning paper shall be entered in the national competition.

IV. Papers for the national contests shall (a) be limited to one paper from each Section or Student Chapter; (b) be submitted, through the Section Secretary (for Associate Members) or Faculty Advisor (for Students) to the Executive Director by June 1; (c) not exceed 2,000 words in length; (d) be written by only one person; and (e) not have previously been published in other than school or Society publications.

V. Associate Member papers shall be judged by the Committee on Younger Members; the Committee on Student Chapters shall review the Student papers. Each committee may nominate a winner from each of the four zones of the Society and of these one may be designated as the national winner. The nominations, if any, shall be presented to the Professional Activities Committee for final action during its meetings in July.

VI. The winners in each zone shall receive an appropriate certificate recognizing their achievements. The national winners will receive appropriate certificates and wall plaques. The papers written by the national winners shall be published in the *Engineering Issues — Journal of Professional Activities* or in *Civil Engineering*.

Associate Members

1940. Allen Jones, Jr.
1941. Don P. Reynolds
1949. S.L. McFarland
1950. Roy G. Cappel
1951. Edgar G. Baugh
1952. Howard L. Payne
1953. Charles William Griffin, Jr.
1954. Alfred E. Waters, II
1955. Merle H. Banta
1956. Robert A. Schaack
1957. Robert S. Braden
1959. Morgan I. Doyne
1960. James M. Abernathy
1961. Donald B. Baldwin
1962. James I. Taylor
1963. T.A. Faulhaber
1964. Francis Pandullo
1965. Paul M. Wright
1966. L. Douglas James
1967. Donald R. Buettner
1968. Francis A. Paul
1969. Theodore Fellinger
1970. Chester Lee Allen
1971. Charles G. Sudduth
1972. Charles R. Schrader
1974. John E. Spitko, Jr.

Students

1940. Harry A. Balmer
1941. Edward Wesp, Jr.
1942. Alfred C. Ingersoll
1947. Lonnie G. Lamon
1948. Frank J. Kersnar
1949. William H. Blair
1950. Gordon L. Laverty
1951. Marion K. Harris
1952. Charles E. Negus, Jr.
1953. Carl Alan Rambow
1954. William H. Blackmer
1955. Jeremiah E. Abbott
1956. James Moser Anderson
1957. Robert Frowein
1958. William G. Benko
1959. L.G. McLaren
1960. Jack Keiser
1961. Robert A. Creed
1962. James R. Wright
1963. David H. Adams
1964. Dabney S. Cradock, III
1965. George E. Hunter
1966. Jerome W. Sargent
1967. Harald G. Biedermann
1968. Charles R. Schrader
1969. Theodore Fellinger
1970. Chester Lee Allen
1971. David Mann
1973. Harold H. Wagle
1974. Terry L. Turnock

The J.C. Stevens Award

This prize was established and endowed in 1943 by John C. Stevens, Past President, ASCE.

I. The prize is designated as The J.C. Stevens Award.

II. It is given to the one adjudged to have submitted the best discussion published by the Society during the twelve-month period ending with June of the year preceding the year of award.

III. The paper and its discussion must be in the field of hydraulics (including fluid mechanics and hydrology).

IV. Only one holding some grade of membership in the American Society of Civil Engineers is eligible for the award, and no one is to receive the prize or any portion thereof more than once.

V. In any year when the excellence of more than one discussion justifies it, the prize committee designates a "second order to merit." A discussion so recognized is considered eligible to compete in the award for the next succeeding year.

VI. Normally the prize consists of books costing not more than $50; however, the Executive Committee of the Hydraulics Division may authorize the Executive Director to exceed that limit if the income from the capital fund is sufficient to justify it.

VII. The recipient may select any books he desires, indicating his choices to the Executive Director of the Society for purchase. In the event that the discussion is submitted by more than one person the prize is to be divided as equally as possible among them.

VIII. In the front of each book is fixed a bookplate setting forth facts regarding the award, signed by the President and Executive Director of the Society.

IX. The selection of the recipient of the award is the responsibility of the Executive Committee of the Hydraulics Division of the Society.

1944. Boris A. Bakhmeteff and Nicholas V. Feodoroff
1945. Thomas R. Camp
1946. John S. McNown
1947. Maurice L. Albertson
1948. Preston T. Bennett
1949. Donald E. Blotcky
1950. William Allan
1951. James S. Holdhusen
1952. Harold R. Henry
1954. W. Douglas Baines
1955. Marion R. Carstens
1956. Serge Leliavsky
1957. Neal E. Minshall
1958. Frederick L. Hotes
1959. Norman H. Brooks
1960. Hans A. Einstein
1961. Daryl B. Simons and Everett V. Richardson
1962. R. Hugh Taylor and John F. Kennedy
1963. Thomas Maddock, Jr. and W.B. Langbein
1964. Herman J. Koloseus
1965. Alvin G. Anderson
1966. John L. French
1967. Richard R. Brock
1968. Ronald W. Jeppson
1969. Fred W. Blaisdell
1970. William W. Sayre
1971. Helmut Kobus
1972. Thomas R. Camp and S. David Graber
1973. Keith D. Stolzenbach and Donald R.F. Harleman
1974. Charles R. Neill

L.F. Harza Latin America Associate Member Awards

A group of awards in the form of entrance fees for Associate Members joining the Society was established in 1945 by L. F. Harza, M. ASCE. These awards were created to encourage mutual understanding between the engineers of the Americas. This fund and the income from it are to be used until exhausted to pay the entrance fees for Associate Membership in the Society, first year's dues, and Society badge for selected qualified engineer graduates of Latin American universities.

I. No more than eight such memberships are to be awarded in any one year nor more than two to graduates from any one country.

II. Candidates for Associate Membership are to be proposed by appropriate faculty heads of the several schools.

III. Selection of candidates is to be made for outstanding scholarship, personality, and interest in Pan American affairs.

IV. To receive consideration by the Society, each candidate proposed is required to submit application for Associate Membership on the standard form, and he must meet all requirements for Associate Membership in the Society.

V. Each candidate who is proposed by his university and who makes application is to be interviewed by one or more members of the Society selected by the Board of Direction.

VI. If, on application, the selected candidate is found by the Board to be qualified for Associate Membership, his entrance fee, his first year's dues, and the cost of his Society badge will be paid from the fund herein described, and he will be notified of his election as an Associate Member.

1948. Rafael Garcia-Montes, of Havana, Cuba.
1950. Julio Escobar-Fernandez, of Bogota, Colombia.
1952. Hernando Cadavid Cardona, of Medellin, Colombia.
1952. Francisco M. Garabis-Lomba, of Hato Rey, Puerto Rico.
1952. Pedro Jimenez Quinones, of Mayaguez, Puerto Rico.
1952. Francisco Domingo Tourreilles, of Montevideo, Uruguay.
1956. Jose Luis Arizti Garay, of Mexico, D. F., Mexico.

Walter L. Huber Civil Engineering Research Prizes

In July 1946, the Board of Direction authorized annual awards on recommendation by the Society's Committee on Research, to stimulate research in civil engineering. In October 1964, Mrs. Alberta Reed Huber endowed these prizes in honor of her husband, Walter L. Huber, Past President, ASCE.

I. Research Prizes are to be awarded to members of the Society in any grade for notable achievements in research related to civil engineering. Preference shall be given to younger members (generally under 40 years of age) of early accomplishment who can be expected to continue fruitful careers in research. No one shall receive a Research Prize more than once.

II. The selection of the nominees for the award is the responsibility of the Executive Committee of the appropriate Technical Division or Research Council not reporting to a Division.

a. Each Division may nominate up to three candidates.

b. Nominations sent to Society Headquarters will be transmitted to the appropriate Division Committee or Research Council.

c. Only those nominations received by ASCE Headquarters by December 31, or other deadline date established for that year will be considered by the Committee on Research.

III. The Society's Committee on Research shall review the Division or Council endorsed nominations.

IV. Each award shall consist of $100 and a suitable certificate.

V. Each year, the number of prizes shall not exceed five. The Committee on Research shall present its nominations for final action by the Board of Direction.

1949. John S. McNown
1955. Lynn S. Beedle, Eivind Hognestad, and Philip F. Morgan
1956. Vinton Walker Bacon, Fred Burggraf, and Chester Paul Siess
1957. Mikael Juul Hvorslev, Bruce G. Johnston, and Lorenz G. Straub
1958. John Wood Clark, Hans A. Einstein, Warren J. Kaufman, Raymond D. Mindlin, and Ivan M. Viest

1959. Charles L. Bretschneider, Norman H. Brooks, Arthur Casagrande, George S. Vincent, and Daniel Frederick
1960. Ray W. Clough, Phil M. Ferguson, Donald R. F. Harleman, Bruno Thurlimann, and David K. Todd
1961. Emmett M. Laursen, William H. Munse, H. Bolton Seed, Anestis Veletsos, and Stanley D. Wilson
1962. Ven Te Chow, William Joel Hall, Alan Hanson Mattock, Robert V. Whitman, and Robert L. Wiegel
1963. Peter S. Eagleson, Robert M. Haythornthwaite, Houssam M. Karara, Thorndike Saville, Jr. and Mete A. Sozen
1964. Steven J. Fenves, Theodore V. Galambos, John F. Kennedy, Perry L. McCarthy, and Emilio Rosenblueth
1965. Herman Bouwer, James K. Mitchell, Joseph Penzien, Gordon G. Robeck, and Rudolph P. Savage
1966. Melvin L. Baron, Louis R. Shaffer, George C. Driscoll, William W. Sayre, and Paul W. Shuldiner
1967. Jack G. Bouwkamp, George Bugliarello, Donald L. Dean, T. Cameron Kenney, and James M. Symons
1968. Alfredo H. S. Ang, Bobby O. Hardin, Marvin E. Jensen, Eduard Naudascher, and Jimmie E. Quon
1969. John W. Fisher, Cyril J. Galvin, Charles C. Ladd, Arthur R. Robinson, and Ronald F. Scott
1970. Kenneth L. Lee, Daniel P. Loucks, Lucien A. Schmit, Marshall R. Thompson, and Jan van Schilfgaarde
1971. Carl A. Cornell, Marvin Gates, Raymond J. Krizek, Norbert R. Morgenstern, and Kam Wu Wong
1972. John A. Hoopes, William B. Ledbetter, Roy E. Olson, Masanobu Shinozuka, and Robert L. Street
1973. Charles G. Culver, James M. Duncan, Lee E. King, John T. Oden, and Chih Ted Yang
1974. Asit K. Biswas, Hugo B. Fischer, Alfred J. Hendron, Jr., Thomas T.C. Hsu, Paul C. Jennings

Moisseiff Award

This prize, established by the Society in April 1947, is a memorial in recognition of the professional accomplishments of Leon S. Moisseiff, M. ASCE, a notable contributor to the science and art of structural design.

Funds have been provided by groups of Mr. Moisseiff's friends to set up the medal, with certificate, and to establish a trust fund to support the yearly awards.

I. The prize is given, in any year, to an important paper published by the Society in the twelve-month period ending with June of the year preceding the year of award (or any year since the last award) dealing with the broad field of structural design, including applied mechanics as well as the theoretical analysis, or constructive improvement, of engineering structures such as bridges and frames, of any structural material.

II. Award of the prize in every year is not mandatory.

III. Papers recognized for award of other Society prizes are disqualified.

IV. All authors, whether members or nonmembers, are eligible to receive the prize.

V. In any year when the excellence of more than one paper justifies it, the prize committee may designate a "second order of merit." A paper so recognized shall be considered eligible to compete in the award for the next succeeding year.

VI. Preparation of a paper by another publisher disqualifies it for this competition.

VII. The selection of the recipient of the award is the responsibility of the Executive Committee of the Structural Division of the Society subject to ratification by the Board of Direction.

VIII. The prize consists of a bronze medal, with an appropriate certificate.

1948. George Winter
1949. Alexander Hrennikoff
1950. Nathan M. Newmark
1951. Frances R. Shanley
1952. Mario G. Salvadori
1953. Arthur W. Anderson, John A. Blume, Henry J. Degenkolb, Harold B. Hammill, Edward M. Knapik, Henry L. Marchand, Henry C. Powers, John E. Rinne, George A. Sedgwick, and Harold O. Sjoberg
1954. Jerome M. Raphael
1955. John M. Biggs
1956. George S. Vincent
1957. David J. Peery
1958. Walter J. Austin, Shahen Yegian, and Tie P. Tung
1959. Alfred L. Parme
1960. Frank Baron and Harold S. Davis
1961. John A. Blume
1962. Frank Baron and Anthony G. Arioto
1963. George S. Vincent
1964. Konrad Basler and Bruno Thurlimann
1965. John H. Wiggins, Jr.
1966. Emilio Resenblueth
1967. Le-Wu Lu
1968. Maxwell G. Lay and Theodore V. Galambos
1969. John A. Blume
1970. Harry H. West and Arthur R. Robinson
1971. Fred Moses and John D. Stevenson
1972. Toshikazu Takeda, Mete A. Sozen, and N. Norby Nielsen
1973. Leroy Z. Emkin and William A. Litle
1974. E. Alfred Picardi, Kanu S. Patel, Robert D. Logcher, Thomas G. Harmon, Robert J. Hansen, Jose M. Roessett, and R. Elangwe Efimba

Rickey Medal

This prize was established in 1947 by Mrs. Rickey in honor of her husband James W. Rickey, M. ASCE, a leader in hydroelectric engineering progress.

Funds were donated to provide for a suitable medal, with certificate, and to establish a permanent fund, the proceeds from which would supply the necessary yearly funds for this medal.

I. The medal is given for a meritorious paper published by the Society during the twelve-month period ending with June of the year preceding the year of award. The schedule of procedure for selection and award follows in general that set up for Society prizes.

II. The award is made to an author, or authors, for a paper in the general field of hydroelectric engineering, including any of its branches, which paper is deemed to constitute a most worthy contribution to the art. Both the value or usefulness of the subject matter as well as the form of presentation are taken into consideration.

III. To fulfill the purposes of the award it is desirable, although not mandatory, to limit the choice to papers not already or otherwise selected for Society prizes during the same year.

IV. Authorship is open to members and nonmembers of the Society.

V. Prepublication of a paper by another publisher disqualifies it for this competition.

VI. Responsibility for canvassing the eligible papers and for recommending the winner, or winners, of the Rickey Medal is vested in the then current Executive Committee of the Power Division, whose choice is subject to ratification by the Board of Direction.

VII. The award comprises for each author a gold medal, a bronze duplicate, and a certificate. It is left to the judgment of the division officers whether, in case the medal is not awarded for a given year, an equivalent cash prize may be added to the following year's award.

VIII. In a year when no paper has been published that merits the award, the Power Division shall nominate for service recognition, a member of the Society who has contributed in an important manner to the science or progress of hydroelectric engineering.

1949. Fred W. Blaisdell
1950. Robert E. Turner
1952. Donald J. Bleifuss
1954. Julian Hinds
1955. Edgar S. Harrison and Carl E. Kindsvater
1956. Adolf A. Meyer
1957. Carlo Semenza and Claudio Marcello

1958. William F. Uhl
1959. James P. Growdon
1960. J. Barry Cooke
1961. Torald Mundal
1962. Calvin V. Davis
1963. J. Edgar Revelle and John N. Pirok
1964. I. Cleveland Steele
1965. Ivor L. Pinkerton and Eric J. Gibson
1966. Ralph W. Spencer, Bruce R. Laverty and Dean A. Barber
1967. Andrew Eberhardt
1968. George R. Rich
1969. Alfred L. Parme
1970. Paul H. Gilbert
1971. J. George Thon, John W. O'Hara, and Clarence H. Whalin
1972. Committee on Hydro Power Project Planning and Design of the Power Division
1974. Wallace L. Chadwick

Ernest E. Howard Award

This award was instituted and endowed in 1954 by Mrs. Howard in honor of her husband, Ernest E. Howard, Past President, ASCE.

I. The award is known as the Ernest E. Howard Award.

II. The award is made annually to a member of the American Society of Civil Engineers who has made a definite contribution to the advancement of structural engineering, either in research, planning, design or construction, including methods and materials, these contributions being made either in the form of papers or other written presentations, or through notable performance or specific actions which have served to advance structural engineering.

III. Not more than one award of the prize is made each year unless the achievement upon which the award is based is considered to be the joint contribution by more than one man, such as the joint authorship of a paper. In such cases duplicate or multiple prizes are tendered only if the balance then available in the award fund is adequate for the purpose.

IV. The annual recommendation as to the recipient of this award shall be made by the Committee on the Ernest E. Howard Award to the Board of Direction. This committee shall consist of three members who shall have overlapping terms of service of three years. The Chairman is the member who, at the time of his appointment, is chairman of the Technical Activities Committee. The other two members are, at the time of appointment, the chairmen who have most recently retired from the Executive Committees of the Structural and Construction Divisions.

V. The Committee on the Ernest E. Howard Award may also designate "second and third order of merit" nominees. These candidates will be reconsidered by the Committee for the next two years. Other remaining candidates will be reconsidered only if responsored after a five-year waiting period.

VI. The award consists of a gold medal with a bronze replica, an engraved certificate, a brochure that describes the accomplishments of the recipient, and cash in the amount of $300.

1956. Ralph E. Boeck
1957. William Ennels Dean, Jr.
1958. Nathan M. Newmark
1959. David B. Steinman
1960. Othmar H. Ammann
1961. Herschel H. Allen
1962. John A. Blume
1963. Lynn S. Beedle and Theodore R. Higgins
1964. Fred N. Severud
1965. Henry J. Brunnier
1966. T. Y. Lin
1967. Henry J. Degenkolb
1968. Chester P. Siess
1969. Thomas C. Kavanagh
1970. Ray W. Clough
1971. Stephenson B. Barnes
1972. George S. Richardson
1973. C. Martin Duke
1974. Bruce G. Johnston

The Thomas A. Middlebrooks Award

The Thomas A. Middlebrooks Award, established by the Society in 1955, is a memorial in recognition of the outstanding professional accomplishments of Thomas A. Middlebrooks, A.M. ASCE. This award is supported by the income from a fund contributed by friends of Mr. Middlebrooks.

I. The award is made to the author, or authors, of a paper published by the Society during the twelve-month period ending with June of the year preceding the year of award which shall be judged worthy of special commendation for its merit as a contribution to geotechnical engineering. Papers by young engineers are given preference. No award is made in years in which no paper of suitable merit is published.

II. The award is not restricted to members of the Society.

III. The nomination of the recipient of the award is the responsibility of the Executive Committee of the Geotechnical Engineering Division of the Society.

IV. In any year when the excellence of more than one paper justifies it, the prize committee may designate a "second order of merit." A paper so recognized is considered eligible to compete in the award for the next succeeding year.

V. The award shall consist of a cash sum to be determined annually by the executive committee of the division, but never to exceed one year's income from the capital sum. The recipient will be given the option of receiving the award, in part or in whole, in the form of books of his own selection. The prize shall also consist of a certificate, and bookplates as required, suitably inscribed with the name of the recipient and the circumstances of the award.

1956. Allen J. Curtis and Frank E. Richart, Jr.
1957. Charles I. Mansur and John A. Focht, Jr.
1958. H. Bolton Seed and Lymon C. Reese
1959. Frank E. Richart, Jr.
1960. Frank E. Richart, Jr.
1961. I.C. Steele and J. Barry Cooke
1962. James K. Mitchell
1963. W.F. Swiger
1964. H. Bolton Seed, Richard J. Woodward, Jr., and Raymond Lundgram
1965. Robert D. Darragh, Jr.
1966. Iraj Noorany and H. Bolton Seed
1967. John Lysmer and Frank E. Richart, Jr.
1968. Raul J. Marsal and Luis Ramirez de Arellano
1969. David D'Appolonia, Elio D'Appolonia, and Richard F. Brissette
1970. James K. Mitchell, Awtar Singh, and Richard Campanella
1971. H. Bolton Seed, Kenneth L. Lee, and I.M. Idriss
1972. Clyde N. Baker, Jr. and Fazlur Khan
1973. James K. Mitchell and William S. Gardner
1974. Aleksandar S. Vesic

Research Fellowship

The ASCE Research Fellowship was established in 1958 by the Board of Direction, for the purpose of aiding in the creation of new knowledge for the benefit and advancement of the science and profession of civil engineering. The grant is made from current income of the Society.

I. The grant, in the amount of $5,000, is made annually.

II. Applicants must be members of the Society in any grade of membership, be citizens of the United States, and have been graduated from an accredited curriculum.

III. Selection for the award of the grant is made by the Committee on Society Fellowships, Scholarships, Grants and Bequests on the basis of the following:

(a) Transcripts of scholastic records.
(b) Ability to conceive and explore original ideas (letters of recommendation will be considered only when they contribute specific information on this ability).
(c) Description of proposed research and its objectives, including a statement from the institution at which the research is to be done that the applicant and proposed research are acceptable to the institution. Preference is given to research that is basic in nature and concept, rather than applied, developmental, or designed to extend or elaborate information. Research requiring an extensive testing program will not be considered.

IV. Each application shall include a statement, in general terms, of the purposes for which the funds are expected to be used.

V. There is no established application form as such. The recipient will be selected on the basis of the quality of his application. Applications will be received until December 1, and awards made by March 15 for a twelve-month period beginning not later than the following October 1. Six copies of the application are required.

VI. The Research Fellow is required to devote full time to his proposed research during the tenure of the fellowship.

VII. The Research Fellow is required to submit a report on his research, suitable for publication, to the Executive Director of the Society on completion of the tenure of his fellowship.

1959. Robert W. Gerstner
1960. Ralph R. Rumer, Jr.
1961. Robert Brown Anderson
1962. Russel C. Jones
1963. John S. Endicott
1964. Joseph F. Lestingi
1965. David J. Ayres
1966. Donald E. Matzzie

Karl Terzaghi Award

This award was established by the Soil Mechanics and Foundations (now the Geotechnical Engineering) Division of the Society by the solicitation of gifts from the many friends and admirers of Karl Terzaghi, Hon. M. ASCE. It was instituted by the Board of Direction on October 10, 1960. Income from the Award Fund will be used only to pay for honoraria and expenses of dies, certificates, and plaques.

I. The Karl Terzaghi Award will be given to an author of outstanding contributions to knowledge in the fields of soil mechanics, subsurface and earthwork engineering, and subsurface and earthwork construction. Contributions which have been published by the American Society of Civil Engineers shall be cited as the principal basis for the Award. However, in exceptional circumstances, and not more than once every eight years, contributions published elsewhere may be cited as a basis for the award.

II. The award will be given without restrictions as to Society membership or nationality.

III. Award recipients will be nominated for Society approval by the Executive Committee of the Geotechnical Engineering Division.

IV. The Award will consist of a plaque and an honorarium of $1000 to be granted at intervals of about two years.

1963. Arthur Casagrande
1965. M. Juul Hvorslev
1968. Willard J. Turnbull
1969. Ralph B. Peck
1971. Laurits Bjerrum
1973. H. Bolton Seed

Edmund Friedman Professional Recognition Award

To recognize the importance of professional attainment in the advancement of "the science and profession of engineering," as defined by the Constitution of the Society, Edmund Friedman, Past President, ASCE, contributed funds to establish the Professional Recognition Award, in 1959, and it was officially instituted by action of the Board of Direction on March 7, 1960. On February 17-18, 1964, the Board of Direction voted to include the name of the donor in the name of the award.

I. The name of the award is the Edmund Friedman Professional Recognition Award.

II. This recognition is awarded annually to a member (except for an Honorary Member) of the American Society of Civil Engineers who is judged to have contributed substantially to the status of the engineering profession by:

(a) Exemplary professional conduct in a specific outstanding instance;

(b) An established reputation for professional service;

(c) Objective and lasting achievement in improving the conditions under which professional engineers serve in public and private practice;

(d) Significant contribution toward improvement of employment conditions among civil engineers;

(e) Significant contribution toward improving the professional aspects of civil engineering education;

(f) Professional guidance of qualified young men who would seek civil engineering as a career; and professional development of young civil engineers in the formative stages of their careers; or,

(g) Other evidence of merit which, in the judgment of the award committee, shall have advanced the Society's professional objectives.

III. The Award shall be made once each year when, in the judgment of the award committee, a suitable candidate is available. Only one recipient will be named in any year; and no recipient may receive the award more than once.

IV. The award committee shall be the Executive Committee of the Administrative Division of PAC. Its recommendation, accompanied by an appropriate citation, shall be reported to the Professional Activities Committee annually.

V. Any individual, or group of individuals may submit nominations, in writing, for this award. Nominations may also originate in the award committee.

VI. At the discretion of nominating bodies or individuals, the same person may be renominated in a subsequent year, provided he is eligible under Rules III and VII. At the discretion of the award committee, meritorious nominees, not selected in a given year, may be held over for consideration in the competition of succeeding years.

VII. Deceased persons are not eligible for nomination to the Award, although an award can be made posthumously to a nominee whose petition has reached the award committee during his lifetime.

VIII. The Award consists of a wall plaque mounted on a walnut shield, shaped like the Society's badge; a duplicate medal; and a formal certificate.

1960. E. Lawrence Chandler
1961. Finley B. Laverty
1962. Lloyd D. Knapp
1963. Donald H. Mattern
1964. Thomas R. Camp
1965. Harold T. Larsen
1966. Edwin C. Franzen
1967. Lawrence A. Elsener
1968. Walter E. Jessup
1969. Alfred C. Ingersoll
1970. Orley O. Phillips
1971. Elmer K. Timby
1972. Leo W. Ruth, Jr.
1973. Fred J. Benson
1974. Billy T. Sumner

Karl Terzaghi Lecture

This lectureship was established by the Soil Mechanics and Foundations (now the Geotechnical Engineering) Division of the Society by the solicitation of gifts from the many friends and admirers of Karl Terzaghi, Hon. M. ASCE. It was instituted by the Board of Direction on October 10, 1960. Income from the Award Fund will be used only to pay for honoraria and certificates.

I. At about yearly intervals and upon recommendation of the Executive Committee of the Geotechnical Engineering Division, the Executive Director will invite a distinguished engineer to deliver a "Terzaghi Lecture" at an appropriate meeting of the Society. The lecturer shall be tendered a certificate and an honorarium of $300.

1963. Ralph B. Peck
1964. Arthur Casagrande
1966. Laurits Bjerrum
1967. H. Bolton Seed
1968. Philip C. Rutledge
1969. Stanley D. Wilson
1970. T. William Lambe
1971. John Lowe III
1972. Bramlette McClelland
1974. F.E. Richart, Jr.

The Theodore von Karman Medal

This award was established and endowed in 1960 by the Engineering Mechanics Division of the Society, with gifts presented by the many friends and admirers of Theodore von Karman, Hon. M. ASCE.

I. The award is known as the Theodore von Karman Medal.

II. The medal is awarded to an individual in recognition of distinguished achievement in engineering mechanics, applicable to any branch of civil engineering.

III. Age, nationality, and Society membership shall not be a consideration in making the award.

IV. The award is normally made every year. It may be omitted at the discretion of the Award Committee. Subject to restrictions imposed by Society policy more than one award may be made in any given year.

V. The Award Committee shall consist of the members of the Advisory Board of the Engineering Mechanics Division whose duty shall be to recommend nominees for formal action by the Board of Direction.

1960. William Prager
1961. R.D. Mindlin
1962. Nathan M. Newmark
1963. Hunter Rouse
1964. Eric Reissner
1965. Warner T. Koiter
1966. Daniel C. Drucker
1967. Maurice A. Biot
1968. Lloyd H. Donnell
1969. Sir Geoffrey Ingram Taylor
1970. Wilhelm Flugge
1971. Alfred M. Freudenthal
1972. Nicholas J. Hoff
1973. Hans H. Bleich
1974. George W. Housner

ASCE Student Chapter Scholarships

Scholarships available for award to members of ASCE Student Chapters in each of the Society's four zones are made possible through the generous bequest of Samuel Fletcher Tapman, M. ASCE, in 1961. The scholarships were established by the Board of Direction in 1961, the first awards being for the academic year 1962-1963.

I. Eligibility:

a. Any member of an ASCE Student Chapter may apply for this scholarship, without restriction as to academic class or marital status, but subject to the requirement that the applicant must be a member in good standing at the time of application and award. Not more than three applications may be submitted from the membership of any one student chapter.

b. The scholarships may be awarded only to those students who will employ them to continue their formal education in a recognized educational institution.

c. Previous holders of these scholarships are eligible to apply in any succeeding competitions for these awards provided the other requirements of eligibility are satisfied.

d. No candidate shall be eligible who holds any other grant not requiring the performance of services which exceeds the tuition and fees costs for the academic year for which the scholarship is awarded.

II. Application:

a. Applications are to be submitted in six copies to the Executive Director, ASCE, for receipt prior to December 1, and must include: (1) a statement by the applicant of the reasons believed to justify an award to him; (2) the applicant's plans for continuing his formal education and a statement outlining how the applicant will finance his education if an award is granted; (3) records of academic performance and standing; and (4) statements of appraisal of the applicant in terms of potential for growth, character, leadership capacity, and interests, from the Student Chapter Faculty Advisor and not less than two other members of the faculty..

III. Selection:

a. One applicant in each Society zone may be selected each year to receive an award.

b. Selection is made by the Society's Committee on Fellowships, Scholarships, Grants and Bequests for recommendation to the Executive Committee.

c. Selection is based on appraisal of the applicants' justification of award, educational plan, academic performance and standing, potential for development, leadership capacity, and financial need.

IV. Awards:

a. The successful applicants will be notified of their selection by the Executive Director, ASCE, on or about March 1.

b. The sum of each scholarship to be awarded for the academic year will be $1,500.

c. The awards will be deposited to the awardee's account with the business office of the institution designated in his application. Equal withdrawals can be made by the awardee for each term of the academic year.

1962. Russell J. deLucia, Richard P. Howard, Roy G. Elmhorst, and William E. Alzheimer
1963. Kenneth E. Arnold, Sidney B. Barnes, Gerald E. Thompson and David Cook
1964. Wilson H. Tang, Dewey N. Corbett, Dwight D. Zeck, and Bruce Bishop
1965. Robert J. Degon, David G. Modlin, H. Dean Bartel, and Gary C. Hart
1966. Alfred M. Corneliuson, Gerald A. Lundeen, Daniel T. Schultes, and William L. McCormack
1967. James L. Greenlee and Stanley M. Klemetson
1968. James F. Marshall, James T. Prewett, Robert D. Sedivy, and Gary L. Windergerst
1969. Thomas H. Derby, Michael Barnett, and Jeffrey A. Layton
1970. Robert M. Czarnecki, Robert W. Troxler, Kenneth L. Bachor, and Steven E. Harris
1971. Rebecca Grant, William T. Boston, David J. Wanninger, and Fred M. Nelson

1972. Stephen R. Staffier, Michael W. Creed, Francis D. Hansen, and Dennis N. Athayde
1973. Maurice R. Masucci, Clarence Fennel, Michael D. Hurst, and Larry A. Myers
1974. Klaus H. Hein, Anthony Janicek, Thomas Basham, and Larry W. Wing

The O.H. Ammann Research Fellowship in Structural Engineering

The O.H. Ammann Fellowship in Structural Engineering was endowed in 1963 by O.H. Ammann, Hon. M. ASCE, for the purpose of encouraging the creation of new knowledge in the field of structural design and construction. Applications are to be submitted in six copies to the Executive Director, ASCE, for receipt prior to December 1.

I. The stipend for the Fellowship shall be $2,000.

II. Applicants must be members of the Society in any grade, or applicants for membership.

III. Each year, in December, a Committee of the Structural Division will recommend a recipient to the Executive Committee of the Society, selected from the several applications submitted in the twelve months preceding the published dead-line date.

IV. Citizens of countries other than the United States may apply if they are eligible under Rule II.

V. During the tenure of his fellowship, the recipient may not work on research projects other than that for which this Fellowship has been granted; but recipients may accept other awards if the conditions of such awards are the same as that for this Fellowship.

VI. Selection for the award of the grant is made on the basis of the following:

(a) Transcripts of scholastic records;

(b) Evidence to indicate that the applicant has the ability to conceive and explore original ideas in the field of structural engineering;

(c) Description of proposed research and its objectives, including a statement from the institution at which the research is to be done that the applicant and proposed research are acceptable to the institution.

VII. Each application shall include a statement, in general terms, of the purposes for which the funds are expected to be used.

VIII. The Research Fellow is required to submit a full report on his completed research to the Executive Director of the Society. The manuscript for such reports must be prepared in conformity with Society publication standards, and subject to review for possible publication.

1964. Daniel M. Brown
1965. John D. Mozer
1966. Pravin M. Shah
1967. John F. Seidensticker
1968. John L. Baumgartner
1969. Mubadda Suidan
1970. Eberhard Haug
1972. Robin K. McGuire
1973. Rick Evans

Civil Government Award

To recognize those members of the engineering profession who have rendered meritorious service in elective or appointive positions in government, William D. Shannon, Past Vice President, ASCE, contributed funds, the annual income of which is to be used for a Civil Government Award. The Award was officially instituted by action of the Board of Direction on October 7-8, 1963.

I. The name of the award is the Civil Government Award.

II. The award shall be made to those members (except for Honorary Members) of ASCE, wherever resident, who have contributed substantially to the status of the engineering profession by meritorious public service in elective or appointive positions in civil government.

III. In the selection of the recipients, primary consideration shall be given to public service which does not require the qualifications of an engineer. The award is intended to recognize service by engineers in such capacities as: mayor, city manager, city councilman, municipal department head, county or special authority official, state governor, member of legislature, state department head, member of congress, cabinet member, federal department administrator, or National President.

IV. The award will not be made to persons holding positions that require the services of a professional engineer or positions that traditionally have been held by engineers; nor will the award be made to persons holding positions filled by competitive civil service examinations.

V. The nominees for this award must be registered professional engineers.

VI. Each Section and Branch shall be entitled to enter the name of one nominee annually. These names should be transmitted to the Executive Director before April 1 of each year for referral to the judging committee. The committee will then submit its official nominee, if any, to the Professional Activities Committee for final action.

VII. Only one person shall receive the award in any year and an awardee may receive the honor only once.

VIII. The awardee shall be presented with an appropriate wall plaque and a certificate.

1964. James K. Carr
1965. George D. Clyde
1966. Robert B. Pease
1967. Vinton W. Bacon
1969. William J. Hedley
1970. W. Scott McDonald
1971. Ben E. Nutter
1972. Kenneth A. Gibson
1973. Milton Pikarsky
1974. JohnW. Frazier

Royce J. Tipton Award

To recognize contributions to the advancement of irrigation and drainage engineering, Royce J. Tipton, past Vice President and F. ASCE, contributed funds for the establishment of an award in this field of endeavor. It was officially instituted by action of the Board of Direction on October 19-20, 1964.

I. The award is known as the Royce J. Tipton Award.

II. The award is made to a member of the American Society of Civil Engineers who has made a definite contribution to the advancement of irrigation and drainage engineering either in teaching, research, planning, design, construction, or management, these contributions being made either in the form of papers or other written presentations, or through notable performance, long years of service, or specific actions which have served to advance the science of irrigation and drainage engineering.

III. Not more than one award of the prize is made each year unless the achievement upon which the award is based is considered to be the joint contribution by more than one man. In such cases duplicate or multiple prizes are tendered only if the balance then available in the award fund is adequate for the purpose.

IV. The Executive Committee of the Irrigation and Drainage Division shall make a recommendation to the Board of Direction for selection of the winner of the award. This recommendation shall be made on or before February 15 of each year. The Executive Committee may solicit nominations from its technical committees or from individual members of the Society for the award.

V. The award shall consist of a plaque together with a certificate suitably inscribed with the name of the recipient and the circumstances of the award.

1965. Orson W. Israelsen
1966. Harry F. Blaney, Sr.
1967. Sidney T. Harding
1968. Dean F. Peterson, Jr.
1969. Harvey O. Banks
1970. Ellis L. Armstrong
1971. William F. Donnan
1972. George D. Clyde
1973. William R. Gianelli
1974. Arthur D. Soderberg

Robert Ridgway Student Chapter Award

To promote excellence among the Student Chapters of the American Society of Civil Engineers, Isabel L. Ridgway endowed this award in honor of her husband, Robert Ridgway, Past President, ASCE. It was officially instituted by action of the Board of Direction on May 17, 1965.

I. The award is known as the Robert Ridgway Student Chapter Award.

II. The award is made annually to the single most outstanding Student Chapter of the American Society of Civil Engineers.

III. The Committee on Student Chapters shall recommend to the President of the Society the winner of the award on the basis of the judging for the certificates of commendation that are presented each year.

IV. The award shall consist of a suitable plaque and copies, for each member of the selected chapter, of the memoir of Past President Ridgway.

V. The award shall be presented to the Chapter at an appropriate meeting by a national officer of the Society.

1966. Virginia Military Institute
1967. University of Utah
1968. Kansas State University
1969. Lamar State College of Technology
1970. South Dakota School of Mines and Technology
1971. University of Nevada
1972. University of Wisconsin-Platteville
1973. Brigham Young University
1974. University of Utah

Civil Engineering History and Heritage Award

To recognize those persons who through their writing, research or other efforts have made outstanding contributions toward a better knowledge of, or appreciation for, the history and heritage of civil engineering, Trent R. Dames, F. ASCE, Past Vice President of ASCE, and member of the first Committee on History and Heritage of American Civil Engineering, and his wife, Phoebe L. Dames, contributed funds the annual income of which is to be used for a Civil Engineering History and Heritage Award. The Award was officially instituted by action of the Board of Direction on October 17-18, 1966.

I. The name of the award is the Civil Engineering History and Heritage Award.

II. The award is made to recognize the recipient's contribution toward a better knowledge of, or appreciation for, the history and heritage of civil engineering.

III. The award is not restricted to members of the Society.

IV. The award may be made annually and may not be made to the same person more than once.

V. Posthumous awards will not be made, except in the case in which the nomination was submitted prior to the nominee's death.

VI. Nominations should be submitted to the Executive Director before March 1 for referral to the Committee on History and Heritage of American Civil Engineering. Any person (or organization) may submit a nomination directly to the Executive Director; however, they are encouraged to submit the nomination through the Section Committee on History and Heritage which has geographical jurisdiction over the locale in which the nominee resides. Nominations may also originate within the Committee on History and Heritage of American Civil Engineering.

VII. The Committee on History and Heritage of American Civil Engineering shall recommend the recipient, if any, to the Board of Direction for final action.

1966. James K. Finch
1968. Ulysses S. Grant, III
1969. Gail A. Hathaway
1970. Stanley B. Hamilton
1971. Carl W. Condit
1972. Charles J. Merdinger
1973. Sara Ruth Watson

ASCE State-of-the-Art of Civil Engineering Award

Because the science and art of civil engineering can cope with the information expansion only if its most gifted practitioners will review and interpret the state-of-the-art for the benefit of their colleagues, in 1966 the professional associates of John D. Winter, M. ASCE, endowed this prize. It is anticipated that a direct benefit of this award will be the scholarly review, evaluation, and documentation of the scientific and technical information needed by the profession.

1. Annually, each technical division of the Society shall encourage an individual, individuals, or committees to prepare papers on the status of knowledge in special areas of interest served by the division.
2. These papers shall be published in the Division Journals.
3. Each Division shall nominate one of its state-of-the-art papers for judging on a Society-wide basis.
4. The author or authors of the nominated papers shall receive suitable letters of acknowledgment.
5. The nominated papers shall be reviewed by the Committee on Society Prizes which shall recommend to the Board of Direction the recipient or recipients of the ASCE State-of-the-Art of Civil Engineering Award.
6. The author or authors of the winning paper shall receive a suitable plaque and a special certificate.

1968. Boris Bresler and James G. MacGregor
1969. Ali Sabzevari and Robert H. Scanlan
1970. William S. Pollard, Jr. and Daniel W. Moore
1971. Bramlette McClelland, John A. Focht, Jr., and William J. Emrich
1972. Flory J. Tamanini
1973. The Task Committee on Structural Safety of the Structural Division's Administrative Committee on Analysis and Design, consisting of Alfredo H.S. Ang (chairman), Mohammed Amin, Tung Au, Colin B. Brown, James O. Bryson, William G. Byers, C. Allin Cornell, Alan G. Davenport, James L. Jorgenson, Niels C. Lind, Fred Moses, Masanobu Shinozuka, Carl J. Turkstra, and James T. Yao
1974. Joint ASCE-ACI Task Committee 426 on Shear and Diagonal Tension of the Committee on Masonry and Reinforced Concrete of the Structural Division, consisting of James G. MacGregor (chairman), Eugene Buth, Alex Cardenas, Vincent R. Cartelli, Edward Cohen, Robert A. Crist, Marvin E. Criswell, H.A.R. de Paiva, Paul F. Fratessa, Peter Gergely, Anand B. Gogate, John M. Hanson, Neil M. Hawkins, Nat W. Krahl, Adrian E. Long, Keith O. O'Donnell, Richard A. Parmelee, Raymond C. Reese, Frederic Roll, Howard P.J. Taylor, and Theodore C. Zsutty

Harland Bartholomew Award

The Harland Bartholomew Award was established by the Urban Planning and Development Division of the Society in recognition of the outstanding professional accomplishments of Harland Bartholomew, Hon. M. ASCE. It was instituted by the Board of Direction in 1968.

The award is made to the person who, during the fiscal year preceding the year of award, shall be judged worthy of special commendation for his contribution to the enhancement of the role of the civil engineer in urban planning and development. The contribution may be in the form of a paper published by the Society or may be his personal efforts and achievements toward that goal.

II. The award is restricted to Fellows, Members, and Associate Members of the Society.

III. The nomination of the recipient of the award is the responsibility of the Executive Committee of the Urban Planning and Development Division of the Society subject to ratification by the Board of Direction.

IV. In any year when the excellence of more than one paper justifies it, the Executive Committee may designate a "second order of merit." A paper so recognized is considered eligible to compete in the award for the next succeeding year.

V. The award shall consist of a wall plaque together with a certificate suitably inscribed with the name of the recipient and the circumstances of the award.

1969. Alan M. Voorhees
1970. Kurt W. Bauer
1971. William H. Claire
1972. Roger H. Gilman
1973. Jack R. Newville
1974. Ladislas Segoe

Samuel Arnold Greeley Award

The Samuel Arnold Greeley Award was instituted in 1968 by the Sanitary Engineering Division (now the Environmental Engineering Division) of the American Society of Civil Engineers, and is endowed by professional associates of Samuel A. Greeley, Past Director and Hon. M. ASCE. The Award is made in accordance with the rules of the Society and the following rules:

I. All original papers dealing with the design, construction, operation or financing of water supply, pollution control, storm drainage or refuse disposal projects published by the Society in the twelve-month period ending with June of the year preceding the year of award are eligible.

II. The eligible papers will be reviewed by a committee of the Environmental Engineering Division, which shall select a nominee for final action by the Board of Direction.

III. The Award is made to the author or authors of the paper that makes the most valuable contribution to the environmental engineering profession. Author or authors shall be members of the Society engaged as principals or employees in the private practice of environmental engineering.

IV. The author or authors of the selected paper will receive an appropriate plaque and a certificate describing the circumstances of the award.

1969. Thomas R. Camp and S. David Graber
1970. Roy E. Ramseier
1971. Robert C. Moore
1973. Ralph G. Berk
1974. James A. Mueller, Thomas J. Mulligan, and Dominic M. DiToro

T.Y. Lin Award

The ASCE Prestressed Concrete Award was endowed in 1968 by T.Y. Lin, F. ASCE, to encourage the preparation of meaningful papers in the designated field of endeavor. The award was instituted by the Board of Direction on May 13-14, 1968. By Board action in October 1969 the name of the award was changed to the "T.Y. Lin Award." The award is made in accordance with the policies of the Society and the following specific rules:

I. All papers written or co-authored by members of ASCE and its Student Chapters that deal with prestressed concrete and which are published in the twelve-month period ending with June of the year preceding the year of award are eligible. Preference shall be given to papers written by younger authors.

II. The American Concrete Institute and the Prestressed Concrete Institute will each be invited to nominate a single paper selected from their respective publications (which meet the foregoing criteria) for judging along with the single paper selected by the Structural Division of ASCE from any ASCE publication.

III. The three papers thus selected will be reviewed by a committee composed of one member each from the American Concrete Institute and the Prestressed Concrete Institute, and one or more members from the Structural Division — one of whom shall be the chairman. This committee shall select a nominee for final recommendation to the Board of Direction.

IV. The author or authors of the selected paper will receive an appropriate wall plaque and a certificate describing the circumstances of the award. In addition, a $500 cash honorarium shall be awarded to the author, or shared equally by the authors. In the case of a paper written jointly by members and nonmembers, the nonmember authors will receive only a certificate.

1969. Robert F. Mast
1970. Howard W. Wahl and Richard J. Kosiba
1971. Arthur R. Anderson and Saad E. Moustafa
1972. Stanley L. Paul
1973. Robert F. Mast and Charles W. Dolan
1974. Raouf Sinno and Howard L. Furr

Wesley W. Horner Award

The Wesley W. Horner Award was instituted in 1968 by the Sanitary Engineering Division (now the Environmental Engineering Division) of the American Society of Civil Engineers, and is endowed by the office partners and family of Wesley W. Horner, Past President of the Society. The Award is made in accordance with the rules of the Society and the following rules:

I. All papers dealing with hydrology, urban drainage, or sewerage that are published by the Society in the twelve-month period ending with June of the year preceding the year of award are eligible.

II. The eligible papers will be reviewed by a committee of the Environmental Engineering Division, which shall select a nominee for final action by the Board of Direction.

III. The award is made to the author or authors of the paper that makes the most valuable contribution to the environmental engineering profession, with preference given to those authors who are in the private practice of engineering.

IV. The author or authors of the selected paper will receive an appropriate plaque and a certificate describing the circumstances of the award.

1969. David M. Greer and Douglas C. Moorhouse
1970. Joseph A. Cotteral, Jr., and Dan P. Norris
1971. Raymond M. Bremner
1974. Charles V. Gibbs, Stuart M. Alexander, and Curtis P. Leiser

Surveying and Mapping Award

This award was established in October 1969 following the solicitation of funds from individual engineers and engineering firms.

I. The award is known as the Surveying and Mapping Award.

II. The award is made annually to a member of the ASCE who has made a definite contribution during the year to the advancement of surveying and mapping either in teaching, writing, research, planning, design, construction, or management, these contributions being made in the form of either papers or other written presentations, or in some instances through notable performance, long years of service, or specific actions which have served to advance surveying and mapping.

III. Not more than one award of the prize is made each year unless the achievement upon which the award is based is considered to be the joint contribution by more than one man, such as the joint authorship of a paper. In such cases duplicate or multiple prizes are tendered only if the balance then available in the award fund is adequate for the purpose.

IV. The Executive Committee of the Surveying and Mapping Division shall make a recommendation to the Board of Direction for selection of the winner of the award. This recommendation shall be made on or before February 15 of each year. The Executive Committee may solicit nominations from its technical committees or from individual members of the Society for the award.

V. The award shall consist of a plaque together with a certificate suitably inscribed with the name of the recipient and the circumstances of the award.

1970. B. Austin Barry
1971. Curtis M. Brown
1972. William A. Radlinski
1973. Philip Kissam
1974. George D. Whitmore

Stephen D. Bechtel Pipeline Engineering Award

To recognize outstanding achievements in pipeline engineering, the Bechtel Foundation donated funds to support this Award in honor of the contributions made by Stephen D. Bechtel, F. ASCE, to the engineering profession for his noteworthy advancements in the design and construction of pipelines throughout the world. The Award was established by the Board of Direction on October 19-20, 1970.

I. The award is made annually to a member of the American Society of Civil Engineers who has made a definite contribution to the advancement of pipeline engineering, either in research, planning, design, or construction. Any of these contributions may be made either in the form of papers or other written presentations, or through outstanding performance or specific and noteworthy actions which have served to advance the art, science, and technology of pipeline engineering.

II. Not more than one award of the prize is made each year unless the achievement upon which the award is based is considered to be the joint contribution by more than one man, such as the joint authorship of a paper. In such cases duplicate or multiple prizes will be awarded. No one shall receive the honor more than once.

III. Each Section and Branch and Technical Division Executive Committee shall be entitled to enter the name of one nominee annually. These names, and supporting documents, will be assembled by the Executive Director and referred to the Executive Committee of the Pipeline Division.

IV. The Executive Committee of the Pipeline Division shall recommend the recipient, if any, to the Board of Direction for final action.

V. The award consists of a plaque and certificate, suitably inscribed with the name of the recipient and the circumstances of the award.

1971. Joseph B. Spangler
1972. Nathaniel Clapp
1973. Eldon V. Hunt
1974. Maynard M. Anderson

Raymond C. Reese Research Prize

To recognize outstanding contributions to the application of structural engineering research, Raymond C. Reese, Hon. M. ASCE, contributed funds to support a structural engineering research prize. The Prize was established by the Board of Direction on April 6-7, 1970.

I. The Raymond C. Reese Research Prize is to be awarded to the author or authors of a paper published by the Society in the twelve-month period ending with June of the year preceding the year of the award that describes a notable achievement in research related to structural engineering and which indicates how the research can be used. The paper should include the results of research (experimental and/or analytical) and, in particular, should indicate and recommend how the research can be applied to design; it is this latter feature that is considered to be most important.

II. No one shall be eligible to receive the prize, or a portion of the prize, more than once.

III. The eligible papers will be reviewed by the Structural Division which shall select the nominee or nominees for final action by the Board of Direction.

IV. The author or authors of the selected paper will receive an appropriate plaque and a certificate describing the circumstances of the award.

1970. James O. Jirsa, Mete A. Sozen, and Chester P. Siess
1971. John F. Wiss, and Otto E. Curth
1972. Munther J. Haddadin, Sheu-Tien Hong, and Alan H. Mattock
1973. Richard A. Parmelee and John H. Wronkiewicz
1974. Peter W. Chen and Leslie E. Robertson

CAN-AM Civil Engineering Amity Award

This Award was established by ASCE on April 24-25, 1972 by the initiative and endowment of James A. Vance, Hon. M. ASCE. The objective of the CAN-AM Civil Engineering Amity Award is to give recognition to those civil engineers who have made outstanding and unusual contributions toward the advancement of professional relationships between the civil engineers of the United States and Canada.

I. The Award is made annually to a member of ASCE for either a specific instance that has had continuing benefit in understanding and good will, or a career of exemplary professional activity that has contributed to the amity of the United States of America and Canada.

II. Each Section and Branch of ASCE shall be entitled to enter the name of one nominee annually. The Engineering Institute of Canada shall also be invited to nominate an individual who meets the criteria for the award. These names, and supporting documents, will be assembled by the Executive Director of ASCE and referred to the judging committee.

III. The annual recommendation as to the recipient, if any, of this award shall be made by the Committee on the CAN-AM Civil Engineering Amity Award to the ASCE Board of Direction. This committee shall consist of three persons appointed by the President of ASCE. Terms of the appointment shall be three years, so designated that one member shall be replaced each year. One member shall be designated Chairman.

IV. Not more than one award of the prize is made each year unless the achievement upon which the award is based is considered to be the joint contribution by more than one man. In such cases duplicate or multiple prizes will be awarded. No one shall receive the honor, or any portion thereof, more than once.

V. The award shall consist of a plaque and a certificate suitably inscribed with the name of the recipient and the circumstances of the award.

VI. The award shall be presented annually, at a national meeting of ASCE. The presentation shall be made at an appropriate general function conducted during the course of such meeting.

1973. L. Austin Wright
1974. Eugene W. Weber

Edmund Friedman Young Engineer Award for Professional Achievement

To recognize the professional contributions of younger members of the Society, Edmund Friedman, Past President ASCE, contributed funds to establish the Edmund Friedman Young Engineer Award for Professional Achievement. The Award was officially instituted by action of the Board of Direction on October 16-18, 1972.

I. The name of the Award is the Edmund Friedman Young Engineer Award for Professional Achievement.

II. This recognition is awarded annually to members of the American Society of Civil Engineers who are less than 32 years of age on February 1 in the year of the award, and who are judged to have attained significant professional achievement, by the degree to which the candidates have shown:

(a) Service to the advancement of the profession;
(b) Evidence of technical competence, high character and integrity;
(c) Leadership in the development of younger member attitudes towards the profession;
(d) Contributions to public service outside of their professional career;
(e) Other evidence of merit, which, in the judgment of the award committee, shall have advanced the Society's objectives.

III. The Award shall be made each year when, in the judgment of the award committee, a sufficient number of suitable candidates are available. No more than five recipients will be named in any year; and no recipient may receive the award more than once.

IV. An individual, or group of individuals, may submit nominations, in writing, for these awards. They are encouraged to submit the nomination through the Sections. Nominations may also originate in the award committee. These nominations shall be submitted to the Executive Director by February 1 of each year for referral to the Committee on Younger Members. The Committee will then submit its nominations, if any, to the Professional Activities Committee through the Member Activities Division Executive Committee for final action.

V. At the discretion of nominating bodies or individuals, a nominee not selected one year may be renominated in a subsequent year, provided he is eligible under Rules II and III. At the discretion of the award committee, meritorious nominees, not selected in a given year, may be held over for consideration in the competition of the following year.

VI. Deceased persons are not eligible for nomination for the Award, although an award can be made posthumously to a nominee whose petition had reached the award committee during his lifetime.

VII. The Award shall consist of a framed certificate inscribed with the name of the recipient and the circumstances of the award, to be presented at a National Meeting.

1973. D. Joseph Hagerty, Gary Parks, and J. Lawrence Von Thun
1974. Eldon A. Cotton, Fred H. Kulhawy, Bruce F. McCollom, Howard Shirmer, Jr., and Robert P. Wadell

Martin S. Kapp Foundation Engineering Award

The Martin S. Kapp Foundation Engineering Award is a memorial in recognition of the outstanding professional accomplishments of Martin S. Kapp, F. ASCE. The award is supported by the income from a fund contributed by the friends and professional associates of Mr. Kapp. The award was established by the Board of Direction on April 7-8, 1973.

I. The Martin S. Kapp Foundation Engineering Award will be given to an individual on the basis of the best example of innovative or outstanding design or construction of foundations, earthworks, retaining structures, or underground construction. Emphasis shall be placed on constructed works where serious difficulties were overcome or where substantial economies were achieved. The example shall have been described in published form available to the entire engineering community.

II. Not more than one award of the prize is made each year unless the achievement upon which the award is based is considered to be the joint contribution by more than one person, such as the joint authorship of a paper. In such cases duplicate or multiple prizes will be awarded. No one shall receive the honor, or any portion thereof, more than once.

III. The award is not restricted to members of the Society.

IV. The recipient of the award — if any — will be nominated for Society approval by the Executive Committee of the Geotechnical Engineering Division.

V. The award consists of a plaque and certificate, suitably inscribed with the name of the recipient and the circumstances of the award.

1973. Anthony J. Tozzoli

Construction Management Award

This award was instituted and endowed by Marvin Gates, F. ASCE and Amerigo Scarpa, F. ASCE. The Award was officially instituted by action of the Board of Direction on October 27-30, 1973.

I. The award is known as the Construction Management Award.

II. The award is made annually to a member of the American Society of Civil Engineers who has made definite contributions in the field of construction management in general and, more particularly, in the application of the theoretical aspects of engineering economics, statistics, probability theory, operations research and related mathematically oriented discriplines to problems of construction management, estimating, cost accounting, planning, scheduling and financing. These contributions being made either in the form of written presentations, or notable performance.

III. Not more than one award of the prize is made each year unless the achievement upon which the award is based is considered to be the joint contribution by more than one person. In such cases, duplicate or multiple prizes are tendered only if the balance then available in the award fund is adequate for the purpose. No one shall receive the Construction Management Award or a portion thereof more than once.

IV. The recommendation as to the recipient of the award shall be made by the Executive Committee of the Construction Division with the advice of previous recipients of the award.

V. The award shall consist of a plaque together with a certificate suitably inscribed with the name of the recipient and the circumstances of the award.

1974. Joseph C. Kellogg

APPENDIX VI

ASCE SECTIONS

To meet the technical and professional needs of members at the local level and to assist them in accomplishing the purposes of the Society, the Board of Direction has authorized, from time to time since 1905, the formation of sections.

Many sections have formed branches to serve members at one or more centers of engineer population in their areas. Several Associate Member forums have been established in sections with large memberships. In addition, groups of sections frequently organize regional councils.

The sections appoint various committees to aid in carrying on the work of the Society and they cooperate with the student chapters in the sections' areas. Newsletters are published periodically by a number of sections to acquaint members with matters of local and national interest.

The following listings show the section's name, the ASCE Zone and District in which the section is located, the year of formation, the number of members residing in the section's area as of 1974, and the branches of the section, with the year of their formation.

AKRON SECTION, Zone II, District 9 (1947); 216 Members.

ALABAMA SECTION, Zone II, District 14 (1931); 813 Members.
Branches: Birmingham (1970); Huntsville (1963); Mobile (1973).

ALASKA SECTION, Zone IV, District 12 (1951); 354 Members.
Branches: Anchorage (1952); Fairbanks (1951); Juneau (1951).

ARIZONA SECTION, Zone IV, District 11 (1925); 699 Members.
Branches: Phoenix (1956); Southern Arizona (1962).

BUFFALO SECTION, Zone I, District 3 (1921); 302 Members.

CENTRAL ILLINOIS SECTION, Zone III, District 8 (1924); 610 Members.
Branches: East (1967); Illinois Valley (1963); West (1967).

CENTRAL OHIO SECTION, Zone II, District 9 (1921); 431 Members.

CENTRAL PENNSYLVANIA SECTION, Zone I, District 4 (1958); 565 Members.

CINCINNATI SECTION, Zone II, District 9 (1920); 318 Members.

CLEVELAND SECTION, Zone II, District 9 (1915); 542 Members.
Branches: Ashland-Mansfield (1961); Youngstown (1960).

COLORADO SECTION, Zone III, District 17 (1909); 1,380 Members.

COLUMBIA SECTION, Zone IV, District 12 (1950); 154 Members.

CONNECTICUT SECTION, Zone I, District 2 (1919); 811 Members.

DAYTON SECTION, Zone II, District 9 (1922); 243 Members.

DELAWARE SECTION, Zone II, District 5 (1953); 251 Members.

DULUTH SECTION, Zone III, District 7 (1917); 73 Members.

FLORIDA SECTION, Zone II, District 10 (1929); 1,297 Members.
Branches: Cape Canaveral (1959); East Central (1958); Gainesville (1949); Jacksonville (1951); Ridge (1968); Tallahassee (1967); West Coast (1951).

GEORGIA SECTION, Zone II, District 10 (1912); 926 Members.
Branches: Albany (1957); Savannah (1951).

HAWAII SECTION, Zone IV, District 11 (1937); 720 Members.

ILLINOIS SECTION, Zone III, District 8 (1916); 2,069 Members.

INDIANA SECTION, Zone II, District 9 (1931); 790 Members.
Branches: Central (1967); Metropolitan-Indianapolis (1966); North Central (1966); Northeastern (1960); Northwestern (1951); Southwestern (1953).

IOWA SECTION, Zone III, District 17 (1920); 413 Members.

ITHACA SECTION, Zone I, District 3 (1932); 169 Members.

KANSAS CITY (MO) SECTION, Zone III, District 16 (1921); 887 Members.

KANSAS SECTION, Zone III, District 16 (1920); 453 Members.
Branch: Wichita (1962).

KENTUCKY SECTION, Zone II, District 9 (1936); 462 Members.

LEHIGH VALLEY SECTION, Zone I, District 4 (1922); 338 Members.

LOS ANGELES SECTION, Zone IV, District 11 (1913); 3,933 Members.
Branches: Desert Area (1953); Orange County (1953); San Bernardino-Riverside (1953); San Luis Obispo (1960); Santa Barbara-Ventura Counties (1953); Southern San Joaquin (1970).

LOUISIANA SECTION, Zone II, District 14 (1914); 997 Members.
Branches: Baton Rouge (1962); New Orleans (1962); Shreveport (1953).

MAINE SECTION, Zone I, District 2 (1950); 312 Members.

MARYLAND SECTION, Zone II, District 5 (1914); 1,085 Members.

MASSACHUSETTS SECTION, Zone I, District 2 (1921); 1,966 Members.
Branch: Western Massachusetts (1962).

METROPOLITAN SECTION, Zone I, District 1 (1920); 4,666 Members.
Branches: Long Island (1972); New Jersey (1962).

MEXICO SECTION, Zone III, District 15 (1949); 149 Members.

MICHIGAN SECTION, Zone III, District 7 (1916); 1,369 Members.
Branches: East Central (1970); Lansing-Jackson (1961); Northwestern (1962); Southeastern (1961); Southwestern (1969); Upper Peninsula (1967); Western Michigan (1961).

MID-MISSOURI SECTION, Zone III, District 16 (1937); 264 Members.

MID-SOUTH SECTION, Zone II, District 14 (1928); 505 Members.
Branches: Little Rock (1949); Memphis (1951).

MINNESOTA SECTION, Zone III, District 7 (1914); 666 Members.

MISSISSIPPI SECTION, Zone II, District 14 (1969); 486 Members.
Branches: Jackson (1949); Vicksburg (1948).

MOHAWK-HUDSON SECTION, Zone I, District 3 (1938); 588 Members.

MONTANA SECTION, Zone IV, District 12 (1945); 255 Members.
Branches: Eastern (1956); Western (1956).

NASHVILLE SECTION, Zone II, District 6 (1921); 336 Members.

NATIONAL CAPITAL SECTION, Zone II, District 5 (1916); 2,561 Members.

NEBRASKA SECTION, Zone III, District 17 (1917); 430 Members.

NEVADA SECTION, Zone IV, District 11 (1963); 284 Members.
Branches: Capital (1964); Southern Nevada (1964); Truckee Meadows (1967).

NEW HAMPSHIRE SECTION, Zone I, District 2 (1958); 225 Members.

NEW MEXICO SECTION, Zone III, District 15 (1929); 361 Members.
Branch: Albuquerque (1964).

NORTH CAROLINA SECTION, Zone II, District 6 (1923); 783 Members.
Branches: Eastern (1962); Northern (1962); Southern (1962); West (1937).

NORTH DAKOTA SECTION, Zone III, District 7 (1973); 107 Members.

OKLAHOMA SECTION, Zone III, District 16 (1920); 483 Members.
Branches: Oklahoma City (1948); Tulsa (1948).

OREGON SECTION, Zone IV, District 12 (1913); 923 Members.
Branch: Willamette (1969).

PANAMA SECTION, (1931); 86 Members.

PHILADELPHIA SECTION, Zone I, District 4 (1913); 1,491 Members.
Branch: Trenton (1953).

PITTSBURGH SECTION, Zone I, District 4 (1918); 1,109 Members.

PUERTO RICO SECTION, Zone I, District 1 (1929); 345 Members.

REPUBLIC OF COLOMBIA SECTION, (1957); 165 Members.

RHODE ISLAND SECTION, Zone I, District 2 (1920); 187 Members.

ROCHESTER SECTION, Zone I, District 3 (1923); 207 Members.

SACRAMENTO SECTION, Zone IV, District 11 (1921); 1,303 Members.
Branches: Central Valley (1953); Marysville (1953); Shasta (1951).

ST. LOUIS SECTION, Zone III, District 16 (1914); 731 Members.

SAN DIEGO SECTION, Zone IV, District 11 (1915); 619 Members.

SAN FRANCISCO SECTION, Zone IV, District 11 (1905); 3,736 Members.
Branches: Fresno (1959); North Coast (1966); Redwood Empire (1958); San Francisco (1971); San Jose (1956).

SEATTLE SECTION, Zone IV, District 12 (1913); 987 Members.

SOUTH CAROLINA SECTION, Zone II, District 10 (1934); 476 Members.
Branches: Central Savannah River Valley (1951); Eastern (1961); Midlands (1963); Northwest (1963).

SOUTH DAKOTA SECTION, Zone III, District 17 (1958); 222 Members.
Branches: Black Hills (1957); Eastern (1959).

SOUTH FLORIDA SECTION, Zone II, District 10 (1927); 653 Members.
Branches: Broward County (1969); Palm Beach (1952).

SOUTHERN IDAHO SECTION, Zone IV, District 12 (1947); 228 Members.

SPOKANE SECTION, Zone IV, District 12 (1914); 259 Members.

SYRACUSE SECTION, Zone I, District 3 (1923); 271 Members.

TACOMA SECTION, Zone IV, District 12 (1930); 313 Members.

TENNESSEE VALLEY SECTION, Zone II, District 6 (1932); 630 Members.
Branches: Chattanooga (1934); Holston (1947); Knoxville (1947).

TEXAS SECTION, Zone III, District 15 (1913); 3,135 Members.
Branches: Austin (1950); Brazos County (1939); Corpus Christi (1950); Dallas (1938); El Paso (1950); Fort Worth (1938); High Plains (1952); Houston (1943); Northeast (1950); San Antonio (1949); San Jacinto (1955); Southeast (1951); West Texas (1952).

TOLEDO SECTION, Zone II, District 9 (1922); 202 Members.

TRI-CITY SECTION, Zone III, District 8 (1940); 166 Members.

UTAH SECTION, Zone IV, District 11 (1916); 371 Members.

VERMONT SECTION, Zone I, District 2 (1960); 186 Members.

VIRGINIA SECTION, Zone II, District 6 (1922); 1,141 Members.
Branches: Blue Ridge (1967); Bull Run (1969); Norfolk (1953); Peninsula (1966); Richmond (1956); Roanoke (1955).

WEST VIRGINIA SECTION, Zone II, District 6 (1937); 372 Members.
Branch: Charleston (1963).

WISCONSIN SECTION, Zone III, District 7 (1923); 921 Members.
Branches: Fox River Valley (1965); Madison (1961).

WYOMING SECTION, Zone III, District 17 (1938); 135 Members.

APPENDIX VII

STUDENT CHAPTERS AND CLUBS

The objective of the ASCE Student Chapter is to help students prepare themselves for entry into the civil engineering profession and the Society.

Student Chapters also help civil engineering students begin those professional contacts and associations which, continued through life, are so valuable to the practicing engineer in serving mankind and the engineering profession more effectively.

Each chapter has a Faculty Advisor, Contact Member and Associate Contact Member, appointed by the Board of Direction upon recommendation of the District Director.

Student Chapters

Name	Place	Date of Organization
Akron, Univ. of	Akron, Ohio	1925
Alabama, Univ. of	University, Ala.	1923
Alaska, Univ. of	Fairbanks, Alaska	1947
Arizona State Univ.	Tempe, Ariz.	1962
Arizona, Univ. of	Tucson, Ariz.	1926
Arkansas, Univ. of	Fayettesville, Ark.	1926
Auburn Univ.	Auburn, Ala.	1921
Bradley Univ.	Peoria, Ill.	1964
Brigham Young Univ.	Provo, Utah	1961
Brown Univ.	Providence, R.I.	1935
Bucknell Univ.	Lewisburg, Pa.	1921
Calif. State Polytechnic Univ.	Pomona, Calif.	1965
Calif. State Univ. at Chico	Chico, Calif.	1969
Calif. State Univ. at Fresno	Fresno, Calif.	1967
Calif. State Univ. at Long Beach	Long Beach, Calif.	1967
Calif. State Univ. at Los Angeles	Los Angeles, Calif.	1964
Calif. State Univ., Sacramento	Sacramento, Calif.	1965
Calif. State Univ., San Diego	San Diego, Calif.	1948
Calif. State Univ., San Francisco	San Francisco, Calif.	1967
Calif. State Univ., San Jose	San Jose, Calif.	1960
Calif., Univ. of at Berkeley	Berkeley, Calif.	1921

Name	Place	Date of Organization
Calif., Univ. of at Davis	Davis, Calif.	1965
Calif., Univ. of at L.A.	Los Angeles, Calif.	1959
Carnegie-Mellon Univ.	Pittsburgh, Pa.	1922
Case Western Reserve Univ.	Cleveland, Ohio	1926
Catholic Univ. of America	Washington, D.C.	1930
Cincinnati, Univ. of	Cincinnati, Ohio	1920
Citadel, The	Charleston, S.C.	1936
Clarkson College of Technology	Potsdam, N.Y.	1929
Clemson Univ.	Clemson, S.C.	1922
Cleveland State Univ.	Cleveland, Ohio	1951
Colorado State Univ.	Fort Collins, Colo.	1940
Colorado, Univ. of	Boulder, Colo.	1920
Columbia Univ.	New York, N.Y.	1927
Connecticut, Univ. of	Storrs, Conn.	1941
Cooper Union, The	New York, N.Y.	1925
Cornell Univ.	Ithaca, N.Y.	1921
Dayton, Univ. of	Dayton, Ohio	1926
Delaware, Univ. of	Newark, Delaware	1932
Denver, Univ. of	Denver, Colo.	1951
Detroit Inst. of Tech.	Detroit, Mich.	1963
Detroit, Univ. of	Detroit, Mich.	1938
Drexel Univ.	Philadelphia, Pa.	1920
Duke Univ.	Durham, N.C.	1933
Florida, Univ. of	Gainesville, Fla.	1926
George Washington Univ.	Washington, D.C.	1923
Georgia Institute of Technology	Atlanta, Ga.	1922
Gonzaga Univ.	Spokane, Wash.	1966
Hawaii, Univ. of	Honolulu, Hawaii	1951
Heald Engineering Coll.	San Francisco, Calif.	1964
Hofstra Univ.	Hempstead, N.Y.	1972
Houston, Univ. of	Houston, Tex.	1953
Howard Univ.	Washington, D.C.	1951
Idaho, Univ. of	Moscow, Idaho	1926
Illinois Inst. of Tech.	Chicago, Ill.	1939
Illinois, Univ. of	Urbana, Ill.	1921
Illinois, Univ. of at Chicago Circle	Chicago, Ill.	1947
Indiana Institute of Technology	Fort Wayne, Ind.	1961
Iowa State Univ.	Ames, Iowa	1920
Iowa, Univ. of	Iowa City, Iowa	1921
Johns Hopkins Univ. Evening College	Baltimore, Md.	1921
Kansas State Univ.	Manhattan, Kan.	1923
Kansas, Univ. of	Lawrence, Kan.	1921
Kentucky, Univ. of	Lexington, Ky.	1921
Lafayette Coll.	Easton, Pa.	1922
Lamar University	Beaumont, Tex.	1962
Lehigh Univ.	Bethlehem, Pa	1922
Louisiana State Univ.	Baton Rouge, La.	1932
Louisiana Tech. Univ	Ruston, La.	1949
Louisville, Univ. of	Louisville, Ky.	1938
Lowell Technological Institute	Lowell, Mass.	1970

Name	Place	Date of Organization
Loyola Univ. of	Los Angeles, Calif.	1966
Maine, Univ. of	Orono, Me.	1921
Manhattan Coll.	New York, N.Y.	1927
Marquette Univ.	Milwaukee, Wis.	1923
Marshall Univ.	Huntington, West Va.	1970
Maryland, Univ. of	College Park, Md.	1936
Massachusetts Inst. of Technology	Cambridge, Mass.	1921
Massachusetts, Univ. of	Amherst, Mass.	1950
Memphis State Univ.	Memphis, Tenn.	1972
Merrimack Coll.	North Andover, Mass.	1961
Miami, Univ. of	Coral Gables, Fla.	1961
Michigan State Univ.	East Lansing, Mich.	1926
Michigan Technological Univ.	Houghton, Mich.	1931
Michigan, Univ. of	Ann Arbor, Mich.	1923
Minnesota, Univ. of	Minneapolis, Minn.	1921
Mississippi State Univ.	State College, Miss.	1928
Mississippi, Univ. of	University, Miss.	1923
Missouri, Univ. of at Columbia	Columbia, Mo.	1922
Missouri, Univ. of at Rolla	Rolla, Mo.	1924
Montana State Univ.	Bozeman, Mont.	1922
Nebraska, Univ. of at Lincoln	Lincoln, Nebr.	1921
Nebraska, Univ. of at Omaha	Omaha, Nebr.	1963
Nevada, Univ. of	Reno, Nev.	1923
Newark Coll. of Engineering	Newark, N.J.	1931
New England Coll.	Henniker, N.H.	1963
New Hampshire, Univ. of	Durham, N.H.	1928
New Mexico State Univ.	University Park, N.M.	1933
New Mexico, Univ. of	Albuquerque, N.M.	1929
New York, City Coll. of the City University	New York, N.Y.	1923
New York, Polytechnic Institute of	Brooklyn, N.Y.	1921
New York, State Univ. at Buffalo	Buffalo, N.Y.	1964
New York Univ.	New York, N.Y.	1921
North Carolina State Univ. at Raleigh	Raleigh, N.C.	1922
North Dakota State Univ.	Fargo, N. Dak.	1932
North Dakota, Univ. of	Grand Forks, N. Dak.	1923
Northeastern Univ.	Boston, Mass.	1940
Northwestern Univ.	Evanston, Ill.	1939
Norwich Univ.	Northfield, Vt.	1937
Notre Dame, Univ of	Notre Dame, Ind.	1943
Ohio Northern Univ.	Ada, Ohio	1926
Ohio State Univ.	Columbus, Ohio	1922
Ohio Univ.	Athens, Ohio	1952
Oklahoma State Univ.	Stillwater, Okla.	1923
Oklahoma, Univ. of	Norman, Okla.	1922
Old Dominion University	Norfolk, Va.	1957
Oregon State Univ.	Corvallis, Oreg.	1921
Pacific, Univ. of the	Stockton, Calif.	1961
Pennsylvania State Univ.	University Park, Pa.	1920
Pennsylvania, Univ. of	Philadelphia, Pa.	1920
Pittsburgh, Univ. of	Pittsburgh, Pa.	1921

Name	Place	Date of Organization
Prairie View Agric. & Mech. Coll.	Prairie View, Tex.	1965
Princeton Univ.	Princeton, N.J.	1926
Puerto Rico, Univ. of	Mayaguez, P.R.	1949
Purdue Univ.	Lafayette, Ind.	1921
Rensselaer Polytechnic Institute	Troy, N.Y.	1920
Rhode Island, Univ. of	Kingston, R.I.	1932
Rice Univ.	Houston, Tex.	1923
Rose-Hulman Institute of Tech.	Terre Haute, Ind.	1927
Rutgers Univ.	New Brunswick, N.J.	1921
Saint Martin's Coll.	Olympia, Wash.	1961
Santa Clara, Univ. of	Santa Clara, Calif.	1938
Seattle Univ.	Seattle, Wash.	1963
South Carolina, Univ. of	Columbia, S.C.	1927
South Dakota School of Mines & Tech.	Rapid City, S. Dak.	1928
South Dakota State Univ.	Brookings, S. Dak.	1933
Southeastern Massachusetts Univ.	North Dartmouth, Mass.	1969
Southern California, Univ. of	Los Angeles, Calif.	1924
Southern Methodist Univ.	Dallas, Tex.	1940
Southern Univ.	Baton Rouge, La.	1971
Southwestern Louisiana, Univ. of	Lafayette, La.	1957
Stanford Univ.	Stanford, Calif.	1920
Stevens Institute of Technology	Hoboken, N.J.	1958
Swarthmore Coll.	Swarthmore, Pa.	1921
Syracuse Univ.	Syracuse, N.Y.	1921
Tennessee State Univ.	Nashville, Tenn.	1967
Tennessee Technological Univ.	Cookeville, Tenn.	1961
Tennessee, Univ. of	Knoxville, Tenn.	1923
Texas A & I Univ.	Kingsville, Tex.	1972
Texas A & M Univ.	College Station, Tex.	1924
Texas Tech. Univ.	Lubbock, Tex.	1933
Texas, Univ. of at Arlington	Arlington, Tex.	1965
Texas, Univ. of, at Austin	Austin, Tex.	1921
Texas, Univ. of at El Paso	El Paso, Tex.	1948
Toledo, Univ. of	Toledo, Ohio	1943
Tri-State College	Angola, Ind.	1965
Tufts Univ.	Medford, Mass.	1929
Tulane Univ.	New Orleans, La.	1933
Union Coll.	Schenectady, N.Y.	1926
U.S. Air Force Academy	Colorado Springs, Colo.	1968
Utah State Univ.	Logan, Utah	1935
Utah, Univ. of	Salt Lake City, Utah	1924
Valparaiso Univ.	Valparaiso, Ind.	1959
Vanderbilt Univ.	Nashville, Tenn.	1925
Vermont, Univ. of	Burlington, Vt.	1937
Villanova Univ.	Villanova, Pa.	1925
Virginia Military Institute	Lexington, Va.	1921
Virginia Polytechnic Institute	Blacksburg, Va.	1922
Virginia, Univ. of	Charlottesville, Va.	1922
Walla Walla Coll.	College Place, Wash.	1972
Washington State University	Pullman, Wash.	1924

Name	Place	Date of Organization
Washington Univ.	St. Louis, Mo.	1920
Washington, Univ. of	Seattle, Wash.	1921
Wayne State Univ.	Detroit, Mich.	1947
West Virginia Inst. of Technology	Montgomery, W.Va.	1963
West Virginia Univ.	Morgantown,, W.Va.	1921
Widener Coll.	Chester, Pa.	1960
Wisconsin, Univ. of	Madison, Wis.	1921
Wisconsin, Univ of	Milwaukee, Wis.	1970
Wisconsin, Univ. of	Platteville, Wis.	1967
Worcester Polytech Inst.	Worcester, Mass.	1922
Wyoming, Univ.	Laramie, Wyo.	1925
Youngstown, University	Youngstown, Ohio	1960

Student Clubs

ASCE now sponsors Student Clubs at undergraduate schools in the United States offering a four-year undergraduate program related to civil engineering. These ASCE Student Clubs are intended to stimulate student interest in civil engineering and to provide Society services to undergraduate students wherever they may be studying.

Federal City College	Washington, D.C.	1972
Florida Technological Univ.	Orlando, Fla.	1972
Franklin Institute of Boston	Boston, Mass.	1972
New Haven, Univ. of	New Haven, Conn.	1972
North Carolina, Univ. of	Charlotte, N.C.	1972
Portland State Univ.	Portland, Ore.	1972
Roger Williams Coll.	Providence, R.I.	1972
Southern Illinois Univ.	Edwardsville, Ill.	1972
U.S. Coast Guard Academy	New London, Conn.	1972

APPENDIX VIII

RECORD OF CONSTITUTIONAL AMENDMENTS

Date	*General Nature of Amendments*
Nov. 5, 1852	Original Constitution adopted.
March 1868	Membership admission procedures.
February 1870	Membership grades; Associate Member grade adopted.
Nov. 5, 1873	Voting privilege extended to Non-Resident members, membership requirements and admission procedure.
Jan. 2, 1878	Membership admission and expulsion procedures.
June 18, 1878	Nominating Committee initiated.
January 1879	First codification; membership requirements and dues.
Feb. 4, 1880	Life Member provisions adopted.
Jan. 5, 1881	Annual Meeting changed from November to January.
Mar. 1, 1882	$10,000 of Development Fund allocated to publications.
Mar. 7, 1883	Amendment procedures modified.
Mar. 4, 1885	Compounding of dues adopted.
Mar. 7, 1888	Membership requirements.
Mar. 5, 1890	Codification adopted.
Mar. 4, 1891	New Constitution; composition of Board; Juniors required to advance grade by age 30.
Oct. 3, 1894	Subscribing Member grade abolished. Secretary elected by Board. Seven Districts established.
Mar. 6, 1895	Membership admission; election procedures.
Oct. 6, 1897	Composition of Board.
Oct. 5, 1898	Office of Auditor abolished.
Oct. 3, 1900	Nominating procedure.
Mar. 4, 1903	Membership admission procedure.
Oct. 7, 1903	Membership admission procedure.
Oct. 7, 1908	Board given authority to elect and transfer members, to appoint special committees (when authorized by business meeting), and to fill Nominating Committee vacancies.
Mar. 1, 1911	Life Member provisions adopted.
Oct. 2, 1912	Nominating procedures.
Mar. 3, 1915	Number of Districts increased from 7 to 13.
Oct. 6, 1920	Nominating procedures.
Oct. 5, 1921	New Constitution; composition of Board; Student Chapters and Local Sections authorized; Board empowered to appoint special committees; nominating procedures.
Mar. 1, 1922	Terms of officers clarified.
Mar. 3, 1926	Membership and explusion procedures.
Mar. 2, 1927	Nominating procedures.
Oct. 5, 1927	Membership admission requirements.

Oct. 3, 1928	Honorary membership requirements.
Mar. 19, 1930	Composition of Board: number of meetings reduced.
Oct. 1, 1930	Membership requirements.
Oct. 16, 1935	Membership grades and admission procedures.
Mar. 20, 1940	Constitution amendment procedures.
Mar. 18, 1942	Age limit and dues for Juniors.
Oct. 7, 1947	Voting privilege extended to Junior grade.
Apr. 19, 1950	New Constitution; membership grades; Junior Members empowered to hold office; Fellowship abolished.
June 14, 1954	Dues increased.
June 6, 1959	Membership grades; Fellow grade adopted.
June 15, 1966	Composition of Board
Feb. 25, 1970	Composition of Board; Board empowered to establish dues.
Nov. 18, 1972	Membership requirements strengthened; Life Membership requirements changed.
Late 1974	Nomination and election of officers; Modification of objectives; Board quorums.

APPENDIX IX

ASCE MEMBERSHIP STATISTICS

Date	*Honorary Members*	*Fellows* [a]	*Members*	*Associate Members*	*Juniors* [b]	*Associates or Affiliates*	*Corresponding Members*	*Total*
Oct. 1853	6	—	48	—	—	—	1	56
Jan. 1871	6	—	210	—	43	—	—	259
May 1876	6	70	381	16	43	—	3	519
July 1880	10	66	453	20	48	—	3	600
May 1885	9	62	665	33	79	—	3	851
Aug. 1890	7	56	1,055	59	200	—	3	1,380
March 1895	9	47	1,190	194	276	72	3	1,791
Feb. 1900	9	38	1,376	439	263	93	3	2,221
Feb. 1905	10	27	1,819	918	350	123	2	3,249
Feb. 1910	8	21	2,582	1,845	690	166	2	5,314
Feb. 1915	5	15	3,408	3,270	840	168	1	7,707
Feb. 1920	3	10	3,990	4,697	506	166	1	9,373
March 1925	13	8	4,960	5,365	771	158 [c]	—	11,275
March 1930	18	7	5,673	6,047	2,329	144	—	14,218
March 1935	18	3	5,683	6,097	3,046	99	—	14,946
Feb. 1940	31	1	5,584	6,336	4,144	70	—	16,166
Feb. 1945	36	1	6,189	7,743	6,485	76	—	20,503
Feb. 1950	39	1	7,514	9,667	10,810	74	—	28,105
March 1955	42	—	18,859	11,358	17,840 [b]	70	—	38,169
March 1960	47	10,933 [a]	15,880	18,007 [b]	—	94	—	44,961
March 1965	51	10,540	20,171	23,142	—	150	—	54,054
July 1970	55	10,351	26,646	27,451	—	192	—	64,695
July 1974	67	9,777	26,015	32,427	—	1,385	—	69,671

[a] The grade of Fellow denoted a contribution to the "Fellowship Fund" until 1950; in 1959 it replaced the term Member to denote the "Advanced Professional Grade."
[b] The "Entrance Grade" was termed Junior until April 1950 when it was changed to Junior Member and then to Associate Member in 1959.
[c] The "Non-Engineer Professional Grade" was termed Associate until 1921 when it was changed to Affiliate.

APPENDIX X

GUIDE TO PROFESSIONAL PRACTICE UNDER THE CODE OF ETHICS

ARTICLE 1: "It shall be considered unprofessional . . . To act for his client or for his employer otherwise than as a faithful agent or trustee."

(1) He shall not undertake any assignment which would create a potential conflict of interest between the engineer and his client or his employer.

(2) He shall not disclose information concerning the business affairs or technical processes of his clients or employer without their consent.

(3) He shall not use information coming to him confidentially in the course of his assignment as a means of making personal profit if such action is adverse to the interest of his client, his employer, or the public.

(4) He shall not divulge any confidential findings of studies or actions of an engineering commission or board of which he is a member, without official consent.

(5) He shall not give professional advice which does not fully reflect his best professional judgment.

(6) He shall not misrepresent his qualifications to a client, to an employer, or to the profession.

(7) He shall not accept an assignment the results of which he will later act upon as a member of a public or quasi-public board.

(8) He shall act with fairness and justice to all parties when administering a construction or other contract.

(9) He shall engage, or advise engaging, experts and specialists, when in his judgment such services are to his client's or employer's best interests.

ARTICLE 2: "It shall be considered unprofessional . . . To accept remuneration for services rendered other than from his client or his employer.

(1) He shall not accept compensation from more than one in-

terested party for the same service, or for services pertaining to the same work under circumstances where there may be a conflict of interest without the consent of all interested parties.

(2) He shall not accept any royalty or commission on any article or process used on the work for which he is responsible, without the consent of his client or employer.

ARTICLE 3: "It shall be considered unprofessional . . . To attempt to supplant another engineer in a particular engagement after definite steps have been taken toward his employment."

(1) He shall not continue to seek employment on a specific engagement after being advised that another engineer has been selected subject to approval of detailed arrangements.

(2) He shall not solicit employment from a client who already has an engineer under contract for the same work. He shall not accept employment from a client who already has an engineer for the same work not yet completed or not yet paid for unless the performance or payment requirements in the contract are being litigated or the contracted engineer's services have been terminated in writing by either party. In case of termination or litigation, the prospective engineer before accepting the assignment shall advise the engineer being terminated or involved in litigation.

(3) He shall not, in the event that another engineer has made a study and report on a specific project, approach the prospective client regarding subsequent phases of the project, unless such contact is initiated by the client.

ARTICLE 4: "It shall be considered unprofessional . . . To attempt to injure, falsely or maliciously, the professional reputation, business, or employment position of another engineer."

This does not remove the moral obligation to expose unethical conduct before the proper authorities. Neither does it preclude a frank but private appraisal of employees or of engineers being considered for employment.

ARTICLE 5: "It shall be considered unprofessional . . . To review the work of another engineer for the same client, except with the knowledge of such engineer, unless such engineer's engagement on the work which is subject to review has been terminated."

The article as stated is believed to be sufficiently explicit.

However, even though the first engineer's services have been terminated, it is a matter of common courtesy to let him know that his work is being reviewed.

ARTICLE 6: "It shall be considered unprofessional . . . To advertise engineering services in self-laudatory language, or in any other manner derogatory to the dignity of the profession."

The following are considered to be permissible:

(1) Professional cards in recognized, dignified publications, and listings in rosters or directories published by responsible organizations, provided that the cards or listings are consistent in size and content, and are in a section of the publication regularly devoted to such professional cards. Information given must be factual, dignified, and free from ostentatious, complimentary, or laudatory implications.

(2) Brochures which factually describe experience, facilities, personnel and capacity to render service, providing they are not misleading with respect to the engineer's participation in projects described.

(3) Display advertising in recognized dignified business and professional publications, providing it is factual, free from ostentation, contains no laudatory expressions or implication and is not misleading with respect to the engineer's extent of participation in projects described.

(4) A statement of his name or the name of his firm and statement of his type of service posted on projects for which he renders services.

(5) Preparation or authorization of descriptive articles for the lay or technical press, which are factual, dignified and free from ostentatious or laudatory implications. Such articles shall not imply anything more than his direct participation in the project described.

(6) Permission by an engineer for his name to be used in commercial advertisements, such as may be published by contractors, material suppliers, etc., only by means of a modest dignified notation acknowledging the engineer's participation in the project described.

ARTICLE 7: "It shall be considered unprofessional . . . To use the advantages of a salaried position to compete unfairly with other engineers."

(1) He shall not engage in outside engineering work to an extent

prejudicial to his salaried position or detrimental to established engineering services, or which would result in a conflict of interest.

(2) If permitted by his employer, his outside activities should preferably be confined to consultation on phases of engineering for which he has special qualifications not inherently available in usual engineering practice. Also, he would not ordinarily establish an office for the purpose of conducting such outside activities.

(3) He shall not use the influence of a salaried position to direct clients to an engineering office in which he has financial interest.

ARTICLE 8: "It shall be considered unprofessional . . . To exert undue influence or to offer, solicit or accept compensation for the purpose of affecting negotiations for an engineering engagement."

(1) He shall not make political contributions for the purpose of influencing the selection of engineers on future engagements.

(2) He shall not give or receive any payments for the purpose of influencing the selection of an engineer for an engineering engagement.

(3) He shall not create obligation on prospective clients or employers through extravagant entertainment, gifts, or similar expenditures.

(4) He shall not engage in "fee splitting" or other distribution of fees for other than services performed and in proportion to the value of such services.

(5) He shall not solicit or accept an engineering engagement, or submit a proposal or contract covering engineering services when payment for such services is contingent upon results supporting a predetermined conclusion or upon a favorable finding with respect to economic feasibility or if the conditions are otherwise such as may adversely affect the engineer's professional judgment and/or the public or the client's interest.

(6) He shall not request, propose or accept an engineering engagement on a contingent fee basis if the contingent basis or the contingent services performed influence the selection of the engineer.

ARTICLE 9: "It shall be considered unprofessional . . . To act in any manner derogatory to the honor, integrity or dignity of the engineering profession."

(1) He shall not be associated in responsibility for work with engineers who do not conform to ethical practices.

(2) He shall express an opinion only when it is founded on adequate knowledge and honest conviction while he is serving as a witness before a court, commission, or other tribunal.

(3) He shall not issue statements, criticisms, or arguments on matters connected with public policy which are inspired or paid for by private interests, unless he indicates on whose behalf he is making the statement.

(4) He shall not endorse products or processes in commercial advertisements.

(5) He shall refrain from expressing publicly an opinion on an engineering subject unless he is informed as to the facts relating thereto.

(6) He shall exercise due restraint in criticizing another engineer's work.

(7) This article appropriately may be considered as a summation of the entire Code. It requires that a member of the Society shall act in accord with high standards of moral conduct under any and all circumstances.

APPENDIX XI

NATIONAL MEETINGS AND CONVENTIONS

Annual meetings and conventions, held between 1881 through 1949 in New York, N.Y., are omitted from this list. They were held each January during that period.

Location	Date
New York, N.Y.	June 1869
New York, N.Y.	June 1870
New York, N.Y.	June 1871
Chicago, Ill.	June 1872
Louisville, Ky.	May 1873
New York, N.Y.	June 1874
Pittsburgh, Pa.	June 1875
Philadelphia, Pa.	June 1876
New Orleans, La.	April 1877
Boston, Mass.	June 1878
Cleveland, Ohio	June 1879
St. Louis, Mo.	May 1880
Montreal, Canada	June 1881
Washington, D.C.	May 1882
Minneapolis, St. Paul,	June 1883
Buffalo, N.Y.	June 1884
Deer Park, Md.	June 1885
Denver, Colo.	July 1886
Hotel Kaaterskill, N.Y.	July 1887
Milwaukee, Wis.	June 1888
Seabright, N.J.	June 1889
Cresson, Pa.	June 1890
Lookout Mountain, Tenn.	May 1891
Old Point Comfort, Va.	June 1892
Chicago, Ill.	July 1893
Niagara Falls, N.Y.	June 1894
Hull (Boston) Mass.	June 1895
San Francisco, Calif.	June 1896
Quebec, Canada	June 1897
Detroit, Mich.	July 1898
Cape May, N.J.	June 1899
London, England	July 1900
Niagara Falls, N.Y.	June 1901
Washington, D.C.	May 1902
Asheville, N.C.	June 1903
St. Louis, Mo.	Oct. 1904
Cleveland, Ohio	June 1905
Thousand Islands, N.Y.	June 1906
Mexico City, Mexico	July 1907
Denver, Colo.	June 1908
Bretton Woods, N.H.	July 1909
Chicago, Ill.	June 1910
Chattanooga, Tenn.	June 1911
Seattle, Wash.	June 1912
Ottawa, Canada	June 1913
New Orleans, La.	Oct. 1913
Baltimore, Md.	June 1914
San Francisco, Calif.	Sept. 1915
Pittsburgh, Pa.	June 1916
	1917*
	1918*
Minneapolis, St. Paul, Minn.	June 1919
Portland, Ore.	Aug. 1920
Houston, Tex.	April 1921
Dayton, Ohio	April 1922
Portsmouth, N.H.	June 1922
San Francisco, Calif.	Oct. 1922
New Orleans, La.	April 1923
Chicago, Ill.	July 1923
Richmond, Va.	Oct. 1923
Atlanta, Ga.	April 1924
Pasadena, Calif.	June 1924
Detroit, Mich.	Oct. 1924
Cincinnati, Ohio	April 1925
Salt Lake City, Utah	July 1925
Montreal, Canada	Oct. 1925
Kansas City, Mo.	April 1926
Seattle, Wash.	July 1926

Place	Month	Year
Philadelphia, Pa.	Oct.	1926
Asheville, N.C.	April	1927
Denver, Colo.	July	1927
Columbus, Ohio	Oct.	1927
Washington, D.C.	April	1928
Buffalo, N.Y.	July	1928
San Diego, Calif.	Oct.	1928
Dallas, Tex.	April	1929
Milwaukee, Wis.	July	1929
Boston, Mass.	Oct.	1929
Sacramento, Calif.	April	1930
Cleveland, Ohio	July	1930
St. Louis, Mo.	Oct.	1930
Norfolk, Va.	April	1931
Tacoma, Wash.	July	1931
St. Paul, Minn.	Oct.	1931
Yellowstone Park, Wyo.	July	1932
Atlantic City, N.J.	Oct.	1932
Chicago, Ill.	June	1933
Vancouver, Canada	July	1934
Los Angeles, Calif.	July	1935
Birmingham, Ala.	Oct.	1935
Hot Springs, Ark.	April	1936
Portland, Ore.	July	1936
Pittsburgh, Pa.	Oct.	1936
San Antonio, Tex.	April	1937
Detroit, Mich.	July	1937
Boston, Mass.	Oct.	1937
Jacksonville, Fla.	April	1938
Salt Lake City, Utah	July	1938
Rochester, N.Y.	Oct.	1938
Chattanooga, Tenn.	April	1939
San Francisco, Calif.	July	1939
Kansas City, Mo.	April	1940
Denver, Colo.	July	1940
Cincinnati, Ohio	Oct.	1940
Baltimore, Md.	April	1941
San Diego, Calif.	July	1941
Chicago, Ill.	Oct.	1941
Roanoke, Va.	April	1942
Minneapolis, Minn.	July	1942
Niagara Falls, Canada	Oct.	1942
Los Angeles, Caiit.	July	1943
....		1944*
....		1945*
Philadelphia, Pa.	April	1946
Spokane, Wash.	July	1946
Kansas City, Mo.	Oct.	1946
Phoenix, Ariz.	April	1947
Duluth, Minn.	July	1947
Jacksonville, Fla.	Oct.	1947
Pittsburgh, Pa.	April	1948
Seattle, Wash.	July	1948
Boston, Mass.	Oct.	1948
Oklahoma City, Okla.	April	1949
Mexico City, Mexico	July	1949
Washington, D.C.	Nov.	1949
Los Angeles, Calif.	April	1950
Toronto, Canada	July	1950
Chicago, Ill.	Oct.	1950†
Houston, Tex.	Feb.	1951
Louisville, Ky.	June	1951
New York, N.Y.	Oct.	1951†
New Orleans, La.	Mar.	1952
Denver, Colo.	June	1952
Chicago, Ill.	Sept.	1952§
San Francisco, Calif.	Mar.	1953
Miami Beach, Fla.	June	1953
New York, N.Y.	Oct.	1953†
Atlanta, Ga.	Feb.	1954
Atlantic City, N.J.	June	1954
New York, N.Y.	Oct.	1954†
San Diego, Calif.	Feb.	1955
St. Louis, Mo.	June	1955
New York, N.Y.	Oct.	1955†
Dallas, Tex.	Feb.	1956
Knoxville, Tenn.	June	1956
Pittsburgh, Pa.	Oct.	1956†
Jackson, Miss.	Feb.	1957
Buffalo, N.Y.	June	1957
New York, N.Y.	Oct.	1957†
Chicago, Ill.	Feb.	1958
Portland, Ore.	June	1958
New York, N.Y.	Oct.	1958†
Los Angeles, Calif.	Feb.	1959
Cleveland, Ohio	May	1959
Washington, D.C.	Oct.	1959†
New Orleans, La.	Mar.	1960
Reno, Nev.	June	1960
Boston, Mass.	Oct.	1960†
Phoenix, Ariz.	April	1961
New York, N.Y.	Oct.	1961†
Houston, Tex.	Feb.	1962
Omaha, Nebr. (WR)	May	1962
Detroit, Mich. (T)	Oct.	1962†
Atlanta, Ga. (E)	Feb.	1963
Milwaukee, Wis. (WR)	May	1963
San Francisco, Calif. (S)	Oct.	1963†
Cincinnati, Ohio (T)	Feb.	1964
Salt Lake City, Utah (E)	May	1964
New York, N.Y. (S)	Oct.	1964†
Mobile, Ala. (WR)	Mar.	1965
Minneapolis, Minn. (T)	May	1965
Kansas City, Mo. (E)	Oct.	1965†
Miami Beach, Fla. (S)	Feb.	1966
Denver, Colo. (WR)	May	1966

Philadelphia, Pa. (T) ....Oct. 1966†
Dallas, Tex. (E) ....Feb. 1967
Seattle, Wash. (S)...........May 1967
New York, N.Y. (WR) ....Oct. 1967†
San Diego, Cal. (T) ....Feb. 1968
Chattanooga, Tenn. (E) ......May 1968
Pittsburgh, Pa. (S) ....Oct. 1968†
New Orleans, La. (WR) .....Feb. 1969
Louisville, Ky. (S) ....April 1969
Washington, D.C. (T)........July 1969
Chicago, Ill. (E) ....Oct. 1969†
Memphis, Tenn. (WR) ....Jan. 1970
Portland, Ore. (S) ....April 1970
Boston, Mass. (T) ....July 1970
New York, N.Y. (E) ....Oct. 1970†
Phoenix, Ariz. (WR) ....Jan. 1971
Baltimore, Md. (S) ....April 1971
Seattle, Wash. (T) ....July 1971
St. Louis, Mo. (E) ....Oct. 1971†
Atlanta, Ga. (WR) ....Jan. 1972
Cleveland, Ohio (S).........April 1972
Milwaukee, Wis. (T).........July 1972
Houston, Tex. (E)............Oct. 1972†
Washington, D.C. (WR) ......Jan 1973
San Francisco, Calif. (S).....April 1973
Tulsa, Okla. (T) ....July 1973
New York, N.Y. (E) ....Oct. 1973†
Los Angeles, Calif (WR)......Jan. 1974
Cincinnati, Ohio (S) ....April 1974
Montreal, Canada (T) ....July 1974
Kansas City, Mo. (E) ....Oct. 1974†

---

*1917-1918 meetings omitted because of war.
*1944-1945 meetings omitted because of war.
†Regarded as annual meeting also.
§Centennial of Engineering Convocation.
WR—Water Resources Engineering; E—Environmental Engineering; T—Transportation Engineering; S—Structural Engineering.

APPENDIX XII

OUTSTANDING CIVIL ENGINEERING ACHIEVEMENT AWARDS

Civil engineering achievement within the United States is recognized annually by selection of the "Outstanding Civil Engineering Achievement." The award is given for the achievement — not to an individual — so that the many engineers who have worked on the project are recognized as having contributed.

1960: St. Lawrence Power and Seaway Project
Award of Merit:
Allegheny County Sewage Disposal System, Pennsylvania
1961: John F. Kennedy International Airport, New York
1962: Intercontinental Ballistic Missile Program
1963: Ohio River Valley Clean Streams Program
Awards of Merit:
George Washington Bridge Expansion Project, New York
Whittier Narrows Water Reclamation Plant, Los Angeles County, California
1964: Glen Canyon Dam, Page, Arizona
1965: Chesapeake Bay Bridge-Tunnel, Norfolk to Delmarva Peninsula
Awards of Merit:
Verrazano-Narrows Bridge, New York
Central District Filtration Plant, Chicago, Illinois
Los Angeles County Flood Control District, California
1966: Launch Complex 39
Award of Merit:
Trans-Sierra Freeway Project, California
1967: Gateway Arch, St. Louis, Missouri
Awards of Merit:
Hanford Nuclear Power Plant, Richland, Washington
Oakland-Alameda County Coliseum Complex, California

1968: San Mateo-Hayward Bridge, California
ASCE Comprehensive Program Special Award:
Greater Pittsburgh, Pennsylvania, Renaissance Program
1969: Oroville Dam & Edward Hyatt Power Plant, California
1970: Armco Steel's Middletown, Ohio, Works
1971: World Trade Center, New York
Awards of Merit:
Northfield Mountain Pumped Storage Project, New England
McClellan-Kerr Arkansas River Navigation System Project
1972: California State Water Project
Awards of Merit:
Mississippi River Pollution Abatement Project
The 200 Billion Electron Volt National Accelerator Laboratory, Batavia, Ill.
1973: Ludington Pumped Storage Project, Michigan
1974: Land Reclamation Program of Fulton County, Illinois
Special Civil Engineering Achievement Award:
The National System of Interstate and Defense Highways

APPENDIX XIII

NATIONAL HISTORIC CIVIL ENGINEERING LANDMARKS

In 1966 the Board of Direction authorized designation of the first ASCE National Historic Civil Engineering Landmark. In this continuing program, Section Committees on the History and Heritage of American Civil Engineering propose, with Section endorsement, a nationally significant civil engineering project to the ASCE Committee on the History and Heritage of American Civil Engineering which then may recommend to the Board of Direction that the project be named by the Society as a National Historic Civil Engineering Landmark. A bronze plaque denoting the designation is supplied for a presentation event.

1966

BOLLMAN TRUSS BRIDGE—An 80-foot double-truss span built by Baltimore engineer Wendel Bollman at Savage, Maryland, in 1869. Only remaining example of a design that facilitated rapid expansion of early American railroads.

1967

BIDWELL BAR SUSPENSION BRIDGE—Typical of the type of suspension bridges constructed during California's gold rush days. It spanned the Feather River approximately 10 miles northeast of Oroville and is the only remaining artifact of its time.

ERIE CANAL—Begun in Rome, N.Y. on July 4, 1817, this 365-mile-long canal cut a swath through New York State from the Hudson River at Albany to the Great Lakes at Buffalo. Called the First School of American Civil Engineering, it was the principal route for emigrants from the East and agricultural products from the West.

MIDDLESEX CANAL—One of the oldest man-made waterways in the U.S. Completed in 1803, the canal stretched 27 miles from Lowell to Charlestown, Massachusetts. Served as model for the Erie Canal. First to prove by low freight rates and expansion of traffic the urgency for widespread internal transportation.

1968

ACEQUIAS OF SAN ANTONIO—One of the earliest recorded uses of an engineered water supply and irrigation system in the country. The first of eight original acequias was placed under construction in 1718, and its remains are still visible on the grounds of the Alamo and on the HemisFair grounds.

CANAL AND LOCKS OF THE POTOWMACK CANAL—The canal and locks are a part of the first extensive system of canal and river navigation works undertaken in the United States. Construction was begun by the Potowmack Company in 1785, and the works operated from about 1799 to 1821. The locks required at Great Falls to overcome the 76-foot difference in elevation are a unique and major feature in the system.

CENTRAL PACIFIC RAILROAD—The western terminus of America's first transcontinental railroad. Construction was begun in Sacramento in 1863 and completed in 1869 at Promontory, Utah. Railroad was organized and located by Theodore Judah, noted civil engineer of that era.

DURANGO-SILVERTON BRANCH of the Denver & Rio Grande Western Railroad—Now a tourist excursion train, the last of the narrow-gauge railroads links the Colorado mining towns of Durango and Silverton. Constructed in 1882, the railroad is a surviving testament to the role of the civil engineering profession in developing the West.

ELLICOTT'S STONE—Andrew Ellicott was commissioned by the Federal Government soon after the Republic was formed to establish an International Boundary. This "stone," located a few miles north of Mobile, Alabama, is the key extant monument from the historic survey that located with remarkable accuracy the 31st parallel border between the U.S. and Spanish West Florida.

JOINING OF THE RAILS OF THE TRANSCONTINENTAL RAILROAD—The linking of the continent by 1,766½ miles of trunk line railroad over mountains and deserts was a turning point in American history. It signaled the opening of the West and the emergence of a unified nation. This event took place at Promontory Point, Utah, on May 10, 1869.

WHEELING SUSPENSION BRIDGE—Oldest existing major suspension bridge. Built in 1849, the bridge was the first long-span wire-cable suspension bridge in the world. It served as a link in the National Highway from Washington, D.C., to the West. Wrecked in a storm in 1854, it was reconstructed in 1856 and is still in service.

1969

ALVORD LAKE BRIDGE—Built in 1889 by New York's Ernest L. Ransome, this bridge—located in San Francisco's Golden Gate Park—is believed to be the oldest (and first to be constructed) concrete-arch bridge with steel reinforcing bars in the U.S.

CHARLESTON-HAMBURG RAILROAD (South Carolina) — World's longest railroad (136 miles) at the time of its construction in 1833. Also the first to operate passenger trains on established schedule, first to use a steam locomotive built in the U.S., first railroad to be completely locomotive powered, and first to carry mail.

1970

ASCUTNEY MILL DAM—Built in 1834 and still in service, at Windsor, Vermont, it is among the very earliest masonry dams of significant size. Made of granite and measuring 120 feet in length with a crest 42 feet above tailwater, it is the structural precursor of today's concrete-gravity dams.

BRIDGEPORT COVERED BRIDGE—Built in 1862, it is the longest single-span (230 feet) covered bridge extant west of the Mississippi River. The design is best described as a Howe truss with arch. Still in service, it originally carried heavy freight between Marysville and Virginia City, California.

CORNISH-WINDSOR COVERED BRIDGE (Vermont)—This two-span covered bridge, with an overall length of 460 feet, is the longest covered bridge existing in the U.S. It is a Towne Lattice timber-truss design of a type widely used on many early timber bridges and later in building construction. It was built in 1866 and is still in use.

FRANKFORD AVENUE BRIDGE—This three-span stone arch bridge over Philadelphia's Pennypack Creek was built in 1697 and has served as an important roadway ever since. It is the first known stone arch to be built in this country and probably the oldest extant bridge in the U.S.

THEODORE ROOSEVELT DAM AND SALT RIVER PROJECT—The first project of the Bureau of Reclamation; the first multi-purpose (irrigation, river regulation, power generation and recreation) project in the United States; and the beginning of the Federal reclamation project throughout the West. When it was completed in 1911, the dam (near Phoenix, Arizona) was one of the highest in the world.

UNION CANAL TUNNEL—The oldest existing transportation tunnel in the U.S., it is part of the canal that connects the Susquehanna and Schuylkill rivers, providing an important transportation route to the west. Built in 1825 under difficult conditions, the tunnel is 729 feet long and 18 feet wide by 14 feet in height. It has been restored to its original condition.

1971

DRUID LAKE DAM—The first major earthfill dam to be constructed in the U.S. Built between 1864 and 1871, it has a number of unique characteristics, especially for its time. First was the challenge of constructing in a ravine a barrier of sufficient height to form a reservoir with as much storage capacity as the site would permit for the rapidly growing city of Baltimore. Then there was the additional challenge of using materials to build an economical dam that would be structurally sound and long lived. It was the forerunner of other earthfill dams that have been built across the U.S.

EADS BRIDGE—Named for its designer and builder, James Buchanan Eads, this bridge (across the Mississippi at St. Louis) involved the first use of pneumatic caissons in the U.S. and represented the deepest submarine construction work done in the world. It was the first bridge to utilize cantilever construction methods entirely and was built without falsework. Completed in 1874, the bridge has unprecedented 500-foot arches, some 200 feet longer than any built previously. In addition, it was the first bridge to make extensive use of steel.

FIRST OWENS RIVER-LOS ANGELES AQUEDUCT—This aqueduct system, unprecedented in size and scope at the time of its completion (1913), was the major arch prototype for the extensive water supply system needed to support the major urban complexes of today. Begun in 1907, the aqueduct provided Los Angeles with an annual flow of 440 cubic feet per second.

MORMON TABERNACLE (at Salt Lake City, Utah)—The first building selected as a Landmark. The greatest engineering challenge was the design and construction of the roof in 1865. This was accomplished with 150-foot lattice arches. Stone and lumber building materials were obtained at great effort from surrounding mountains for the railroad that could bring metal building components from the industrialized centers of the East had not been completed. Today, few changes have been made in the original construction, and the roof remains structurally sound.

OLD BETHLEHEM WATERWORKS (Pennsylvania)—The first known pumping system providing drinking and washing water in the North American Colonies. The existing building is dated 1761, but it was preceded by an experimental frame building dated 1754. A wooden waterwheel, driven by the flow of Monocacy Creek, powered wooden pumps that lifted the water through wooden pipes to the top of a hill where it was distributed by gravity. The system served as a model for Latrobe's Washington Square Pump House in Philadelphia.

1972

BROOKLYN BRIDGE—The longest suspension bridge in the world, when it was completed in 1883, and the first to use steel cables and trusses. It was designed by John A. Roebling and built under the supervision of his son, Washington, both of whom were civil engineers.

CABIN JOHN AQUEDUCT—The longest stone-masonry arch in the world until 1903. This structure at Cabin John, Maryland, is still serving the basic purpose for which it was built, providing water supply for Washington, D.C., as well as carrying modern traffic loads on a major highway.

CHESBROUGH'S CHICAGO WATER SUPPLY SYSTEM—The water tower and pumping station stand as symbols of the daring engineering achievement of building, in 1869, a 2-mile tunnel under Lake Michigan with intake crib to provide a safe potable water supply for the citizens of Chicago. These buildings have also been recognized as a memorial to the victims of the Chicago Fire in 1871.

GUNNISON TUNNEL—The key to the first major transmountain irrigation system in the U.S. When completed in 1909, the 5.1-mile Gunnison Tunnel (in Colorado) was the longest irrigation tunnel in the U.S. The 30,582-foot-long tunnel initially supplied irrigation water to 146,000 acres of vegetable-and fruit-growing cropland.

MIAMI CONSERVANCY DISTRICT—The first regionally coordinated flood-control system in the U.S. embodying retention reservoirs for controlled release of flood waters. The actual project consisted of the construction of five dams, levee and channel improvements at nine villages and towns, and relocation of four railroad lines and of many highways and wire lines, the removal of one village, the lowering of water and gas mains, and many minor works. Since its completion in 1922, the protected Miami Valley (Ohio) has not been damaged by flooding.

ROEBLING'S DELAWARE AQUEDUCT—John Augustus Roebling's earliest (1848) still-standing suspension bridge (at Lakawaxen, Pennsylvania) and perhaps the oldest existing cable suspension bridge in the world that retains its original principal elements. One of the nation's most significant engineering relics and the earliest extant work of the man who is rightfully acknowledged the "father of the modern suspension bridge."

1973

BUFFALO BILL DAM (at Cody, Wyoming)—This was the first major dam to be designed and built using the trial-load analysis technique and, at the time of its completion (1910), was the highest dam in the world and the only dam with a height/width ratio greater than one.

CHEESMAN DAM—When completed in 1905, the Cheesman Dam was the world's highest gravity-arch stone masonry dam. It was the first major dam in the U.S. to incorporate the gravity-arch concept in its design and is the key structure in the Denver water supply system.

EMBUDO, NEW MEXICO, STREAM-GAUGING STATION—The first stream-gauging system ever undertaken, this project led to the development of techniques that have been used extensively to collect essential data not only for hydraulics projects but also for land use and urban planning. The project was established in 1888 and is still in use.

INGALLS BUILDING—The first reinforced concrete skyscraper in the world. A 16-story (210-feet) structure that demonstrated for the first time the safety and economy of reinforced concrete frames for high-rise construction, and was a vital stimulus to the use of reinforced concrete in fireproof construction. It was built at Cincinnati, Ohio, in 1902-03 at a cost of $400,000.

PELTON IMPULSE WATER WHEEL—The monument site (at Camptonville, California) of the first successful impulse water wheel. This significant development by Lester Allen Pelton (1829-1908) is used throughout the world for creating water power. Its high efficiency was due to the use of the first splitter-type bucket used on a water wheel. This method was the key to tapping the vast water power of the American West.

STARRUCCA VIADUCT (at Lanesboro, Pennsylvania)—This was the key structure in the New York and Erie Railroad that was among the earliest major links between the Eastern seaboard and the Midwest. It was constructed, in 1848, in record time, and was among the first, if not the first, important engineering work to utilize structural concrete.

1974

MILWAUKEE METROPOLITAN SEWAGE TREATMENT PLANT—This plant is unique in that it is America's earliest large-scale activated sludge type municipal sewage treatment plant. From its successful operations many other municipalities have adopted its system of efficient environmental recycling. When it began operation in 1926, it was a major improvement over contemporary methods and was a major advance in the history of municipal sanitary engineering.

INDEX